Glossary of Metalworking TERMS

By

Richard P. Pohanish

Industrial Press, Inc.
New York

Library of Congress Cataloging in Publication Data

Pohanish, Richard P.
  Glossary of metalworking terms / by Richard P. Pohanish
436p.cm.
  ISBN 0-8311-3241-8
    1. Metal-working--Dictionaries. I. Title

TS204.P64 2003
671'.03--dc21   2003040610

Industrial Press Inc.
200 Madison Avenue
New York, NY 10016-4078

First Paperback Edition, April 2005

Sponsoring Editor: John Carleo
Cover Design: Janet Romano

Printed in the United States of America

10   9   8   7   6   5   4   3   2   1

Preface

Dr. Samuel Johnson, compiler of the first *Dictionary of the English Language*, once remarked that people need to be reminded more, and informed less. This clever observation may explain the enduring need for references works. While working on *Machinery's Handbook Pocket Companion*, the author realized that no modern glossary for terms related to the fabrication and use of metals could be found. A few impressive dictionaries were found, but these were quite old and are scholarly works covering the nomenclature of professional metallurgists, engineers, and industrial technologists.

By its very nature a *glossary* is a collection of selected specialized terms in a field of knowledge, which is easy to use and with definitions that are geared to the interests of an expected audience. Thus, the readership for this *Glossary of Metalworking Terms* is one seeking a source of review information: students and professionals in any of the subdivisions of engineering, science, and related fields of technology; secondary school, public, college, military, and industrial libraries; and for management personnel in a wide variety of industrial areas, including sales and purchasing. In other words, the main interest of many potential users may be other than metals, metal finishing, or metallurgy, but they have a need to be conversant with terms related to these fields. It is hoped that this modest book will provide an accessible and transportable reference source that will meet their needs.

When compiling a reference of such broad coverage, it is necessary to seek the advice and expertise of colleagues, to all of whom the author is deeply grateful. In particular the author would like to thank Franz Gruss for allowing him the use of his technical book collection and John Carleo of Industrial Press for his encouragement and support for this project.

Last but not least, the author realizes that treatment of some terms may be considered superficial by some standards; nevertheless, any comments from the readers for additions in the next edition will be greatly appreciated. Please send all suggestions in writing to the author's attention c/o Industrial Press, Inc.

Key to Abbreviations

b.c.c. - body-centered cubic (crystal structure)

bp - boiling point

°C - degrees of temperature in Celsius (Centigrade)

ca - about

CAS - Chemical Abstracts Service (number)

cc - cubic centimeter

DOT - U.S. Department of Transportation

EPA - U.S. Environmental Protection Agency

°F - degrees in temperature in Fahrenheit

f.c.c. - face-centered cubic (crystal structure)

h.c.p. -hexagonal close-packed (crystal structure)

kg - kilogram

lb - pound

mp - melting point

OSHA - U.S. Occupational Safety and Health Administration

% - percent(age)

> -greater than

< - less than

$\leq$ - equal to or less than

$\geq$ - equal to or more than

A

A: (1) Symbol for ampere; (2) an unofficial symbol for the element argon; (3) symbol for acceleration; (4) symbol for amplitude; (5) symbol for atomic weight; (6) symbol for area. See also alpha.

Å: Symbol for angstrom unit.

A, α: See alpha.

A₁: The eutectoid temperature of steel.

AA: (1) Abbreviation for arithmetical average; (2) Abbreviation for Aluminum Association.

AAI: Abbreviation for Aluminum Association, Inc.

AAR: Abbreviation for Association of American Railroads.

ABMA: Abbreviation for American Bearing Manufacturers Association.

ABFMA: Abbreviation for Anti-Friction Bearing Manufacturers Association.

abherent: Any substance that prevents adhesion of a material to itself or to another material, such as mold release agents.

abrasion: The process of wearing away material by mechanical action, friction, shearing; or, using abrasive materials in actions such as erosion, grinding, honing, lapping, rubbing, superfinishing, or polishing.

abrasion resistance: The ability of a material to resist wear by friction or shearing action.

1

abrasive: A manufactured or natural very hard, tough material capable of wearing away another material softer than itself. Used for grinding and improving a surface by friction, as in polishing, buffing, cleaning. Emery (alumina), pumice, silica (sand), rouge (iron oxide), and diamond dust are examples of natural abrasives. Aluminum oxide [Al_2O_3], silicon carbide [SiC], cerium oxide, fused alumina, and boron carbide [B_4C] are examples of *artificial abrasives*. Sandpaper, steel shot, glass beads, and steel wool are also examples of abrasives. Abrasives can vary in hardness from diamond (hardest), silicon, carbides, emery, feldspar, to rouge (softest). See also Mohs hardness scale. See also abrasion.

abrasive belt: A belt that is continuous and abrasive coated. Used for removing stock, light cleaning up of metal surfaces, grinding welds, deburring, breaking and polishing hole edges, and finish grinding of sheet steel. The types of belts that are used may be coated with aluminum oxide [Al_2O_3], the most common coating used for stock removal and finishing of all alloy steels, high carbon steel, and tough bronzes. Abrasive belts may also be coated with silicon carbide [SiC] for use on hard, brittle, and low-tensile strength metals such as aluminum and cast iron.

abrasive cutting: Generally means severing a smaller piece from a length of material using an abrasive cutoff wheel at relatively high speed. See also abrasive cutoff wheel.

abrasive cutoff wheel: A thin grinding wheel made with resinoid, rubber, shellac, and fiber bond, used for cutting steel, brass, and aluminum bars and tubes of all shapes and hardness, ceramics, plastics, insulating materials, glass, and cemented carbides. Abrasive wheels are commonly available in four types of bonds: resinoid, rubber, shellac, and fiber or fabric reinforced. In general, resinoid bonded cutoff wheels are used for dry cutting where burrs and some burn are not objectionable and rubber-bonded wheels are used for wet abrasive cutting on practically all materials where cuts are to be smooth, clean, and free from burrs. Shellac-bonded wheels have a soft, free cutting quality that makes them particularly useful in the tool room where tool steels are to be cut without discoloration. Fiber-reinforced bonded wheels are able to withstand severe flexing and side pressures; Fabric-reinforced bonded wheels, which are highly resistant to breakage caused by extreme side pressures, are fast cutting and have a low rate of wear. Wheels are available in various diameters, from less than 1 in. up to 30 in. and thicknesses that range from 0.003 to about 5/32 in. See also cutoff wheel.

abrasive cutting machine: There are four basic types of abrasive cutting machines: chop stroke, oscillating stroke, horizontal stroke, and work rotating. Each of these four types may be designed for dry cutting or for wet cutting (includes submerged cutting, whereby the work is immersed in a coolant). Circular saws may be used on certain models for materials that are adaptable for saw cutting.

abrasive dressing stick: An abrasive stick, usually hand held in a holder, used to dress the face of smaller-size grinding wheels. Because it also shears the grains of the grinding wheel, it is often used for roughing or preshaping, prior to final dressing with a diamond dresser. Abrasive dressing sticks are made of hard material such as silicon carbide [SiC] or boron carbide [B_4C] grains with a hard bond such a fluorocarbon polymer. See also grinding wheel dresser.

abrasive wear: This occurs when a hard rough surface slides against a softer one, ploughing a series of grooves and removing material; it also occurs when abrasive particles are introduced between sliding surfaces, or when a part is moved through an abrasive medium. Erosion abrasion takes place when abrasive particles impact on surfaces. These particles may be suspended in liquids, carried by air, or flow of their own weight, such as sand particles down a chute.

abrasive wheel: See grinding wheel.

abrasive wire bandsaw: A cutting using a cutting blade made from a small diameter wire containing hard, abrasive particles of diamond dust, alumina, or cubic boron nitride bonded to its surface.

absolute dimensioning system: A numerical control (N/C) system in which all tool movements are measured and programmed from the fixed datum or reference point.

abs.: Abbreviation for absolute.

abscissa: The distance between points measured along a horizontal scale; the coordinate of the distance of a point from a point of origin, measured along the *x*-axis.

absolute: (1) The temperature scale used by scientists and engineers. The temperature ($0°K = -459.72°F = -273.15°C$) at which molecular movement ceases and the volume of an ideal gas would become zero. Referred to as absolute zero. Also known as the Kelvin (°K) scale; (2) a term used to indicate something that is exact, perfect, or free from impurities.

Ac_1, Ac_2, Ac_3, Ac_4: Symbols used to describe the transformation temperature (temperatures of phase changes at equilibrium conditions) during the heating of iron and steels. Ac_1, austenite begins to form; Ac_2, magnetic transformation of *alpha (α)-beta(β)* ferrite; Ac_3, transformation of ferrite to austenite is completed; Ac_4, austenite transformed to *delta (δ)*-ferrite.

AC41A alloy: A die- and sand-casting zinc alloy containing 95.96% zinc, 4% aluminum, 1% copper, and 0.04% magnesium.

ACC: Abbreviation for the American Copper Council.

Accrolon® 9039: A proprietary product of Accro-Seal Corporation for a bearing material made from an engineering thermoplastic that is self-lubricating and having excellent resistance to abrasion, wear, and fatigue.

Accrolube®: A proprietary product of Accro-Seal Corporation for a grease containing Teflon®, used to reduce wear between metal surfaces.

accuracy: The degree of conformity of an indicated value to the correct value or an accepted standard value.

ACerS: Abbreviation for the American Ceramic Society.

acetylene (ethyne): A colorless gaseous hydrocarbon compound HC≡CH with an ether-like odor and made, originally, from the action of water on calcium carbide. Acetylene burns with an intense flame. Used for cutting metals, and in welding and soldering. Acetylene requires careful handling and storage. A highly reactive compound, acetylene may form explosive mixtures with silver, mercury, copper, and copper alloys. It is highly flammable and explodes readily.

acicular: Having a crystalline structure (e.g. acicular martensite) that is elongated and needle-shaped. This is the shape of "metal whiskers."

acid: Any organic or inorganic chemical compound containing hydrogen replaceable by metals, and having a pH of zero up to 7, the neutral point. Strong acids in the pH range of zero to 2 are corrosive and will cause chemical burns to the skin, eyes, and mucous membranes. Handle with care. Acids turn litmus red. An acid is termed "organic" when the molecule contains carbon. Inorganic acids are known as mineral acids.

Acid Aid 5LXS-IH®: A proprietary product of Crown Technology, Inc., used for increasing steel pickling rates.

Acid Aid X®: A proprietary product of Crown Technology, Inc., used for metal cleaning and pickling.

acid embrittlement: See hydrogen embrittlement.

acid gas: A gas that forms an acid when dissolved in water.

acid lead or acid lead alloy: See hard lead.

acid lining: A term used in the steel-making industry to describe the silica brick furnace lining.

acid solution: A solution having a pH value less than 7.0.

acid steel: Steel produced in a furnace with an acid hearth and lining, and buried under an acidic slag from substances such as silica. This process requires raw materials that have low phosphorus and sulfur content. Considered superior to basic steel.

acme tap: A tool used for cutting internal screw threads. These threads have an included angle of space of 29°.

Acme thread: A screw thread, either outside (male) or inside (female), that has an included angle of 29°. Acme threads have a section that is between the square and V. They are used extensively for feed screws, to cause movement, or to generate power transmission through the rotation of the male thread within the female thread. This thread is easier to machine than a square thread and was developed to carry heavy loads without causing excessive bursting pressure in the hole or nut. Acme threads are designed as *general purpose* or as *stub*. The stub thread is shallower in height and wider at the crest and root of the thread. It is generally used when the deeper and narrower general purpose thread would weaken the remaining wall of the workpiece.

ACS: Abbreviation for both American Carbon Society and American Chemical Society.

activated carbon: A granular or powdered form of carbon that is highly porous and absorbent; produced by heating (carbonizing) vegetable material in the absence of air; used for absorbing contaminants, purifying gases and vapors, and clarifying liquids. Also known as activated charcoal.

activated charcoal: See activated carbon.

activation: The changing of the passive surface of a metal to a chemically active state. Contrast with passivation.

active spline length, involute spline (L_a): The length of spline that contacts the mating spline. On sliding splines, it exceeds the length of engagement.

actual fit: The actual fit between two mating parts is the relation existing between them with respect to the amount of clearance or interference that is present when they are assembled. Fits are of three general types: clearance fit, transition fit, and interference fit.

actual size: The measured size.

actuator: A mechanical piece such as a cam or arm used to activate a device, or set it in motion.

acute angle: If the inclination of the arms of an angle are less than a right angle (90°), the angle is called *acute*.

acute triangle: A triangle in which all the angles are less than 90°. If one of the angles is larger than 90°, the triangle is called an obtuse-angled triangle. Both acute and obtuse-angled triangles are known under the common name of oblique-angled triangles. The sum of the three angles in every triangle is 180°.

adapters: Devices used to mount or hold cutters of various types and sizes on the spindle of a milling machine.

addendum: (1) *gear:* A term describing the height of tooth above pitch circle or the radial distance between the pitch circle and the top of the tooth; that circle which passes through the top of the tooth of a gear; (2) *thread:* A term describing the distance from the pitch diameter to the crest of the thread.

additives, cutting oils: A substance used to increase the effectiveness of basic mineral cutting oils, additives are of two main types, *polar* and *chemical*. Polar additives are capable of ionizing. They include *animal fats and oils,* which develop unpleasant odors and bacteria, *vegetable oils* such as castor oil and coconut oil, and *marine oils* derived from the fatty tissues of fish and whales and which have a fishy odor. Chemical additives can be synthetic and semisynthetic. They are compounded from chlorine, sulfur, and phosphorus, and provide solid film lubrication through chemical reaction with the surface of the workpiece. See also soluble oils.

adherend: An object which is held to another object by an adhesive.

adhesive: Any substance, organic or inorganic, that is capable of fastening or bonding by means of surface attachment. The bond durability depends on the strength of the adhesive to the base material or substrate (adhesion) and the strength within the adhesive (cohesion). Adhesives can be classified as structural or nonstructural. In general, an adhesive can be considered structural when it is capable of supporting heavy loads, and nonstructural when it is incapable of supporting heavy loads. Surface preparation is very important; the presence of oil, grease, mold-release agents, or even a fingerprint can destroy a good bond. Surface preparation such as chemical etching or mechanical roughening may be needed to improve joint strength on some materials. A good source of further information is the Research Center for Adhesives and Coatings at Case Western Reserve University in Cleveland, Ohio.

adhesive sealant: An adhesive that performs both bonding and sealing functions.

adjustable angle plate: Also know as sine angle plates, this tool is precision made and has provisions for setting them with micrometer or rectangular gage blocks. A sine bar is frequently used when an adjustable angle plate is not available. See also angle plate.

adjustable caliper snap gage: See caliper snap gage.

adjustable hook rule: See hook rule.

adjustable reamer: Also known as an expansion reamer. A type of hand reamer containing multiple (perhaps five or more) carbon- or high-speed steel blades carried in collars mounted on a central threaded shank. The shank can be adjusted, usually by means of a cone adjuster, to control the cutting edges uniformly throughout their length and vary the diameter over a certain limited range. These reamers are available in standard sizes and they may be expanded to the smallest diameter of the next-size reamer. When dull or broken, the blades are not re-ground, but replaced. Adjustable reamers should never be expanded beyond their specified limits; such an action may put the reamer out of round and cause cutting edge breakage.

adjustable tap: A tool for cutting the thread of an internal screw, made with slots in the shank (tap body) for holding separate blades or chasers that can be adjusted to control the tap's cutting edge.

adjustable tap wrench: A tool containing two opposing handles and a V-shaped opening in the center. One of the handles is adjustable, making it possible to hold the square ends of taps and reamers of various sizes in the V-shaped opening. *See also* tap wrench.

admiralty brass: Also known as admiralty metal. Alloys nominally containing 70-73% copper, 0.75-1.20% tin, and 26-29% zinc; may also contain a small amount (0.01-0.05%) arsenic to increase resistance to dezincification. It has good corrosion resistance to sea water, dilute acids, dilute bases. A type of alpha brass used for cold working.

admiralty brass, inhibited: Alloys resistant to marine corrosion, containing 71% copper, 28% zinc, and 1% tin.

admiralty gun metal: An alloy resistant to marine corrosion, containing 88% copper, 10% tin, and 2% zinc. See also gun metal.

advance cutting tool material: Materials able to withstand extremely harsh condition of elevated cutting tool speeds and temperatures. These materials include ceramic, cermet, polycrystalline diamond (PCD), and cubic boron nitride (CBN) substrates.

advance metal: Alloys nominally containing 56-60% copper (Cu) and 40-44% nickel (Ni) used for precision electrical instrumentation.

advance per revolution: The distance a cutting tool will advance in one revolution. Calculated by dividing the feed rate by the revolution per minute (rpm).

AEC: Abbreviation for Aluminum Extruders Council.

Aero®: A registered trademark of Cytec Industries for a group of chemical products used for case-hardening, metal heat-treating, catalysts metallurgical additives.

aerosol: Fine liquid or solid particles suspended in air (dispersed in a gas), the particle size often being in the 0.01 to 100 microns range. Natural aerosols include smoke (solid particles) and fog (liquid particles). Man-made aerosols are manufactured by filling a valved container, usually a can, with a suspension (e.g., paint, lubricants) in a gas under pressure.

AESF: Abbreviation for American Electroplaters and Surface Finishers Society.

AFBMA: Abbreviation for Anti-Friction Bearing Manufacturers Association.

affinity, chemical: The selective tendency of one element to react or combine with another to form a new substance.

afterblow: In the Bessemer process for steel making, continuing the blast air flow to remove phosphorus following carbon removal.

AFS: Abbreviation for American Foundrymen's Society.

Ag: Symbol for the element silver. From the Latin, *argentum.*

AG40A alloy: A die-casting zinc alloy containing 95.96% zinc, 4% aluminum, and 0.04% magnesium.

age hardening: Also known as precipitation hardening. A spontaneous increase in hardness and strength as a result of grain structure in a wide range of alloys that are aged or allowed to "rest" at room temperature for a few days following quenching, rapid cooling, or cold working.

aging: (1) A time-temperature-dependent change in the properties of ferrous alloys that occurs more slowly at ambient (room) temperature and more rapidly at slightly elevated temperatures following hot working or heat treating processes, or after cold working operations. Except for strain aging and age softening, it is the result of precipitation from a solid solution of one or more compounds whose solubility decreases with decreasing temperature. For each alloy susceptible to aging, there is a unique range of time-temperature combinations to which it will respond. (2) aging metal: rusting, pitting, and scaling due to corrosion. Avoided by use of alloys containing noncorrosive metals (e.g., stainless steel) or by treating the surface of the base metal by cladding or plating, or other coatings or treatments.

AGMI: Abbreviation for American Gear Manufacturers Association.

AIChE: Abbreviation for American Institute of Chemical Engineers.

AIMCAL: Abbreviation for Association of Industrial Metalizers, Coaters and Laminators.

air: Another name for the atmosphere, the mixture of gases that surround the earth. At sea level the composition of air is 78% nitrogen, 20.9% oxygen, 0.94 argon, and 0.033 carbon dioxide. Additionally, there are trace amounts of other gases. The composition of any given sample can vary with the altitude at which it was taken.

air compressor: A mechanical apparatus usually driven by an electric motor, gasoline or steam for supplying air or other gasses under increased pressure.

air gage: An instrument for measuring internal and external dimensions by measuring the amount of air that escapes between a work surface and a gaging spindle which is not in contact the surface being measured. For external work surface, a snap-type spindle or other device is used. The greater the clearance, the higher will be the velocity of escaping air.

air-hardening steel: A steel that contains sufficient carbon and other alloying elements to harden fully during cooling in air or other gaseous mediums from a temperature above its austenitising temperature (transformation range). The term should be restricted to steels that are capable of being hardened by cooling in air in fairly large sections, about 2 in. or more in diameter. Same as self-hardening steel.

AISC: Abbreviation for American Institute of Steel Construction.

AISE: Abbreviation for Association of Iron and Steel Engineers.

AISI: Abbreviation for American Iron and Steel Institute.

AISI/SAE steel designations: A four digit number system used to designate the chemical composition of carbon and alloy steels. This system describes plain carbon and low-to-medium alloy content, used primarily in machine parts. The first two numerals designate either plain carbon or the alloy grouping and quantity, and the last two numerals give the mean carbon content in hundredths of a percent. The first digit for carbon steels is 1. Plain carbon steels are designated 10xx. Plain carbon steels containing 0.75% carbon are designated 1075, etc.

Additionally letters are added between the first and last pairs of digits to designate special qualities and additives as follows:

B: indicates the presence of boron in amounts of 0.0005-0.003% for enhanced depth-hardening.

L: indicates the presence of lead in amounts of 0.15-0.35% for enhanced machinability.

M: indicates merchant quality steel, for alloy steels .

E: indicates electric furnace steel, for alloy steels.

H: indicates hardenability requirements, for alloy steels.

Carbon Steels	Description
10xx	Nonresulfurized, 1.00 Mn maximum.
11xx	Resulfurized
12xx	Rephosphorized and resulfurized
15xx	Nonresulfurized, >1.00 Mn maximum

Alloy Steels	Description
13xx	1.75 Mn
40xx	0.20 or 0.25 Mb or 0.25 Mb and 0.042 S
41xx	0.50, 0.80, 0r 0.95 Cr and 0.12, 0.20. or 0.30 Mb
43xx	1.83 Ni, 0.50-0.80 Cr, and 0.25 Mb
46xx	0.85 or 1.83 Ni and 0.20 or 0.25 Mb
47xx	1.05 Ni, 0.45 Cr, 0.20 or 0.35 Mb
48xx	3.50 Ni and 0.25 Mb
51xx	0.80, 0.88. 0.93, 0.95, or 1.00 Cr
51xxx	1.03 Cr
52xxx	1.45 Cr
61xxx	0.60 or 0.95 Cr and 0.13 or 0.15 V (min.)
86xx	0.55 Ni, 0.50 Cr, and 0.20 Mb
87xx	0.55 Ni, 0.50 Cr, and 0.25 Mb
88xx	0.55 Ni, 0.50 Cr, and 0.35 Mb
92xx	2.00 Si or 1.40 Si and 0.70 Cr
50Bxx	0.28 or 0.50 Cr
51Bxx	0.80 Cr
81Bxx	0.30 Ni, 0.45 Cr, and 0.12 Mb
94Bxx	0.45 Ni, 0.40 Cr, and 0.12 Mb

Cr=chromium, Mb=molybdenum, Ni=nickel, S=sulfur, Si=silicon, V=vanadium

Al: Symbol for aluminum. This symbol often appears in capitals, such as AL, which is erroneous.

A-L 4750: An alloy nominally containing 50-53% iron and 47-50% nickel, having high permeability at low field strength and high electrical resistance.

alclad: Composite sheet produced by bonding either corrosion-resistant aluminum alloy of the duralumin type with high purity aluminum, thereby combining the strength of the alloy with the corrosion resistance of the pure metal, resulting in a structurally-stronger aluminum alloy.

Alfenol®: A family of iron-aluminum alloys.

12 Alfenol®: An alloy nominally containing 88% iron and 12% aluminum.

16 Alfenol®: An alloy nominally containing 84% iron and 16% aluminum.

Alger metal: A casting alloy nominally containing 89.7-90% tin, 10% antimony, 0-0.3% copper.

algicide: A chemical agent, such as copper sulfate, added to water-based cutting fluids to destroy algae.

Algiers metal: (1) Another name for Alger metal; (2) a name for a casting alloy nominally containing 94-95% tin, 5% copper, 0.5% antimony. Used for small bells.

aligning bearing: A bearing which, by virtue of its shape, is capable of considerable misalignment.

alignment variation, involute spline: The variation of the effective spline axis with respect to the reference axis.

alkali: (1) A term loosely applied to the hydroxides and carbonates of the alkali metals and alkaline earth metals (alkali metals), as well as the bicarbonate and hydroxide of ammonium. Any acid-destroying compound having a pH of 7-14. Strong alkalis (or bases) in the pH range of 12 to 14 are considered corrosive and will cause serious chemical burns to the skin, eyes, and mucous membranes. Alkalis turn litmus blue and neutralize acids to form salts. Widely used alkali substances include sodium carbonate [$CO_3 \cdot 2Na$] and potassium carbonate [$CO_3 \cdot 2K$], which are mild alkalies, and sodium hydroxide [NaOH] and potassium hydroxide [KOH], which are strong caustic alkalies; (2) a term referring to Group 1A of the Periodic Table of Elements.

alkali metals: Any one of the metal elements in group 1A of the periodic table: lithium, sodium, potassium, rubidium, cesium, and francium. These are the most strongly electropositive of the metals. All are corrosive in the presence of moisture and should be handled with care. They react vigorously, at times violently, with water. The density increases and the melting and boiling points become lower with increasing atomic weight. Cesium has the greatest atomic weight of the group.

alkaline earth metals: A term applied to the oxides of the metals of the alkaline earth group: calcium, barium, strontium, radium (Group IIA of the Periodic Table of Elements). Magnesium is sometimes included. In general, they are extrudable, malleable, and machinable. All are insoluble in water and form strong bases.

Allen screw: Specialized set screws, socket-head screws, and machine bolts containing a hollow head with an hexagonal-shaped hole that is used for tightening and loosening the screw and made to accept, and be adjusted by, an allen wrench.

Allen's metal: A bearing ally containing 55% copper, 40% lead, 5% tin, and small amounts of sulfur that may replace a portion of the lead.

Allen wrench: Also known as hex wrenches or hollow-setscrew wrench. Available in assorted sizes, these L-shaped hand wrenches are made of hexagonal-shaped bar stock of hardened steel, used to fit into hollow-headed Allen screws (safety set-screws), socket-head screws, and fasteners.

allotropic forms: Forms of substances, called allotropes, that differ in physical properties, but have the same chemical composition. Examples of well-known allotropes include carbon (coal, diamond, graphite, carbon black, charcoal), tin (gray and white), phosphorus (red and white), and iron (alpha iron, beta iron, etc). Uranium has three crystalline forms, manganese four. While most allotropes are those of crystalline metals, gaseous and liquid allotropes also exist. For example, diatomic oxygen (O_2) and ozone (O_3) are allotropes of molecular oxygen.

allotropy: Also known as polymorphism. The existence of certain elements in two or more forms which are identical in chemical composition, but significantly different in physical or chemical properties due to crystalline structure, density, etc. A common example is carbon, occurring as coal, diamond, carbon black, charcoal, graphite, and several amorphous forms. Various metals also have several allotropic forms that are often designated by Greek letters, e.g., α-, γ-, and Δ-iron.

allowance: (1) The amount of acceptable clearance, or the desired difference in dimensions, between mating parts. The minimum working clearance or maximum interference prescribed difference used to achieve various classes of fits between different parts. Allowances may be positive (sliding fit) or negative (force fit); (2) in screw threads, the prescribed difference between the design (maximum material) size and the basic size. It is numerically equal to the absolute value of the ISO term *fundamental deviation*. Not to be confused with tolerance.

alloy: A substance formed by melting any combination of metals together, or the mixture of one or more metals with non-metallic elements, as in cast iron or carbon steels. One metal is usually in much larger proportion than the others. The melting point of an alloy is constant for a definite composition and is commonly lower than any of the components. The properties of an alloy are usually different from those of its components. An alloying element is used to improve the properties of an alloy. The major classes of alloys are ferrous and nonferrous, depending on whether or not iron is a component. Where mercury is an ingredient, the product is called an amalgum.

1040 Alloy: An alloy nominally containing 72% nickel, 14% copper, 11% iron, 3% molybdenum, having high permeability at low field strength and high electrical resistance.

alloy cast iron: Describes castings containing alloying elements such as chromium, copper, molybdenum, nickel, and manganese in adequate amounts to appreciably change the physical properties, adding strength or special properties such as higher corrosion resistance, wear resistance, or heat resistance. Machinable alloy cast irons having tensile strengths up to 70,000 pounds per square inch or even higher may be produced. Used for automotive cylinders, pistons, piston rings, and brake drums; for application where the casting must resist scaling at high temperatures.

alloy, fusible: A low-melting point alloy melting in the range of approximately 50-260°C, usually containing bismuth, cadmium, lead, tin, or indium.

alloying element: A chemical element added to a metal to effect changes in physical properties and produce or improve a desired quality such as toughness, hardness, corrosion resistance, and tensile strength.

alloy steel: Any steel containing carbon and one or more alloying elements such as chromium, nickel, molybdenum, tungsten, or vanadium. These steels may also contain commonly accepted amounts of manganese, silicon, sulfur, and phosphorus which impart some distinctive mechanical or physical properties not found in plain carbon steels. Alloy steels comprise not only those grades that exceed the element content limits for carbon steel, but also any grade to which different elements than used for carbon steel are added, within specific ranges or specific minimums, to enhance mechanical properties, fabricating characteristics, or any other attribute of the steel. By this definition, alloy steels encompass all steels other than carbon steels; however, by convention, steels containing over 3.99% chromium are considered "special types" of alloy steel; these include the stainless steels and many of the tool steels. In a technical sense, the term alloy steel is reserved for those steels that contain a modest amount of alloying elements (about 1- 4%) and generally depend on thermal treatments to develop specific mechanical properties. Alloy steels are always killed, but special deoxidation or melting practices, including vacuum, may be specified for special critical applications. Alloy steels generally require additional care throughout their manufacture because they are more sensitive to thermal and mechanical operations. See also steel.

Almelec®: A proprietary trade name for an alloy containing 98.5% aluminum, 0.7% magnesium, 0.5% silicon, and 0.3% iron.

alnico: Generic name for carbon free ferrous alloys containing up to 15% aluminum, 20% nickel, 25% nickel, and small amounts of titanium, having outstanding properties of a permanent magnet.

aloin: A chemical used in electroplating.

Alox®: A registered trademark of Alox Corp. for a line of chemicals used as corrosion inhibitors, film-forming rust preventives, lubricity agents, and metalworking fluids.

Aloxite®: A registered trademark for a pure, crystalline, granular aluminum oxide [Al_2O_3] produced in an electric furnace. Used as an abrasive for grinding high tensile strength steel, as a reagent for measuring the carbon content in steel, a filtering material, and as a basic refractory material.

Aloxlube®: A registered trademark of the Alox Corporation for a series of chemicals used in extreme-pressure (EP) and lubricity additives. Used for making metalworking and other industrial fluids such as hypoid and other gear oils, cutting and grinding fluids, slideway lubricants, hydraulic fluids, and stamping and drawing compounds.

alpha (A or α): The first letter of the Greek alphabet used as a symbol (α) for angle, coefficient of linear expansion, spherical optical rotation, thermal diffusibility, angular acceleration, thermal coefficient of resistance, and electrical dissociation. In metallurgy it stands for the major allotropic form of a metal or alloy.

alpha brass: Alloys nominally containing 63-70% copper and 30-37% zinc, used for cold working. See also aluminum brass, admiralty brass, and gilding metal.

alpha iron: One of the four solid phases of pure iron. It is a soft, magnetic, incapable of dissolving carbon, and does not occur in twin crystals when viewed under the microscope; the major allotropic form of iron which is stable below 1,640°F/910°C. In iron-base carbon alloys, the transition temperature varies with the carbon content.

Alrak process: A chemical process used to form a corrosion-resistant oxide film on the surface of aluminum and aluminum alloys by immersion in a boiling solution of 5% soda ash [Na_2CO_3] and 1% sodium chromate(IV) [$CrO_4 \cdot 2Na$].

Alrok process: See Alrak process.

alternating current: An electric current that reverses its direction at regular recurring intervals.

alum: A widely used name for aluminum sulfate $[Al_2(SO_4)_3]$.

Alumail®: A proprietary trade name of Bayer AG for a range of metal surface coating products.

Aluman®: An aluminum alloy nominally containing 85-88% aluminum, 10-12% zinc, and 2% copper.

Alumel®: A trademarked name for a series of electrical, high-resistance nickel alloys containing approximately 94-98% nickel, 1-2% aluminum, 0.5-1% silicon, 2.0-2.5% manganese, and 0.5% iron. When used as thermocouples, this alloy may contain up to 2% chromium.

Alumilite®: A proprietary trade name for a surface coating used on aluminum products.

alumina: Aluminum oxide $[Al_2O_3]$, α-corundum. The natural aluminum oxide film formed on aluminum. See aluminum oxide.

aluminate: A term indicating a material or chemical compound containing aluminum as the major component.

aluminium: U.K. spelling of the metallic element aluminum (Al).

aluminizing: The process for improving the oxidation resistance by bonding an aluminum or aluminum alloy coating on a metal by hot dipping or spraying, or diffusion coating.

aluminum: CAS number: 7429-90-5. The most abundant metal in the earth's crust. Pure aluminum weighs about one third as much as iron and has a comparatively low tensile strength, about 13,000 psi. Used as an alloying material with copper, chromium, magnesium, manganese, nickel, lead, bismuth, iron, titanium, tin, or zinc. A strong deoxidizer used in steel alloys and improves the soundness of ingots and castings. Also present in steels made specially for nitriding. In powder form aluminum is flammable and highly reactive with oxidizers and many other materials.

Symbol: Al
Physical state: Silvery solid
Periodic Table Group: IIIA
Atomic number: 13
Atomic weight: 26.98154
Valence: 3

Density (g/cc): 2.703
Melting point: 1,220°F/660°C
Boiling point: 4,545°F/ 2,507°C
Source/ores: Bauxite, boehmite, gibbsite
Oxides: Al_2O_3
Crystal structure: f.c.c.
Brinell hardness: 27 (cold rolled 25%)

aluminum alloys: Aluminum containing variable amounts including bismuth, chromium, copper, lead, magnesium, manganese, nickel, silicon, tin, or zinc. Using these elements a wide variety of alloys may be obtained for specific qualities such as corrosion resistance, die-casting or working, heat-treatability, electrical transmission, and possessing a broad spectrum of strength, machinability, and weldability.

Aluminum Association (AAI): 900 19 St. N.W., Suite 300, Washington, DC 20006. Telephone: 202/862-5100. FAX: 202/862-5164. WEB: http://www.aluminum.org An industry association of producers of aluminum.

aluminum-base grease: A grease prepared from aluminum soap (aluminum oleate, aluminum palminate, aluminum resinate, aluminum stearate) and a lubricating oil.

aluminum brass: Alloys containing 59-76% copper, 21.50-40% zinc, and 0.3-2.50% aluminum, and sometimes small amounts of iron. Also an alpha brass containing 76% copper, 22% zinc and 2% aluminum.

aluminum bronze: A name for various alloys that are termed "light" and "heavy." Light alloys generally contain 82-89% aluminum and 10-18% copper. Heavy alloys contain from 81-95% copper, 9-11% aluminum, 0-4% nickel, and small amounts of iron, manganese, and/or tin. Properties include high strength, hardness, and ductility, as well as resistance to seawater, most chemicals, shock, and fatigue. The heavy alloys have moderate to difficult machinability. See also gold bronze. Aluminum bronze, 5% contains 95% copper and 5% aluminum. Aluminum bronze (1) (known as 9% aluminum bronze) contains 91% copper, 9% aluminum. Aluminum bronze (2) contains 81.5% copper, 9.5% aluminum, 5% nickel, 2.5% iron, and 1% manganese. Aluminum bronze (3) contains 91% copper, 7% aluminum, and 2% iron. Following is a description of some casting alloys: Alloy 9A, 88% copper, 3% iron, 9% aluminum. Alloy 9B, 89% copper, 1% iron, 10% aluminum. Alloy 9C, 85% copper, 4% iron, 11% aluminum. Alloy 9D, 81% copper, 4% iron, 11% aluminum, 4% nickel. Propeller Bronze, 82% copper, 4% iron, 98% aluminum, 4% nickel, 1% manganese.

Aluminum Extruders Council (AEC): 1000 N. Rand Road, Suite 214, Wauconda IL 60084. Telephone 847/526-2010. FAX: 847/526-2010. E-mail: @aec.com WEB: http://www.aec.org

aluminum iron brass: Alloys nominally containing 61.2% copper, 35.3% zinc, 2.3% aluminum, and 1.2% iron.

aluminum iron bronze: Alloys nominally containing 85-90% copper, 6-10% aluminum, and 3-7% iron.

aluminum magnesium bronze: Alloys nominally containing 90-94% copper, 5-10% aluminum, and small amounts (about 0.5%) of magnesium.

aluminum manganese alloy: Alloys nominally containing 97-98% aluminum and 2-3% manganese.

aluminum manganese bronze: Alloys nominally containing 89% copper, 9.6% aluminum, and 1-2% manganese.

aluminum nickel alloy: Alloys nominally containing aluminum and nickel in varying amounts.

X *aluminum oxide:* [Al_2O_3] generally called *alumina, emery,* or *corundum.* The natural oxide that forms a protective coating on aluminum metal (alumina). Possessing very definite qualities of hardness, toughness, and type of fracture. Aluminum oxide is used extensively as an abrasive for snagging work in foundries, OD and centerless grinding on bench and floorstand grinders, sharpening of circular saw blades, and cutting-off. Natural aluminum oxide is called corundum and usually contains variable amounts of impurities. Also known by the following chemical and proprietary trade names: Alcan AA-100; Alcan C-70, 71, 72 and 73; Alcoa F1; Alexite; Almite; Alon; Alon C; Aloxite; Alufrit; Alumina; α-Alumina; β-Alumina; γ-Alumina; Aluminite 37; Aluminum oxide; α-Aluminum oxide; β-Aluminum oxide; γ-Aluminum oxide; Aluminum oxide c; Aluminum sesquioxide; Aluminum trioxide; Alumite; Alundum; Alundum 600; Bauxite; Bayerite; Boehmite; Brasivol; C-1; Catapal S; Compalox; Conopal; Corundum; D 201; Dialuminum trioxide; Diaspore dirubin; Dispal alumina; Dispal M; Dotment 324; Dotment 358; Dural; Dycron; Emery; Exolon XW 60; F 360 (alumina); Faserton; Fasertonerde; G2 (oxide); Gibbsite; Hypalox II; KA101; Lucalox; Ludox CL; Microgrit WCA; PS-1 (alumina); Purdox; Realox *Also see* emery.

aluminum silicon alloy: An alloy nominally containing 86.88% aluminum, 12.74% silicon, 0.34% iron, 0.02% zinc, 0.010% copper, and small amounts (about 0.005%) of titanium and manganese.

aluminum silicon bronze: An alloy nominally containing 91% copper, 7% aluminum, and 2% silicon.

aluminum stearate: [$Al(C_{18}H_{35}O_2)_3$] A white powder or gray mass used in lubricants and cutting compounds.

aluminum-titanium alloy (1:1): An alloy containing 50% aluminum and 50% titanium.

aluminum tin bronze: An alloy nominally containing 85% copper, 10% tin, 5% aluminum, and 2% zinc.

Alundum®: A registered trademark for a pure, crystalline, granular aluminum oxide [Al_2O_3] Made by the fusion of bauxite in the electric furnace. Used for various grinding operations involving hard materials such as steel, as a reagent for measuring the carbon content in steel, as a filtering material, and as a basic refractory material.

amalgum: A semiliquid or solid mixture or alloy of mercury with one or more alloy, metal, or nonmetal, including cesium, gold, lithium, potassium, silver, sodium, tin, and zinc.

Amborite®: A registered trademark of the DeBeers Corporation for polycrystalline cubic nitride, used as a production abrasive.

Anti-Friction Bearing Manufacturers Association (AFBMA): formerly called American Bearing Manufacturers Association. 2025 M Street, NW, Washington DC 20036-2422. Telephone: 202/367-2422. FAX: 202/367-1155. Web: http://www.abma.dc.org

American Carbon Society (ACS): Penn State university, 205 Hosler Building, University Park PA. Telephone: 814/863-0594. FAX: 814/865-3248. Web: http://www.ems.psu.edu/carbon. Sponsors the Biennial American Carbon Conferences and the international journal *Carbon*. Devoted to the physics, chemistry, research, education, and dissemination of technology of organic crystals, polymers, chars, graphite, and carbon materials.

American Ceramic Society (ACerS): PO Box 6136, Westerville OH 43086-6136. Telephone: 614/890-4700. FAX: 614/899-6109. Web: http://www.acers.org Devoted to the scientific and educational advancement of theory and practice of technology related to glass, cerametallics, cements, refractories, nuclear ceramics, electronics, structural clay products, and white wares.

American Chemical Society (ACS): 1155 16th Street, Washington, DC 20036. Telephone 202/872-4600; 800/227-5558. FAX: 202/872-4615. Web: http://www.acs.org The professional society for chemists in the U.S. Publishes the weekly scientific publications, *Chemical Abstracts* and *Chemical & Engineering News.*

American Copper Council (ACC): 2 South End Ave., Suite 4C, New York, NY 10280. Telephone: 212/945-4990. E-mail: marcabb@aol.com

American Electroplaters and Surface Finishers Society (AESF): 12644 Research Parkway
Orlando, FL 32826. Telephone: 407/281-6441. FAX: 407/281-6446. Web: http://www.aesf.org

American Foundrymen's Society (AFS): 505 State Road, Des Plains, Illinois 60016. Telephone: 708/824-0181; 800/537-4237. FAX: 708/824-7848. Web: http://www.afsinc.org A technical society devoted to research, education, and dispersion of technology related to manufacture and utilization of castings.

American Gear Manufacturing Association (AGMA): 1500 King. St., Alexandria, VA 22314-2730. Telephone: 703/684-0211. FAX: 703-684-0242.

American Institute of Steel Construction (AISC): 1 Wacker Drive, Chicago, IL 60601-2001. Telephone: 312/670-2400. FAX: 312/670-5403. Web: http://www.aisc.org An industry association of fabricated structural steel manufacturers. Provides design information and publishes standards.

American Iron and Steel Institute (AISI): 1101 17th Street, N.W., Suite 1300, Washington, DC 20036. Telephone: 202/452-7100. FAX: 202/463-6573. Web: http://www.steel.org An industry association of iron and steel producers. Publishes various steel products manuals.

American National Standards Institute (ANSI): 11 W 42nd Street, New York, NY 10036. Telephone: 212/642-4900. FAX: 398-0023. Web: http://www.ansi.org

A federation of trade associations, professional groups, technical societies and consumer organizations which make up the clearinghouse for voluntary standards on the national level in the United States. ANSI is the U.S. member of the International Organization (ISO) for Standardization and the International Electrotechnical Commission (IEC). It is a nonprofit organization that publishes National Standards in cooperation with technical and engineering societies, trade associations, and government agencies.

American National Standard pipe thread: See American Standard Pipe Thread.

American Oil Chemists Society (AOCS): 2211 Bradley Avenue, PO Box 3489, Champaign, IL 61826-3489. Telephone: 217/359-2344. FAX: 217/351-8091. Web: http://www.aocs.org A technical society devoted to the advancement and diffusion of knowledge related to chemistry, physics, extraction, refining, safety, packaging, quality control, and applications for the use of animal, marine, and regular oils and fats.

American Petroleum Institute (API): 1220 L Street, NW, Washington, D.C. 20005. Telephone: 202/682-8000. FAX: 202/682-8029. Web: http://www.api.org A professional association of the petroleum industry devoted to research, education, and dissemination of technology and standards involved with welding, storage tanks, pipelines including welding and maintenance.

American Society of Mechanical Engineers (ASME): 3 Park Avenue, New York, NY 10019-5990. Telephone: 212/705-7722. FAX: 212/705-7674. Web: http://www.asme.org

American Society for Metals (ASM): See ASM International.

American Society for Nondestructive Testing, Inc. (ASNT): 1711 Arlingate Lane, PO Box 28518, Columbus, OH 43228-0518. Telephone: 614/274-6003. FAX: 614/274-6899. Web: http://www.asnt.org An engineering society devoted to the scientific and educational advancement of theory and practice of nondestructive test methods used to improve product quality and reliability.

American Society for Quality Control, Inc. (ASQ): 611 E. Wisconsin Avenue, P.O. Box 3005, Milwaukee, WI 53201-3005. Telephone: 414/272-8575. FAX: 414/272-1734. Web: http://www.asq.org An engineering society devoted to the creation, promotion, and diffusion of knowledge related to quality control science and its application to industrial products.

American Society for Testing and Materials (ASTM): 100 Barr Harbor Drive, West Conshohocken PA 19428-2959. Telephone: 610/832-9585. FAX: 610/832-9555. Web: http://www.astm.org The society operates via more than 125 main technical committees that function in prescribed fields under regulations that ensure balanced representation among producers, users, and general-interest participants. Organized in 1898 and first chartered in 1902. A scientific and technical organization for standards, materials, products, and systems. It is the world's largest source of voluntary consensus standards.

American Society of Tool Engineers (ASTE): See Society of Manufacturing Engineers (SME).

American Standard Dryseal Pipe Thread (NPTF): Also called American National Standard dryseal pipe thread. The external thread is tapered, while the internal thread may be straight or tapered. The latter is the stronger choice. Used for liquid- or gas-pressure joints where the use of pipe compound or sealing compound is unsuitable.

American Standard Pipe Thread: Also called American National Standard pipe thread. The most commonly used system of pipe threads. The system contains two types of pipe threads: (1) American Standard straight pipe thread (NPS); (2) American Standard taper pipe thread (NPT) and a variation of NPT called American Standard dryseal pipe thread (NPTF).

American Standard Straight Pipe Thread (NPS): Also called American National Standard straight pipe thread A non-tapered pipe thread. The diameter of the external and internal threaded surface is uniform over the length of the effective threads. A pipe compound is normally used with this kind of thread. Not generally used for water pipes.

American Standard tapered pipe thread (NPT): Also called American National Standard tapered pipe thread. A pipe thread with threads that are tapered 3/4 in (62.48 mm) per foot of length from the small end of the thread toward the large end. Both the external and internal threads are tapered and the farther the tapered threads are screwed together, the joint becomes tighter and more. The angle between the sides of the thread is 60°. Used to provide low-pressure seal against leakage of liquid and gasses. A pipe compound is normally used with this kind of thread.

American Tin Trade Association: P.O. Box 53, Richboro PA, 18954. Telephone: 215/504-9725. E-mail: americantintrade@mailcity.com

American Welding Society (AWS): 550 NW LeJeune Road, Miami, FL 33126. Telephone: 305/443-9353. FAX: 305/443-7559. Web: http://www.aws.org A nonprofit technical society devoted to advancing the art and science of welding. The AWS publishes codes and standards concerning all phases of welding. Publishes *The Welding Journal.*

American Zinc Association: 1112 16th St. N.W., Suite 240, Washington, DC 20036. Telephone: 202/835-0164. FAX: 202/835-0155. Web: http://www.zinc.org

amines: Rust inhibiting chemical agents added to cutting fluids. Chemical compounds derived from ammonia and having the chemical suffix -NH_2.

ammeter: An instrument used for measuring electric current in amperes.

ammonium chloride: [NH_4Cl]. A white crystalline solid or powder used as a soldering flux, and in electroplating. Also known as sal ammoniac.

ammonium fluosilicate: [$(NH_4)_2SiF_6$] A white crystalline powder used in electroplating.

ammonium hexachloroplatinate: [$Cl_6Pt \cdot 2H_4N$] Yellow powder or orange-red crystalline solid used in electroplating.

ammonium persulfate: [$(NH_4)_2S_2O_8$] A white crystalline solid used in electroplating.

ammonium silicomolybdate: [$(NH_4)4SiMo_{12}O_{40} \cdot xH_2O$] A yellow crystalline solid used as an additive in electroplating.

ammonium soap: A mixture of finely powdered rosin and a strong ammonia solution used as a flux for copper soldering.

ammonium sulfamate: [$NH_4OSO_2NH_2$] a white solid used in electroplating.

ammonium thiocyanate: [NH_4SCN] A colorless solid used in zinc coating and electroplating.

ammonium thiosulfate: [$(NH_4)2S_2O_3$] A white crystalline solid used as a brightener in silver electroplating baths and in cleaning compounds for zinc-based, die-cast metals.

amorphous: Noncrystalline; a solid substance devoid of regular atomic structure or definite geometrical shape.

amorphous carbon: See carbon.

amp.: Abbreviation for ampere (A).

Ampco®: A registered trademark of Ampco Metal Inc. for a series of alloys used for bushings, bearings, gears, slides, etc., containing 79.75-92.5% copper. 6-15% aluminum, 1.5-5.25% iron.

Ampcoloy®: A registered trademark of Ampco Metal Inc. for a line of industrial copper alloys including low-iron-aluminum bronzes, nickel-aluminum bronzes, tin bronzes, manganese bronzes, lead bronzes, beryllium-copper, and high-conductivity alloys.

Ampco-Trode®: A registered trademark of Ampco Metal, Inc. for a line of arc-welding electrodes and filler rods made of aluminum-bronze.

ampere: The constant current that produces a force of 2×10^{-7} N/m length between two parallel conductors placed 1 m apart in a vacuum.

Ampvar®: A proprietary name of Atlas Minerals and Chemical Co for a synthetic-resin metal conditioner used to prepare metal surfaces prior to the application of corrosion-proof coatings.

p-anisaldehyde: [$C_8H_8O_2$] A colorless oil used in electroplating.

angle gage blocks: Precision tools made from hardened tool steel that is ground and lapped to precision tolerances, used to accurately measure and inspect angles. Like rectangular gage blocks, these angle blocks can be wrung together in various combinations. Complete sets contain 16 blocks measuring 4 in. on the base with 5/8 in. thickness that can be used to accurately measure 356,400 angles ranging from 1 second up to 99 degrees. See also gage blocks, rectangular.

angle of cutter entry: An angle, typically referred to as being either negative or positive, determined by the position of the cutter centerline relative to the edge of the workpiece.

angle of drill point: The commercial standard for angle of drill point: 118° included angle, 12° to 15° lip clearance. The angle of chisel point 125° to 135° is best suited for drills engaged in average classes of work.

angle of thread: The angle included between the flanks of the thread measured in an axial plane.

angle plate: An L-shaped tool having two adjacent faces accurately ground at right angles (90°) to each other, used with clamps to hold a workpiece for machining, layout, and inspection of work. See also adjustable angle plate.

angstrom (Å): A linear unit of wavelength; 10^{-8} mm, or one-250 millionth inch. Also 1 ten-thousandth micron or 1 ten billionth of a meter. The angstrom is defined in terms of the wavelength of the red line of cadmium (6438.4696 Å) in air @ 59°F/15°C. Used to describe distances between atoms, wavelengths of short-wave radiation, etc. Occasionally still used in science.

angular gear: A kind of gear having beveled teeth used for transmitting rotary motion at an angle; made for shafts whose center lines meet, but are not at right angles to each other. See also bevel gear.

anhydrous: A term meaning "without water."

anion: A negatively charged ion of an electrolyte which tends to move toward and collect at the anode (positive pole) when subjected to electric potential.

anisotropic: A crystallography term meaning that the physical properties of a material depend on the direction of measurement. Various degrees of anisotropy exist, depending on the amount of symmetry of the material or component shape. For example, cast metals and extruded plastics tend to be isotropic so that samples cut in any direction within a cast body tend to have the same physical properties. However, rolled metals tend to develop crystal orientation in the direction of rolling so that they have different mechanical properties in the rolling and transverse-to-rolling directions. Extruded plastics film also may have different properties in the extruding and transverse directions so that these materials are oriented biaxially and are anisotropic. Composite materials that have fiber reinforcements carefully oriented in the direction of applied loads, surrounded by a plastics matrix, have a high degree of property orientation with direction at various points in the structure and are anisotropic. In addition to describing mechanical properties, anisotropy is also used in referring to the way a material shrinks in the mold. Anisotropic shrinkage is important in molding crystalline and glass-fiber-reinforced materials for which shrinkage values are usually listed for the flow direction and the cross-flow direction. These values are of most concern to the tool designer and molder, but the existence of anisotropy and its severity must be considered when a material is chosen for a part having tight tolerances.

anisotropy: The characteristic of exhibiting different values of physical properties in different directions with respect to a fixed reference system in the material.

annealing: A heat-treatment process that decreases the hardness and brittleness of metal (and other crystalline materials) by relieving internal stresses by recrystallization. The work is heated and held at a definite high temperature and then allowed to cool at a relatively slow rate. In addition to removing hardness, the process may also result in grain refinement and improved mechanical properties. The annealing temperature and the cooling rate depend on the material being treated and the particular purpose and effects that are desired in the finished material. Certain more specific heat treatment of iron-based alloys covered by the term annealing are black annealing, blue annealing, box annealing, bright annealing, cycle annealing, flame annealing, full annealing, intermediate annealing, isothermal annealing, quench annealing, recrystallization annealing, graphitizing, malleabilizing, process annealing, and spheroidizing. The object of annealing is to alter mechanical or physical properties, remove strains, to produce a definite microstructure in the metal and to render it soft enough for machining.

annular: Ring shaped.

anode: The positive electrode of an electrolytic cell, to which negatively charged ions travel when an electric current is passed through the cell. In a battery, the anion is the negative electrode (usually zinc) from which current is drawn and at which oxygen is generated.

anode mud or slime: An insoluble residue formed on the anode of refining or plating baths. During the electrolytic refining of copper, for example, this residue may contain platinum, silver, gold, or other relatively inert metals or rare elements that can be profitably recovered.

anodic coating: See anodizing.

anodic oxidation: An electrochemical process in which objects such as aluminum are given a thin, corrosion-resistant surface film of oxide by making them the anodes in an oxidizing electrolyte.

anodize: To apply a process of anodic oxidation, usually by connecting as an anode the object being processed, in an oxidizing electrolyte.

anodizing: Also know as anodic oxidation. A thin corrosion-resistant oxide film will form naturally on some metal surfaces on exposure of the metal to air. Anodizing is the process of treating the surfaces of aluminum, magnesium, and certain of their alloys with stable and artificially thickened films of oxides. The metal being coated acts as the anode in an electrolytic bath of sulfuric, oxalic, and chromic acids. As the current is passed, oxygen evolves and thickens the oxide coating on the treated metal to about 0.001 in., depending on resistance. These oxidation coatings are hard, provide good electrical insulating properties, and can be dyed to obtain a wide range of attractive colored finishes, including black. Anodized coatings can also be used as preparatory treatments to electroplating. Plating metals such as cadmium, copper, iron, nickel, and silver have been successfully deposited over anodic oxide coatings.

ANSI: Abbreviation for American National Standards Institute.

antifreeze: A substance, such as ethylene glycol, that is added to water to prevent freezing.

antifriction alloy: Any soft alloys nominally containing antimony, arsenic, copper, lead, lead-silver, tin, or zinc that are antifriction and nonseizing, used especially for bearings. See also Babbit metal.

antifriction bearing: A term use to describe ball and roller bearings to distinguish them from plain, sliding bearings, and other nonrolling element bearings. This kind of bearing has closer tolerances than do plain bearings and is used where high speeds, precision, and heavy loads are encountered.

antimonial alloys: Alloys containing antimony. Sb is used to harden alloys for bearings (nominally 75-80% antimony, 5-25% copper, and 5% tin), and to harden lead alloys for use in bearings, batteries, and chemical plants.

antimonial lead: Alloys nominally containing approximately 1-28% antimony. Used for cable sheathing, storage battery grids, electrical applications, and pipes for industrial purposes where high strength and corrosion resistance is required. See also hard lead.

antimonial tin solder: Alloys nominally containing 95% tin, 5% antimony. Used for the joining of copper alloys, joints in copper tubing, and electrical equipment.

antimony: A bright, hard, brittle, silvery metalloid. Used as a hardening alloy in lead, bearing metals, solder, Britannia metal, pewter, and type metal. Antimony and many of its compounds are highly toxic. Antimony is resistant to attack by ammonia, carbon dioxide, and hydrofluoric acid.

Symbol: Sb (from stibium)	**Density (g/cc):** 6.62
Physical state: Silvery solid	**Specific heat:** 0.0503
Periodic Table Group: VA	**Melting point:** 1,168°F/631°C
Atomic number: 51	**Boiling point:** 2,975°F/1,635°C
Atomic weight: 121.75	**Source/ores:** Stibnite, ullmanite, valentinite
Valence: 3,4,5	**Crystal structure:** Rhombohedral (gray)
	Brinell hardness: 30-58

antistat: A material that prevents the formation of static electricity.

antimony lead: See antimonial lead and hard lead.

anvil: A block of cast steel or iron, with an upper horizontal surface of hardened steel on which metals are forged or hammered. A square hole is usually provided to hold blacksmith's tools such as fuller blocks, hardies, etc.

AOD process: The process of injecting an argon and oxygen mixture into molten steel to reduce carbon impurities.

API: Abbreviation for American Petroleum Institute.

API gravity: A scale generally used in the petroleum and lubrication industry and established by the American Petroleum Institute. This scale is based on the unit called the API degree, and is defined in terms of specific gravity as follows: [(141.5/specific gravity @ 60°F) – 131.5]; e.g., Arabia oil (light = 34°; heavy = 27°).

Applix®: A registered trademark of Union Butterfield for spiral point taps for stainless steel.

apprentice: One who learns a trade or art by practical experience under skilled workers.

approach ratio: In gear design, the ratio of the arc of approach to the arc of action.

apron: A protective or covering plate used to enclose a mechanism, as in the apron of a lathe.

aq: The standard abbreviation for aqueous.

Aquadag®: A registered trademark for a lubricant made from a colloidal suspension of graphite in water.

aqua regia: A yellow, fuming, corrosive mixture normally containing (by volume) three parts of hydrochloric acid, and one part of nitric acid [HNO_3]. A solvent for noble metals.

aqueous: Watery or water-like. A solution in which a solvent is water. The standard abbreviation is aq.

aqueous fluids: Fluids containing water.

Ar: The symbol for the chemical element argon.

Aradite®: A registered trademark for an epoxy resin used for castings and adhesives.

arbor: (1) A machining term for a shaft, spindle, or bar used for holding, supporting, and driving cutting tools for milling, grinding, drill press, bandsawing, or other machining operations. An arbor frequently has a taper shank fitting the spindle of a machine tool. An arbor is not the same as a mandrel (a workholding device), and the terms should not be used interchangeably; (2) a foundry term describing reinforcements in sand molds made from metal.

arbor press: A machine used for driving and removing arbors, shafts, or mandrels from workpieces.

arc: (1) A section or portion of a curved line; (2) an electric current jumping between two electrodes.

arc blow: The deflection of an electric arc from its normal path due to magnetic forces.

architectural bronze: An alloys nominally containing 57% copper, 40% zinc, and 3% lead.

arc of action: In gear design, the arc of the pitch circle through which a tooth travels from the first point of contact with the mating tooth to the point where contact ceases.

arc of approach: Arc of the pitch circle through which a tooth travels from the first point of contact with the mating tooth to the pitch point.

arc of recess: In gear design, the arc of the pitch circle through which a tooth travels from its contact with the mating tooth at the pitch point to the point where its contact ceases.

arc voltage: The voltage across the welding arc.

arc welding: A term describing group of surface fusion processes that use electric energy from an arc between electrode and workpiece to produce intensive heat. Direct current is used and the piece being welded is usually the positive terminal; the welding rod is the negative. Various processes use carbon electrode, metal inert gas (MIG), or tungsten inert gas (TIG).

AREA: Abbreviation for American Railway Engineering Association.

argenite: Also known as silver glance. An ore of silver.

argentalium: An alloy of aluminum containing antimony.

argentum: Latin for silver, and source of the chemical symbol for silver, Ag.

Argentine metal: A casting alloy nominally containing 85% tin, 15% antimony.

arithmetical average (AA): The mean between an aggregate of values. See roughness average.

Armaloy®: A proprietary name for a nonferrous cast alloy used as brazed tips on tool shanks, as removable tool bits, as inserts in toolholders and milling cutters.

Armco® 48: An alloy nominally containing 50-53% iron and 47-50% nickel, having high permeability at low field strength and high electrical resistance.

Armco® ingot iron: A registered trademark for commercially very pure iron containing less than 0.1% total impurities.

arsenic: A metalloid element. Toxic and carcinogenic. Used as an alloying element, especially with lead and copper to impart specific physical properties; in copper to improve oxidation resistance, creep resistance, and high-temperature strength; in brasses to improve resistance to dezincification; in lead alloys to add hardness, etc.

Symbol: As
Physical state: Gray solid
Periodic Table Group: VA
Atomic number: 33
Atomic weight: 74.92
Valence: 2,3,5

Density (g/cc): 5.78 (α- metallic)
Melting point: 1,503°F/817°C
Boiling point: Sublimes ca.1,135°F/613°C
Oxides: As_2O_3, As_2O_5
Crystal structure: Rhombohedral (α); hex. (β)
Mohs hardness: 3.5

arsenical bronze: Alloys nominally containing 80% copper, 10% tin, 9.25% lead, and 0.75% arsenic.

arsenical copper: A copper alloy nominally containing up to 0.5% arsenic to improve oxidation resistance and creep strength.

arsenical lead: See lead, arsenical.

arsenical lead babbitt: A lead-base babbitt metal containing up to 3% arsenic. Examples include: SAE 15 containing 83% lead, 15% antimony, 1% tin, and 1% arsenic. "G" Babbitt containing 83.5% lead, 12.75% antimony, 0.75% tin, and 3% arsenic.

artificial abrasives: Also known as synthetic abrasives or manufactured abrasives. With the exception of diamond (Knoop hardness about 7,000), man-made abrasives are harder and have greater shock resistance than natural abrasives. These include aluminum oxide [Al_2O_3] (Knoop hardness about 2,000), boron carbide [B_4C](Knoop harness about 2800), silicon carbide [SiC] (Knoop hardness about 2,500), cerium oxide, and fused alumina.

As: Symbol for arsenic.

ASA: Abbreviation for American National Standards Institute.

ASCII: Pronounced "ass-key." An acronym for American Standard Code for Information Interchange. One of the binary coded decimal systems used to write N/C programs.

ASD: Abbreviation for Association of Steel Distributors.

ASHRE: Abbreviation for the American Society of Heating, Air Conditioning and Refrigerating Engineers.

ASM International (ASM): 9639 Kinsman, Materials park, Novelty, OH 44072. Telephone: 440/338-5151; 800/336-5152. FAX: 440/338-4634. Web: http://www.asm.intl.org

ASME: Abbreviation for the American Society of Mechanical Engineers.

ASNT: Abbreviation for American Society for Nondestructive Testing, Inc.

asphyxiant gas: A gas that may not be toxic but can cause unconsciousness and/or death by replacing air and thus depriving an organism of oxygen. Asphyxiant gases include acetylene, argon, carbon dioxide, neon, nitrogen, nitrous oxide (laughing gas), helium, hydrogen, fluorocarbon-114, butane, butene, β-butylene, ethane, ethylene, methane, liquefied petroleum (LPG), and other hydrocarbon gases.

ASQC: Abbreviation for American Society for Quality Control, Inc.

Association of American Railroads (AAR): 50 F Street, NW, Washington, D.C. 20001. Telephone: 202/639-2100. FAX: 202/639-2989. WEB: http://www.aar.org An industry association of railroads. Among other things, it publishes specifications for rolling stock and welding qualifications.

Association of Industrial Metalizers, Coaters and Laminators (AIMCAL): 5005

Association of Iron and Steel Engineers (AISE) 39

Rockside Rd., Ste. 600, Cleveland, OH 44131; Telephone: 216/573-3773; FAX: 216/573-3783. WEB: http:// www.aimcal.org

Association of Iron and Steel Engineers (AISE): 3 Gateway Center, Suite 1900, Pittsburgh PA 15222. Telephone: 412/281-6323. FAS: 412/281-4657. WEB: http://www.aise.org

Association of Steel Distributors (ASD): 401 N. Michigan Avenue, Chigago IL 60611. Telephone: 312/644-6610. FAX: 312/527-6705. E-mail: asd@ssba.com WEB: http://www.steeldistributors.org/asd/.

ASTM: Abbreviation for American Society for Testing and Materials.

Astroloy®: A nickel-based cobalt alloy containing 47-59% nickel, 17-20% cobalt, 13-17% chromium, 4.5-5.7% molybdenum, 3.7-4.7% aluminum, 3-4% titanium, 0-1 iron, and 0-0.1% carbon.

atm: Abbreviation for atmosphere.

atmosphere: (1) The air or gases surrounding the earth; (2) the standard barometric pressure of 14.696 pounds per square inch; the pressure exerted by the air at sea level that will support a column of mercury (Hg) 760 mm high (about 30 in.) at 32°F/0°C at sea level. The standard abbreviation is atm. See also air.

atom: The smallest chemically indestructible particle of an element that can enter into chemical combination, and that remains during chemical reaction. All chemical compounds are formed of atoms, made up of particles of matter called protons, neutrons, and electrons. The difference between compounds is attributable to the nature, number, arrangement of their constituent atoms, and packing fraction (an energy concept indicating stability of the forces binding nuclear particles).

atomic-hydrogen welding: An electric arc welding process using intense heat (flame temperature of 7,232-9,032°F/4,000-5,000°C) generated from an arc between two tungsten or other suitable electrodes in a hydrogen atmosphere. The use of pressure and filler metal is optional. The industrial use of this method has decreased in favor of inert gas-welding techniques, and also due to the dangers related to hydrogen embrittlement.

atomic number: The number of free unit positive charges (protons) carried in the nucleus of an atom.

atomic weight: The official international standard for atomic weight is the average weight (or relative atomic mass) per atom of an element, compared with 1/12 of the mass of an atom of the nuclide ^{12}C (the 12 isotope of carbon) taken at precisely 12.000.

atomization: The breaking up or dispersion of molten metals into small particles by using compressed gas and used in the production of powder metals and meal coating.

ATR alloy: Alloys nominally containing 99% zirconium, 0.5% copper and 0.5% molybdenum used in reactors.

attribute gage: A simple gage that is used to confirm that a dimension of a part or workpiece is within preset limits.

Au: Symbol for the element gold. From the Latin, aurum.

Auer metal: An alloy nominally containing 65% misch metal and 35% iron.

auric compounds: Gold compounds.

auriferous: Containing gold.

austempering: A process for heat-treating ferrous alloys that have been previously heated above the transformation range [austenitizing temperature (Ac_3)]. Following heating the alloy is held at uniform temperature in a quenching bath medium having a suitably high rate of heat abstraction, and maintaining the alloy, until the transformation is converted into austenite. The temperature for austenite transformation is below that of pearlite formation and above that of martensite formation, and is chosen based on desired properties. Austempering has been applied chiefly to steels having 0.60% or more carbon content with or without additional low-alloy content, and to pieces of small diameter or section, usually under 1 in., but varying with the composition of the steel. Case-hardened parts may also be austempered.

austenite: An interstitial solid solution of one or more elements in which face-centered cubic iron (α-iron) acts as the solvent. Unless otherwise designated, the solute is generally assumed to be carbon. Austenite is ductile, non-magnetic, and highly resistant to many forms of corrosion.

austenitic alloys: Alloys of iron, chromium, and nickel having an austenitic structure at room temperature after slow cooling in normal air, noted for their corrosion resistance.

austenitic manganese steel: Alloys nominally containing 85.8-86.8% iron, 12-13% manganese, and 1.2% carbon. Some commercial manganese alloys contain 10-14% manganese and 1-1.4% carbon and may also contain up to 0.3% silicon, 0.08% sulfur, and 0.8% phosphorus.

austenitizing: A process of heating a ferrous alloy to a temperature that changes the structure to austenite. Heating a ferrous alloy into the transformation range (Ac_1) results in partial austenitizing or above the transformation range (Ac_3) results in complete austenitizing. When used without qualification, the term implies complete austenitizing.

Australian gold: An alloy of 91.67% gold and 8.33% silver used in coinage.

autofrettage: Prestressing a hollow metal cylinder by the use of momentary internal pressure exceeding the yield strength.

autogenous welding: A type of fusion welding in which metals are joined without compression or hammering, and without the use of welding rod or flux.

autoignition temperature: The minimum temperature at which a substance will ignite spontaneously, or cause self-sustained combustion in the absence of any heated element, spark, or flame. The closer the autoignition temperature is to room temperature, the greater the risk of fire. An example would be white phosphorus with an autoignition temperature of 86°F/30°C. This chemical ignites spontaneously on exposure to air.

automatic centerless grinding: A centerless grinding machine equipped with a magazine, gravity chute, or hopper feed, provided the shape of the part will permit using these feed mechanisms, used for the grinding of relatively small parts. See also centerless grinding.

automatic center punch: A metal punch that contains an internal spring-controlled hammer. When sufficient pressure is applied to the handle of the punch by hand (as no hammer blow is used) the spring is compressed until the striker is released and the point leaves its mark on the metal workpiece. The force of the striker can be adjusted; and, when set, all marks made at that setting will be uniform in size. Used for making a mark where a fine degree of accuracy is required.

auxillary anode: In electroplating, a supplementary anode used to get better plate distribution and improve uniformity of plating thickness.

average: The computed or estimated median figure between two or more values (such as height, weight, or pressure).

avg.: Abbreviation for average.

Avogadro's number: The number of molecules in 1 mole (gram-molecule) of a substance. (Equal to 6.0247 x 10^{23}). Based on Avogadro's Law: *equal volumes of gases under like conditions of pressure and temperature contain the same number of molecules* formulated by the Italian scientist, Amadeo Avogadro (1776-1856).

AWI: Abbreviation for the Australian Welding Institute.

awl: A hand tool for making holes or marking the surface of a workpiece. Depending on intended use, the blade and point of an awl may have different shapes.

AWS: Abbreviation for American Welding Society.

axial internal clearance: The measured maximum possible movement parallel to the bearing axis of the inner ring in relation to the outer ring.

axial load bearing: A bearing in which the load acts in the direction of the axis of rotation.

axial pitch: In gear design, the distance measured parallel to the center line of the gear.

axial plane: In a pair of gears it is the plane that contains the two axes. In a single gear, it may be any plane containing the axis and a given point.

axis: (1) An imaginary line passing through the center of an object, around which all parts of the object can rotate and are symmetrical; (2) a fixed line, real or imaginary, along which distances are measured or to which positions of an object are referred; (3) a drill term used to describe the imaginary straight line which forms the longitudinal centerline of the drill; (4) a file term used to describe an imaginary line extending the entire length of a file equidistant from faces and edges.

axis of thread: In screw thread design, a term used to describe an axis coincident with the axis of its pitch cylinder or cone.

B: Symbol for the element boron.

Ba: Symbol for the element barium.

Babbit metal: A general term for a group of soft alloys widely used as cast, machined, or preformed bimetallic bearings in the form of a thin coating on a steel base. The main types contain antimony, lead, lead-silver, lead-bronze, cadmium, tin, or arsenic, and small percentages of copper and bismuth. High grade alloys consist of tin, copper and antimony; low grade alloy is made from lead, copper, and antimony. Trade names include Bearite®, Bearium®. Discovered in 1839 by Isaac Babbitt, a Boston goldsmith, they have good bonding characteristics with the substrate metal, maintain oil films on their surfaces, and are nonseizing and antifriction. See also arsenical babbitt.

back, file: The convex side of a file having the same or similar cross-section as a half-round file.

backlash: (1) Wear in gear, screw mechanism, or other moving parts that can result in lost motion, slippage, or *end play* causing vibration; (2) a possibly dangerous reaction involving the sudden release of potential energy of a body in motion when the body stops, usually causing the body to quickly reverse directions. *Down milling* on milling machines require a backlash eliminator.

back rest: In turning and grinding, a tool used to support slender work.

back taper: The slight decrease in diameter from point to back in the body of the drill.

bactericides: Pesticides or chemical agents that are destructive to bacteria and used to retard or control their growth in cutting fluids.

baddeleyite: A natural, chemical resistant, zirconium oxide (ZrO_2). Used for corrosion prevention, and a natural source of zirconium.

Bahnmetall: A general name for lead-base alloys used as bearing metals and typically containing 98.45% lead, 0.7% calcium, 0.6% sodium, 0.2% aluminum, and 0.05% silicon, lithium, or nickel.

bainite: A decomposition product of austenite consisting of an aggregate of ferrite and carbide. In general, it forms at temperatures lower than those where very fine pearlite forms and higher than that where martensite begins to form on cooling. Its appearance is feathery if formed in the upper part of the temperature range; acicular (needlelike), resembling tempered martensite, if formed in the lower part.

baking: A low-temperature heat-treatment used to remove entrained gases from a metal. Used particularly to reduce the hydrogen content of pickled or electroplated high-strength steels.

balance, dynamic: See dynamic balancing.

balance, static: See static balancing.

ball-and-ring method: A method for determining the melting point of a material. The aperture of a metal ring is filled with the material to be tested, and a metal ball is placed on it for weight. The material is then heated to the temperature at which the material softens and is pushed out of the ring by the weight of the ball. It is used for materials that soften before melting, but having no absolute melting point.

ball bearing: An antifriction bearing that contains loose hardened steel balls that convert sliding friction into rolling friction. The balls fit between a race and journal.

ball-bearing steels: A chrome-steel alloy nominally containing 97-97.5% iron, 1-1.5% chromium, 1% carbon, and 0.5% manganese or 86-87.6% iron, 12% chromium, and 0.4-2.0% carbon. Depending on the alloy formula, these steels are oil quenched from 1,490-1,742°F/810-950°C and tempered at 302-392°F/150-200°C.

ball end mill: A shank-type end milling tool used for milling pockets, fillets, and slots with rounded bottoms, and die-making. These mills have end-cutting teeth that can be used for plunge milling or longitudinal milling. Two-fluted and four-fluted ball end mills with end- and center-cutting lips are also available, and used for similar operations.

ball-peen hammer: The most commonly used machinist's hammer. The head is made of tool steel having one face that is slightly convex (called the flat face), and the other a ball face or peen. The flat face is used for driving center punches, chisels, and for various other general purposes, the peen (or rounded) end is used for purposes such as flattening rivet heads. Available in many sizes from 1 oz (28.35 grams) to 3 lb. (1.36 kg), the 16 oz. (453.6 grams) size is generally preferred for general shop work.

ball-pein hammer: See ball-peen hammers.

ball reamer: A specialized, hemispherical reamer used for finishing the cup-like, receiving recess of a ball joint. See also reamer.

band polisher: A polishing machine having a continuous band with abrasive materials bonded to the surface.

bandsaw: A vertical or horizontal sawing machine having a continuous blade (called a band) of various widths and cutting teeth configurations. The horizontal bandsaw is specially designed for stock cutoff and is available as portable. The vertical bandsaw can be used for straight-line cuts, angular cuts, and contour cuts; consequently, it is also called a *contour bandsaw*. A fine pitch is used for sawing soft metals or large thicknesses and a finer pitch should be used for sawing thinner or harder metals. Also available are blades without teeth, containing a sharp knife-like or scalloped edge, blades containing short sections of file blade used for power filing, and blades containing abrasives used for band polishing. The blade without teeth is called a knife bandsaw and is used to cut soft, seal-type material such as rubber, textiles, cork, and some plastics. Also available to cut soft materials and light metals is a spiral blade, available in various diameters from 0.020 (0.51 mm) to 0.074 in (1.88 mm). The band is driven by an electric motor and travels over multiple wheels or pulleys. See also electrobandsaw.

barberite: Alloys nominally containing 88.5% copper, 5% nickel, 5% tin, 1.5% silicon. Highly resistant to seawater, moist sulfurous atmospheres, and sulfuric acid [H_2SO_4] in concentrations up to about 60%.

barium: An alkaline earth element. Toxic. An extrudable and machinable metal, used as a deoxidizer for copper. Soluble compounds are poisonous.

Symbol: Ba	**Density (g/cc):** 3.5
Physical state: Silvery solid	**Specific heat:** 0.068
Periodic Table Group: IIA	**Melting point:** 1,334°F/729°C
Atomic number: 56	**Boiling point:** 2,084°F/1140°C
Atomic weight: 137.327	**Source/ores:** Barite (baryte), witherite
Valence: 2	**Oxides:** BaO
	Crystal structure: b.c.c.

barium carbonate: [$BaCO_3$] A white powder used in case hardening baths, and ferrite manufacturing.

barium cyanide: [$Ba(CN)_2$] A white crystalline powder used in electroplating and metallurgy. Highly poisonous.

barium hypophosphite: [$BaH_4(PO_2)_2$] A white, crystalline powder used in nickel electroplating.

Barium XA®: A registered trademark for a product used by manufacturers of high quality tool steels. Eliminates chain-type occlusions and degasifies the steel.

bark: The decarburized layer just beneath the scale that results from heating steel in an oxidizing atmosphere.

bar peeling: A heavy machine method of feeding bars continuously through a rotating head equipped with three or more carbide cutting tools. The peeling operation produces bar stock of consistent dimensional accuracy and good surface.

base: A substance that reacts with acids to form salts and water. All bases create solutions having a pH of more than 7.0, the neutral point, and may be corrosive to skin and other human tissue. Strong bases (or alkali substances) in the pH range of 12 to 14 are considered corrosive and will cause chemical burns to the skin, eyes and mucous membranes. Alkalis turn litmus blue. Widely used industrial alkali substances include sodium carbonate [$CO_3 \cdot 2Na$], potassium hydroxide [KOH] and potassium carbonate [$CO_3 \cdot 2K$].

base circle, gear: The circle from which an involute tooth curve is generated or developed.

base diameter, involute spline (D_b): The diameter of the base circle.

base helix angle, gear: The angle, at the base cylinder of an involute gear, that the tooth makes with the gear axis.

base metal: (1) The primary metallic element of an alloy to which other elements are added (e.g., copper in brass); (2) the original core metal to which coating of plating or cladding is applied; (3) in welding, the metal composition of the pieces to be joined.

base pitch, gear: In an involute gear it is the pitch on the base circle or along the line of action. Corresponding sides of involute teeth are parallel curves, and the base pitch is the constant and fundamental distance between them along a common normal in a plane of rotation. The normal base pitch is the base pitch in the normal plane, and the axial base pitch is the base pitch in the axial plane.

basic dimension: A numerical value used to describe the theoretical or nominal perfect size, form, orientation, or location of a feature or datum target from which all permissible variations are established by tolerances on other dimensions, in notes, or in feature control frames.

basic hole system: A system of fits in which the design size of the hole is the basic size and the allowance, if any, is applied to the shaft.

basic oxygen process (BOP): A process for speeding up the chemical reaction for steelmaking by using pure oxygen under high pressure; the oxygen is injected into the molten iron ore, pig iron, and scrap iron mixture. Also known as the basic oxygen converter process.

basic profile (of thread): A thread design term used to describe the cyclical outline, in an axial plane, of the permanently established boundary between the provinces of the external and internal threads. All deviations are with respect to this boundary.

basic shaft system: A system of fits in which the design size of the shaft is the basic size and the allowance, if any, is applied to the hole.

basic size: (1) The theoretical or nominal standard size from which all variations are made; (2) that size from which the limits of size are derived by the application of allowances and tolerances; (3) the size to which limits of deviation are assigned. The basic size is the same for both members of a fit.

basic space width, involute spline: The basic space width for 30° pressure angle splines; half the circular pitch. The basic space width for 37.5° and 45° pressure angle splines, however, is greater than half the circular pitch. The teeth are proportioned so that the external tooth, at its base, has about the same thickness as the internal tooth at the form diameter. This proportioning results in greater minor diameters than those of comparable involute splines of 30° pressure angle.

basis brass: See brass.

basis metal: A term in electroplating that describes any one of a number of metals being plated; the cathode. See also base metal.

bastard cut: A grade of file coarseness between coarse and second cut of American pattern files and rasps. *See also* bastard file.

bastard file: A file between coarse and second cut having approximately 30 teeth per inch; used for fast removal of work material. By comparison, a coarse or rough-cut file has approximately 20 teeth per inch and a second-cut file has approximately 40 teeth per inch. The name is a misnomer in that the file was invented by an Englishman named Barsted. When English workers came to the United States and requested a Barsted file, Americans thought this was the English pronunciation for "bastard." (Raabe)

bauxite: A native aggregate of aluminum-bearing minerals in which the aluminum occurs largely as hydrated aluminum oxide $[Al_2O_3 \cdot 2H_2O]$. A gray-red, claylike mineral, the purest form of aluminum oxide found in nature. Properties: Mohs hardness 1-3; Density 2-2.55.

BBIM: Abbreviation for Brass and Bronze Ingot Manufacturers Association.

Be: Symbol for the element beryllium.

bead: A reinforcing narrow ridge formed in a sheet-metal workpiece or part.

bearing: The support in which a moving part such as a shaft, journal, or pin revolves. May be plain, ball, or roller. Ball and roller bearings are called anti-friction and contain a number of balls or rollers interposed between the shaft and the support. Although the terms bearing and bushing are often used synonymously, a bearing is a full-round or semicircular (half-bearing), cylindrical part containing a special material or lining alloy that is bonded to a steel, bronze, or similar supporting backing material. Bushings are full-round cylindrical parts made from one special alloy intended to provide a cushion between moving metal parts or other specialized use.

bearing bronze: Copper-tin alloys nominally containing 5-20% tin, 0-20% lead and small amounts of phosphorus, the remainder being copper. This material has anti-frictional properties making it useful for bearings, worm wheels, axles, etc.

bearing metal: A general term for the various alloys used for their anti-frictional properties in bearings. See also bearing bronze, Babbitt metal.

bearing scraper: A specially shaped, hardened steel, hand tool that is slender, point, and V-shaped with two cutting edges. Used to remove sharp internal edges, burrs, and for scraping the surface of cylindrical bearings.

Bearite®: A registered trademark for a series of lead-base alloys nominally containing 17% antimony and small amounts of bismuth and copper. Used for bearings.

Bearium®: A registered trademark for high-lead bronze alloys nominally containing 18-28% lead and 10% tin, used for bearings and similar items.

Beilby layer: The atomic layer having a thickness of about 30Å, formed on a metal when it is polished.

beldongrite: Native manganese iron oxide.

bell metal: A bronze, or copper-based alloy, that generally contains three or four parts of copper, 15-40% tin, and possibly some iron, lead, or zinc, and is used for making bells and similar musical instruments.

bell-mouthed slot or groove: A condition where the ends of the slot or groove become gradually wider than the center.

bench vise: The most commonly used general-purpose vise. The bench vise is bolted securely to a workbench and usually has a swivel base.

bend test: A standard test in which metals are bent over a form containing a specific radius, through a specific angle, for a specific number of cycles until visible cracks appear. This test is used to determine the relative ductility of certain sheet metals such as sheet, strip, plate, or wire, tubes, rivets, and for determining the toughness of malleable cast iron and other metals.

Benedict metal: A nickel silver alloy nominally containing 55-59% copper, 20% zinc, 10-12% nickel, 9-10% lead, 2-3% tin. See also nickel brass, leaded.

beneficiation: An extractive metallurgical process used to separate and concentrate the high-grade ore or other valuable constituents of ores from impurities (gangue) in preparation for further processing (smelting). The steps may include calcination or physical separation by flotation, screening, washing, milling, or magnetic separation.

Bengough-Stuart process: A process of anodizing aluminum using a 3% chromic acid [CrO_3] solution as an oxidizing agent, producing a gray surface finish.

beryllium: A rare metal element. A carcinogen (OSHA). An alloying element in nickel and copper. Highly reactive but corrosion and oxidation resistant at room temperature. No reaction to hydrogen at any temperature. The lightest known structural metal. Brazing and welding are difficult. Highly permeable to x-rays. Used in space and missile technology as a heat shield and structural material, making non-sparking tools, and as an alloying element with copper, aluminum, and magnesium.

Symbol: Be
Physical state: White solid
Periodic Table Group: IIA
Atomic number: 4
Atomic weight: 9.012
Valence: 2

Density (g/cc): 1.848 @ 68°F/20°C
Specific heat: 0.475
Melting point: 2,332°F/1,277°C
Boiling point: 5,378°F/2,970°C
Source/ores: Beryl, bertrandite
Oxides: BeO, Be(OH)$_2$
Crystal structure: h.c.p. (α); b.c.c. (β)
Brinell Harness: 75-85

beryllium aluminum alloy: An alloy containing 62% beryllium and 38% aluminum.

beryllium bronze: Alloys nominally containing 96.5-97.5% copper, 2.5-2.6% beryllium, and up to 1.1% nickel. It has moderate machinability.

beryllium copper: Alloys nominally containing 97.9% copper, 0.2% nickel or cobalt, and 1.9% beryllium.

beryllium potassium sulfate: [$BeSO_4K_2SO_4$] Shiny crystalline solid used in electroplating of metals including chromium and silver.

Bessemer process: A process for making steel by pouring molten cast iron in a specially-designed, refractory-lined vessel or converter and blowing a stream of air through the molten pig iron to oxidize the carbon, silicon, and manganese.

Bessemer steel: Steel produced by the Bessemer process.

beta iron: One of the four solid phases of pure iron. The secondary allotropic modification of iron. It is intensely hard and brittle, weakly magnetic at the higher points of its temperature (1,648°F/923°C) and non-magnetic at the lower point (1,416°F/769°C).

Betterton-Kroll process: A metallurgical process for producing high quality, almost pure (99.995%) bismuth and purifying desilverized lead. Metallic magnesium or calcium is mixed with molten lead, which reacts with the bismuth and separates and floats to the surface as scum layer. This floating material is skimmed off and the excess calcium and magnesium are removed from the lead with chlorine gas.

Betts process: An electrolytic process for the refining and extraction of lead by using an electrolyte of lead fluosilicate and fluosilicic acid with a small amount of gelatin. The resulting impurities are treated to recover antimony lead, silver, gold, bismuth, etc.

bevel: (1) Any flat surface not at right angle (90°) to the rest of the piece; (2) another name for a bevel square.

bevel gear: A gear having beveled teeth used for transmitting rotary motion at an angle. Bevel gears connect shafts which are not parallel to each other, and whose center lines meet each other at right angles. Bevel gears may be made for shafts whose center lines meet, but are not at right angles to each other; in this case the gears are sometime called *angular gears*. When the shaft center lines do not intersect each other, *hypoid gears* (offset bevel gears) may be used

bevel protractor: An instrument that contains the features of a bevel and a protractor that can be used and adapted to all classes of work where angles are to be accurately laid out. Depending on the need for degrees of accuracy, bevel protractors can be constructed with or without a vernier dial.

bevel square: A tool, similar in appearance to a try square or steel square, but having a blade hinged to the stock, which allows it to be moved and set to any desired angle in its own plane.

bevel vernier protractor: An instrument for laying out angles with a fine degree of accuracy. See also bevel protractor.

BHN: Abbreviation for Brinell harness number.

Bi: Symbol for the element bismuth.

bilateral tolerance: A tolerance in which variation is permitted in both directions (partly plus and partly minus) from the specified dimension or design size.

bilateral tolerance system: A design plan that uses only bilateral tolerances.

billet: A solid block of material usually semifinished and in the form of cylinder or rectangular prism that has been cast or hot worked by forging, rolling, or extrusion preparatory to some finishing process.

binary alloy: A metal containing two principal metals, exclusive of impurities.

biocide: A general name for any substance that is added to cutting fluids to kill or inhibit the growth of contaminants and microorganisms such as bacteria, molds, slimes, or fungi. See also slimicide.

Birmabright®: A registered trademark for a range of aluminum alloys nominally containing 1-7% magnesium.

Birmingham platinum: A white brass containing 65-80% zinc and 20-45% copper.

bis(hydroxyethyl)butynediol ether: A dark brown liquid used in electroplating and as a corrosion inhibitor.

bismanol: [MnBi] An alloy or compound of bismuth and manganese produced by powder metallurgy methods, used as a permanent magnet. A product of the U.S. Naval Ordnance Laboratory.

bismuth: Used as an additive in tempering baths; to improve machinability of steel; making low-melting alloys and solders; a coolant for nuclear reactors. The powder is flammable.

Symbol: Bi	**Density (g/cc):** 9.78
Physical state: Silver-white; slight pink solid	**Specific heat:** 0.0303
	Melting point: 520°F/271°C
Periodic Table Group: VA	**Boiling point:** 2,840°F/1560°C±5
Atomic number: 83	**Source/ores:** Bismite, bismuthinite, bismutite
Atomic weight: 208.98	**Oxides:** Bi_2O_3, Bi_2O_5
Valence: 2,3,4,5	**Crystal Structure:** Rhombohedral

bismuthinite: Also known as bismuth glance. [Bi_2S_3] an ore of bismuth contains 81.2% bismuth, 18.8% sulfur and may contain copper or iron.

black annealing: A process of box annealing iron-base alloy sheet, strip, or wire after hot rolling, shearing and pickling. The process does not impart a black color to the product if properly done, The name originated in the appearance of the hot-rolled material before pickling and annealing.

black crest thread: A thread whose crest displays an unfinished cast, rolled, or forged surface.

blacking holes: Irregular shaped defects from casting usually exposed during machining operations.

black metal: In electroplating, black deposits from certain metals, e.g., platinum.

black oxide: A black, thin, corrosion-resistant oxide coating on metal produced by immersing the metal in hot oxidizing salts or salt solutions.

blade pitch: The distance between the centers or other corresponding points on one saw tooth to the adjacent tooth.

blank: Any piece of material (in sheet metalworking, usually a flat sheet of metal) of any desired size and shape that has been cut by a blanking and ready to be made into a finished form by subsequent press and forming operation such as stamping, piercing and extrusion.

blank development: The procedure of determining the size and shape of the resultant flat pattern used for a blank.

blank, file: A file in any stage of manufacture before being cut.

blanking: Cutting or punching out pre-designed flat shapes from metal sheets. This is nearly always the first operation in producing the finished article. Blanking may be combined with other operations in one tool, all the work being performed by a single stroke of the press. See also press, punching.

blast furnace: A vertical coke-fired furnace used for smelting metallic ores, e.g., iron ore.

blast-furnace gas: By-product gas from smelting iron ore in blast furnaces, typically containing 57% nitrogen, 26% carbon monoxide [CO], 13% carbon dioxide, and 3.7% hydrogen.

blasting: See sand blasting.

blind hole: A hole that has been drilled into, but not completely through, the workpiece.

blind hole tapping: A process of cutting threads to a specified depth. This is usually done in a blind hole. Often called bottom-hole tapping.

blister: A metal defect, on or near the surface, resulting from the expansion of gas in subsurface pockets during solidification. Very small blisters are called "pinheads" or "pepperblisters."

blister copper: A crude copper produced by oxidizing impure copper ores or matte. Its blistered appearance is caused by the expansion and release of gases in subsurface pockets, including SO_2. Electrolysis is used for subsequent refining.

block tin: An alloy of tin containing antimony, arsenic, cobalt, iron, and lead.

bloom: (1) A term used to describe the fluorescence of lubricating oils; (2) a hot-rolled piece of steel, rectangular in cross section, made from an ingot and produced on a blooming mill, or made by forging. For iron and steel, the width is not more than twice the thickness, and the cross-sectional area is usually not less than $36in^2$.

blowpipe: A metal tube that is finely tapered to a point and used to blow a controlled amount of air into a flame for soldering and jewelry making.

blue annealing: The process of softening iron-base alloys by heating in an open furnace to a temperature within the transformation range, between A_{c1} and $A_{c3,}$ and then cooled air. Used for alloys in the form of steel plate and hot-rolled sheet and for helical spring cold-coiled from wire steel. The formation of a bluish oxide mill scale on the surface is incidental of the process.

blue brittleness: Brittleness exhibited by some steels heated within the temperature range that produces blue oxide films (i.e., 300-350°C/572-662°F) and subsequently deformed at room temperature. Killed steels are virtually free of this kind of brittleness.

blueing: See bluing.

bluing: A process of heating and tempering the surface of hot-rolled sheet in an open furnace to a temperature within the transformation range and then cooling in air. A thin blue oxide film is formed on initially scale-free surface, as a means of improving appearance and a measure of resistance to corrosion, especially when oiled. Bluing in air reduces the hardness and imparts toughness and springiness, this term is also used to denote a heat treatment of springs after fabrication, to reduce the internal stress created by coiling and forming.

blunt, file: A file whose cross-sectional dimensions from point to tang remain unchanged.

blunt start thread: The removal of the incomplete thread at the starting end of the thread. This is a feature of threaded parts that are repeatedly assembled by hand, such as hose couplings and thread plug gages, to prevent cutting of hands and crossing of threads. It was formerly known as Higbee cut.

bort: Also spelled boart. Low-grade industrial diamonds used as abrasives in metal cutting and grinding wheels.

body: That portion of a drill extending from the shank or neck to the outer corners of the cutting lips.

body-centered cube: A *unit* or *cell* of space lattice (crystalline) structure having 9 atoms; one located at each corner of the eight corners of the cubic unit and one at the center of the cube. Metals with this arrangement include niobium (columbium), molybdenum, tungsten, and vanadium. α-Iron, or ferrite, also has this lattice arrangement at temperatures below the hardening temperature range; however, at this range, it is converted to face-centered cubic.

body-centered tetragonal: A *unit* or *cell* of space lattice (crystalline) structure having 9 atoms; one located at each corner of the eight corners of the oblong "cube"and one at the center of the oblong "cube." This arrangement is almost identical to the body-centered cubic unit except for the fact that the sides are not all square, having one elongated axis. Martensite has this space lattice arrangement.

body clearance diameter: A drill term describing that portion of the land that has been cut away so it will not bind against the walls of the hold.

body diameter clearance: A drill term use to describe that portion of the land that has been cut away so it will not rub against the wall of the hole.

boiler plate: A mild steel alloy nominally containing 0.15 -0.18% carbon.

boiler scale: A rocklike deposit occurring on the walls and tubes in boilers where hard water has been used. Depending on the content of the water, this scale consists largely of calcium carbonate, calcium sulfate, or similar materials. This scale tends to decrease the boiler's rate of heat transfer, cause increased heating costs, and shortened boiler life. Boiler feed water should be softened to chemically remove calcium and magnesium ions before use. Scale may be removed by treatment with ammonium bicarbonate solution or other commercially available products.

bonded abrasive: Natural or manufactured abrasive materials that are reduced to particles or grains and mixed with suitable bonding agent such as sodium silicate, clay, fluorocarbon polymers, rubber, shellac, resins, or combinations of these materials. The grains are compressed, sometime with the addition of heat, into useful shapes such as wheels that are rotated by a power-driven shaft. The bonding material used in a grinding wheel is normally identified by a letter code as part of the standard marking system used by wheel's manufacturers.

bonderizing: A process that adds a complex phosphate conversion coating on steels by immersion in a hot acid bath containing phosphates of zinc and manganese. The process provides corrosion resistance and good paint adhesion.

bone ash: An ash made by calcining bones and reducing them to ashes, composed principally of tribasic calcium phosphate, but containing small percentages of calcium carbonate, magnesium phosphate, and calcium fluoride. Synthetic products are also available. Used for cleaning and polishing metals, coating molds for copper wire, etc.

borates: Corrosion inhibiting chemical agents added to cutting fluids.

borax: A commercial name for sodium borate [$B_4O_7 \cdot 2Na$] A white crystalline solid used as a metal flux in welding and brazing.

Borazon ®: A registered trademark of the General Electric Company for its polycrystalline cubic boron nitride (PCBN) superabrasive

Borcher's metal: A group of heat- and corrosion-resistant alloys variably containing 34-65% nickel, 30-32% chromium, 34-35% cobalt and small amounts of molybdenum and/or silver or gold. One alloy contains 65.9% nickel, 32.3% chromium, and 1.8% molybdenum.

bore: The width or diameter of a hole.

bore, bearing: The area of the bearing that makes contact with the bearing seat on the shaft.

boric acid: [H_3BO_3] A colorless or white powder used as a welding flux, for brazing copper, and in electroplating.

boring: The process of internal finishing that includes enlarging and finishing off a previously drilled or cored hole to accurate size (generally the hole is drilled undersize by 1/16 in to 1/8 in). The process uses a single-point, cutting tool called a boring tool or boring bar; it travels along the inside of the work as it revolves. Boring can be accomplished on lathes and with boring heads on drill presses and milling machines, or on specially built boring machines. The workpiece must be clamped securely to the machine table and boring must be done at low rpm with automatic feed.

boron: CAS 7440-42-8. A hard, nonmetallic element. Alloying element with various metals, and deoxidizing agent for steels and copper. Used to make special-purpose alloys, oxygen scavenger for copper and other metals, boron-coated tungsten wires, high-temperature brazing alloys, cementation of iron. Neutron absorber in nuclear power reactors.

Symbol: B
Physical state: dark brown powder or yellow crystals
Periodic Table Group: IIIA
Atomic number: 5
Atomic weight: 10.81
Valence: 3

Density (g/cc): 2.54 (crystalline); 2.45 (amorphous)
Melting point: 4,172°F/2,300°C
Boiling point: 6,616°F/3,658°C
Source/ores: Borax, kernite, colemanite, ulexite.
Oxides: B_2O_3
Mohs hardness: 9.3

boron alloy: A uniformly dispersed mixture of boron with another metal or metals. See also ferroboron.

boron carbide: [B_4C] Hard, black crystals. A very hard, black, crystalline solid used as an abrasive on paper and in grinding wheels.

bottoming tap: A tap used for cutting internal threads or "tapping" blind holes (hole that go only part way into the work) to the bottom of the hole. Bottoming taps may have full threads to the end or have one to two threads at the front end of the land that are tapered so that the tap will start easily by distributing the cutting action over several teeth as the threads are gradually cut. See also chamfer taps.

box annealing: A process of annealing which, to minimize surface oxidation, is carried out in a closed sealed metal box or pot, with or without packing material. The charge is usually heated slowly to a temperature below, but somewhat above or within, the transformation temperature range and subsequently cooled slowly. This process is also known as close annealing or pot annealing.

Br: Symbol for the element bromine.

brale indenter: The diamond pointed indenter used in Rockwell-C hardness testing.

brass: A copper-zinc alloy of varying proportions. Depending on use, brasses may contain small percentages of other elements including about 1-2% lead, and sometimes aluminum, manganese, silicon, iron, or tin. Aluminum and tin improve strength and corrosion resistance. Manganese and iron help to refine the metal's grain. The most common brass for general use contains nominally 63-67% copper and 33-37% zinc. Also known as basis brass, ordinary brass, or gold button brass. See also iron brass, leaded brass, low-zinc brass, forging brass, naval brass, red brass, yellow brass, cartridge brass, Muntz metal, ronia metal.

Brass and Bronze Ingot Manufacturers Association (BBIM): 200 Michigan Avenue, Suite 1100, Chicago IL 60604-2404. Telpehone: 312/372-4000. FAX: 312/939-5617. E-mail pbb@defrees.com

brass, iron: See iron brass.

brazing: A welding method for joining two or more metals by using nonferrous metals or metallic alloys, in the form of filler rods, applied as a thin layer at temperatures above 800°F/427°C, and drawn into the space between the edges of the metals to be joined. Brazing by immersion in a molten salt or metal bath is called *dip brazing.* The term soldering or soft soldering is used for a similar process when the temperature is lower than the arbitrary value of 800°F/427°C.

breaking out: Describes the removal of a casting from a mold.

breakout: The uneven separation of material breaking off as the drill exits the workpiece.

bright annealing: An annealing process in a controlled furnace atmosphere which inhibits or reduces surface oxidation, thereby preventing discoloration of the bright surface.

bright plate: An electroplated finish that requires little or no polishing to produce a smooth and gleaming finish.

brighteners: Organic chemical compounds that are added to a nickel electroplating solution of the Watts type [nickel sulfate and nickel chloride (6:1) plus boric acid], used to cover surface irregularities of the substrate and yield mirror brightness to the finish. One or more brighteners may be added to the plating formulation.

Brinell hardness test: A measure of the relative hardness of the smooth surface of a metallic material, obtained by measuring the resistance it offers to the depth of indentation of a standard 10 mm (0.394 in) hard steel or carbide ball at a standard load of 3,000 kg (6600 lb) for a period of 15 seconds for steel and 30 seconds for nonferrous metals. The hardness is computed by an expression whereby the value obtained is directly proportional to the applied pressure, and inversely proportional to the depth of the penetration on the metal. In other words, the greater the distance of penetration, the softer the work, and the higher the Brinell harness number (BHN).

Britannia metal: A tin-base alloy or pewter normally containing 91-92% tin, 1-3% copper and 5-10% antimony. See also pewter.

British Standards Institution (BSI): Headquartered at British Standards House, 2 Park Street, London, W.1., England. A nonprofit body devoted to the drafting and publication of improvements, standardization, and simplification of engineering and industrial material in the England.

British thermal unit (Btu): The amount of heat required to raise the temperature of one pound of water one Fahrenheit degree (usually from 39 to 40°F/3.8-4.4°C). This is equivalent to 1,055 joules or about 252 gram calories. A *therm* is usually 100,000 Btu, but is sometimes used to refer to other units.

brittle: Breaking easily and suddenly with a comparatively smooth fracture. The brittleness of castings and malleable work is reduced by annealing.

brittle fracture: Fracture in metals, especially structural steels, by rapid cracking with little or no appreciable plastic deformation.

brittleness: (1) The property of a material to fracture or break without appreciable macroscopic plastic deformation; (2) a lack of ductility; (3) the measure of how easily a metal fractures when subjected to outside forces, such as bending or low temperatures.

brittle point: The temperature (slightly above the transition point) at which a material shatters when pressure is applied.

brittle temperature range: The temperature range between 400° and 700°F/ 204° to 371°C.

broach: A slightly tapered, bar-shaped cutting tool containing a series of chisel-like cutting edges or teeth on its surface, each of which is progressively higher and/or differ in shape from the starting end. Each removes a small amount of material as it moves through the workpiece. The design of the broach allows it to rough, semifinish, and finish internal or flat and external surfaces, regular or irregular, in a single pass. For some operations the broach may be more cost effective to use than a milling cutter; it needs sharpening only about 1/16th times as often. The same name is sometimes used for a small reamer used by jewelers. See also broaching.

broaching: An operation for machining metal with a hardened, multi-toothed form tool called a broach which is either pushed or pulled in a straight line through, or along, the surface of a workpiece. Because the teeth are progressively larger, each tooth removes successively larger amounts of metal. Broaching may be applied in machining holes of complex shapes (called internal broaching) and also to many flat or other outside surfaces (called external broaching). Internal broaching is used to form either symmetrical or irregular holes, keys, grooves, or slots in machine parts, especially when the size or shape of the opening, or its length in proportion to diameter or width, make other machining processes impracticable. It is usually a one-step operation that is rapid and accurate, leaving a smooth surface. See also broach.

bromine: CAS number: 7726-95-6. A dense, deep red liquid. A nonmetallic element. Used to harden metals. Toxic.

Symbol: Br	**Density (g/cc):** 3.19
Physical state: Red liquid	**Specific heat:** 0.107
Periodic Table Group: VIIA	**Melting point:** 18.9°F/-7.3°C
Atomic number: 35	**Boiling point:** 138°F/59°C
Atomic weight: 79.904	**Sources/ores:** Seawater, natural brines
Valence: 1,3,4,5,7	**Crystal structure:** Orthorhombic

bronze: Alloys of copper and tin in varying proportions. It often contains zinc and occasionally contains lead. Bronze can contain 73-95% copper, 1-20% tin, 0-18% zinc, and 0-18% lead. The alloy is made more brittle by increasing the proportion of tin above 5%. Generally, bronze weighs about 0.32 lb/in^3 with a melting point about 1675°F/913°C. There is a great variety of special bronzes containing other ingredients such as aluminum (aluminum bronze), phosphorus (phosphor bronze), silicon (silicon bronze), and manganese (manganese bronze).

bronze, aluminum: See aluminum bronze.

bronze, commercial: See commercial bronze.

bronzes: A general term used to describe various copper-rich alloys, other than brass, that may contain tin. See also bronze.

bronze welding: A method of brazing by using an arc welder to join cast iron. This method minimizes the risk of cracking the cast iron from thermal stress because the temperature used to deposit the filler metal is about 1,472°F/800°C instead of 2,102°F/1,150°C to melt the iron.

bronze wire: An alloy containing 98.75% copper, 1.2% tin, and 0.05% phosphorus.

bronzing solder: Alloys nominally containing 50/50% copper and zinc.

BSI: Abbreviation for the British Standards Institution.

Btu: Abbreviation for British thermal unit.

BUE: Abbreviation for built-up edge.

buffer: A chemical reagent containing both a weak acid and its conjugate weak base which is added to a solution to minimize the change in pH when acid or alkali are added to the solution.

buffing: The process of smoothing a metal objects by holding and pressing them against a fabric disk or belt imbedded with loosely applied buffing compound, running at high speed. The buffing compound consists of fine abrasive particles held in a composition of wax or similar binding materials, and often formed into hand-held sticks or cylinders. See also polishing.

build-up: In electroplating, the excessive metal deposition on corners of components, caused by relatively high current densities in these areas. In metal cutting, see build-up edge.

built-up edge (BUE): A metal cutting term used to describe a type of tool wear problem caused, more or less, by the formation, adhesion, and build-up of layers of chip material to the top rake surface of a cutting tool during a machining operation. BUE is largely a temperature and, therefore, a cutting speed-related phenomenon normally associated with the cutting of ductile materials with high speed tools without the use of cutting fluid at ordinary cutting speeds. However, it can also be the result of high pressure, high frictional resistance, edge flagging, and other wear. Surface finish is often the first to suffer as the BUE grows, but if this type of wear is allowed to continue, it may harden the work surface and cause shortened tool life from rapid edge break-down, and possible fracture.

bulk density: Mass of powdered or granulated solid material per unit of volume.

bullion: Bulk quantities of precious metals such as gold, silver, or gold-silver alloy as produced in refining.

bumper: A vibrating machine used to blend and compact sand in mold making

burning: (1) A term applied to permanent damage of metal or alloy by exposure to high temperature (close to the melting point) causing embrittlement, loss of impact strength, ductility, and some loss of tensile strength The damage may involve melting of some constituent or penetration by, and reaction of the metal with, a gas such as oxygen, or by loss of component elements of the metal from the crystal boundaries. The original properties are lost and cannot be restored without remelting; (2) in grinding, when the work becomes hot enough that discoloration or a change in the microstructure is caused; (3) may be used to describe cutting with oxy-acetylene.

burning of workpiece: See burning

burnishing: A metal finishing process for improving the surfaces of metals by using frictional contact between the workpiece and hardened steel tools, including small steel balls or metal pads. The surface is plastically smeared, producing a smooth, bright, and lustrous surface.

burnt deposit: In electroplating, a ragged deposit usually caused by uncontrolled or excessive current density.

burr: (1) A sharp, thin, usually jagged sliver of metal left on a workpiece as the tool from a machine or punch operation exits the cut; (2) a rotary tool with a serrated surface used to clean up interior surfaces of dies.

bushing: A protective liner used to provide a cushion or bearing between moving parts such as revolving shafts. See linear bushings, press fit bushings, renewable bushings, pilot bushings, rotary bushings. See also bearing for comparison.

Butoxyne 497®: A registered trademark of International Specialty Products for a liquid nickel brightener used in electroplating, copper pickling inhibitor, and corrosion inhibitor. International Specialty Products 1361 Alps Rd. Wayne, NJ 07470.

butt joint: (1) A joint between two components or plates, lying in the same plane. The ends of the plates to be joined abut squarely against each other. By comparison, a lap joint has the plates to be joined overlapping each other; (2) one of two kinds of riveted joint. In the butt-joint, the plates being joined are in the same plane and are joined by means of a cover plate or butt strap, which is riveted to both plates by one or more rows of rivets. The other rivet joint is the lap joint.

by-product: Any material, other than the principal product, generated as a consequence of an industrial process.

BZN ®: A registered trademark of the General Electric Company for polycrystalline cubic boron nitride (PCBN) superabrasive products.

C

C: (1) symbol for Celsius or Centigrade, a unit of temperature, as °C; (2) symbol for the element carbon.

C-22®: A registered trademark of Haynes International, Inc. for a nickel-based alloy having high strength, excellent resistance to corrosion, ease of welding, and fabrication.

Ca: Symbol for the element calcium.

CAD: Abbreviation for computer-aided design.

Cadalyte®: A registered trademark of DuPont for a series of compounds for cadmium electroplating.

CAD/CAM: Abbreviation for computer-aided design/computer-aided manufacturing.

CADD: Abbreviation for computer-aided drawing and design.

cadmium: Metallic element. CAS number: 7440-43-9. An alloying element. Lowers melting point of certain alloys when used in low percentages. Used for protective coatings on metals, especially copper and steel (in electroplating); bearing metals such as babbitt, and low-melting alloys such as solders; brazing alloys; a deoxidizer in aluminum, nickel and silver alloys; control rods for reactors. Corrosion resistance is poor in industrial atmospheres; tarnishes in moist air. Highly toxic; carcinogenic, and the powder is flammable.

Symbol: Cd	**Density (g/cc):** 8.65
Physical state: Silvery solid	**Specific heat:** 0.0549
Periodic Table Group: IIB	**Melting point:** 610°F/321°C

Atomic number: 48
Atomic weight: 112.41
Valence: 2

Boiling point: 1,409°F/765°C
Source/ores: Greenockite; also found in many lead and copper ores containing zinc.
Oxides: CdO
Crystal structure: close-packed hexagonal
Mohs hardness: 2.0

cadmium acetate: A colorless crystalline solid used in electroplating.

cadmium bronze: Copper alloys containing small amounts (0.5-1.2%) of cadmium.

cadmium chloride: $[CdCl_2]$ A white odorless crystalline solid used in electroplating.

cadmium copper: A copper alloy nominally containing 99.0 - 99.5% copper and 0.1 - 1.0% cadmium. It has little reduction in electrical conductivity and 50% greater tensile strength than pure copper.

Cadmium Council, Inc: 12110 Sunset Hills Rd., Ste. 110, Reston, VA 22090. Telephone: 703/709-1400.

cadmium cyanide: $[Cd(CN)_2]$ an electrolyte for cadmium and copper plating.

cadmium hydroxide: $[Cd(OH)_2]$ A chemical used for cadmium plating.

cadmium iodide: $[CdI_2]$ A white crystalline solid used in electroplating.

cadmium oxide: $[CdO]$ Brown to yellow-brown crystalline solid or powder used in electroplating; a component of silver alloys.

cadmium pigment: A family of pigments used in high-gloss baking enamels often used as automotive finishes and other metal applications. Red shades come from cadmium selenide; yellow and orange come from cadmium sulfide. See also Cadmolith®.

cadmium sulfate: [CdSO₄] A colorless crystalline solid used in electrodeposition of cadmium, copper, and nickel.

Cadmolith®: A registered trademark of SCM Corp. for a series of yellow and red cadmium pigments extended with barium sulfate and termed cadmium lithopone (cadmium selenide lithopone) pigments. Used in automotive finishes, baking enamels for metals.

CAE: Abbreviation for computer-aided engineering.

caesium: See cesium.

cal: Abbreviation for calorie.

calaverite: [AuTe₂] An important source of gold. Pale bronze-yellow color or tin-white; exposure causes tarnishing to bronze-yellow. Contains 40-43% gold, 1-3% silver. Mohs hardness 2.5.

calcination: Roasting a substance to a temperature below its melting point, but high enough to cause evolution of carbon dioxide and effect dissociation (by eliminating volatile constituents such as bound water and other impurities). Also, to induce phase changes in metal ores. See also roasting, smelting.

calcite: Natural calcium carbonate [CaCO₃). The essential ingredient of chalk, limestone, and marble. Mohs hardness: 3.

calcium: An alkaline-earth, metallic element. CAS number: 7440-70-2. The fifth most abundant element in the earth's crust. Oxidizes in air, forming a protective, bluish-gray film. May be machined, extruded, or drawn. Used as an alloying element for aluminum, copper, lead; as an deoxidizer for ferrous and non-ferrous alloys; decarburizaton and desulfurization of iron and its alloys; as a reducing agent in preparation of rare earth minerals, chromium metal powder, uranium. Reacts with air and moisture. Must be handled under dry, inert gas or vacuum to avoid contamination.

Symbol: Ca	**Density (g/cc):** 1.57
Physical state: Silvery solid	**Specific heat:** 0.157
Periodic Table Group: IIA	**Melting point:** 1,553°F/845°C
Atomic number: 20	**Boiling point:** 2,703°F/1,484°C
Atomic weight: 40.08	**Source/ores:** calcite, limestone, dolomite, flourspar
Valence: 2	**Oxides:** CaO (lime)

calcium boride: A deoxidizer used in making copper.

calcium lead: See lead, calcium.

calcium silicide: [CaSi and $CaSi_2$] A refractory powder used in cast iron.

calcium silicon alloy: An alloy containing 30% calcium; it may ignite spontaneously in air.

calcium stearate: [$Ca(C_{18}H_{35}O_2)_2$] A white powder used as a mold release agent.

caliper: A tool designed for measuring diameters that are not required to be extremely accurate. Calipers are available for different jobs and in differing sizes. Outside calipers have legs with a large curve inward.

calipers: An instrument used for testing the work where the accuracy of a micrometer is not required and for measuring the dimensions of work pieces, especially internal and external diameters of cylindrical pieces. The caliper consists of two pieces of steel that are curved and are hinged together with a tight joint at one end, the distance between the points representing the measurement taken. In general, calipers are used to measure distances between or over surfaces, or for comparing distances with standards, such as those on a graduated rule. Their size is measured by the greatest distance they can be opened between the two points. Never use calipers on work that is moving or revolving.

There are different classes of calipers including *outside*, *inside*, *hermaphrodite*, and *foundry* and *forging*. There are also different forms of calipers including *firm joint* and *spring joint*. Outside and inside calipers are self-defining. Hermaphrodite calipers have one short divider leg with a sharp point (which may be adjustable) and one caliper leg; they are used to scribe center lines on a shaft, or to mark out lines on a surface parallel with the edge. Firm joint calipers have a friction joint between the legs that are joined with an adjustable friction screw. Spring joint calipers have a curved spring between the two legs and an adjusting screw and nut. The legs are forced together evenly against the tension of the curved spring.

caliper gage: Similar to a snap gage, for measuring external members. See caliper snap gage.

caliper snap gage: A tool for repetitively checking work where one dimension (e.g., a shaft diameter) has to be checked frequently. The caliper snap gage is a C-shaped frame with two flat faces, one on each jaw, ground so that the distance between them permits the gage to fit smoothly onto a part that is the correct size. Adjustable caliper snap gages are also available. These also have a C-Shaped frame, but in place of the usual jaws, four cylindrical jaws are inserted in holes in the frame, and endwise movements of these by means of a screw provide the necessary adjustments. See also snap gage.

Calloy®: Alloys nominally containing approximately 90% aluminum and 10% calcium, used as a steel deoxidizer for "killed steel."

calorie (cal): The metric system unit for measuring quantity of heat. The amount of heat required to raise one gram of water one degree centigrade, equivalent to about 0.003968 Btu. More common is the kilogram calorie, also know as kilocalorie (abbreviated K-cal or kg-cal = 1000 g-cal). Equivalent to about 3.97 Btu.

calorizing: The process for improving the oxidation resistance of mild and low-alloy steels by heating them in aluminum powder at 1,832°F/1,000°C. The heated aluminum powder forms an alloy [Al_2O_3] with the steel surface and produces a thin, tightly-bonded coating.

calsintering: A method used to treat fly ash (a mixture of alumina, silica, unburned carbon, and various metallic oxides) to recover alumina (aluminum oxide [Al_2O_3]) that can be converted to aluminum by conventional methods. A possible alternative source of aluminum.

cam: A disc containing irregularly shaped lobes that are mounted on a shaft (camshaft), used to activate the opening and shutting of mechanical devices such as automotive valves.

CAM: Abbreviation for computer-aided manufacturing.

camber: (1) Deflection from a straight line; (2) deviation from edge straightness usually referring to the greatest deviation of side edge from a straight line; (3) used to describe the crown in rolls where the center diameter has been made slightly greater to compensate for deflection under load (the rolling pressure).

Canadian Standards Association (CSA): A national association of technical committees providing national standards for Canada. Address: 178 Rexdale Boulevard, Rexdale, Ontario, Canada M9W 1R3.

Canadian Welding Bureau (CWB): A division of the Canadian Standards Association. It provides codes and standards for all phases of welding. Address: 254 Merton Street, Toronto, Ontario, Canada M4S1A9.

Canadian Welding Society, Inc. (CWS): The national welding society of Canada. Address: 6 Milvan Drive, Weston, Ontario, Canada M9L1Z2.

Canadian Welding Development Institute (CWDI): A division of the Canadian Standards Association. Address: 254 Merton Street, Toronto, Ontario, Canada M4S1A9.

canning: A concave, dish-like deformity in a flat metal surface, sometimes called "oil canning."

cantilever: A beam with one fixed end and loaded at the other.

cape chisel: A sturdy, narrow cold chisel made from hexagonal- or octagonal-shaped tool steel, and having a cape or flare for the widest flat at the cutting edge, used for cutting keyways, narrow slots, chipping grooves, and similar work.

carat: (1) One twenty-fourth of a part; a term used to express the degree of purity of gold. Pure gold is 24 carat or 1000 "fine." Gold alloy is described by the number of parts of gold contained in 24 parts of the alloy. Thus, 14 carat gold contains 14/24 gold or 58.33% gold, or 583.3% "fine." Also spelled karat; (2) a unit of weight equal to 200 mg, the gem weight used for precious stones. See also karat.

carbide: A term used to describe hard metals that are binary compounds of carbon with one or more other elements, produced by powder metallurgy in which the powders are pressed in molds and heated (sintered) at high temperature. The most familiar carbides are those of calcium, boron, iron (cementite), silicon, tantalum, titanium, tungsten, vanadium, and zirconium. See also coated carbide, cemented carbide.

Carbide Industry Classification System: A classification system recognized by many manufacturers, used in grouping machining applications for cemented-carbide tools:
Cast iron and nonferrous materials: C-1: Rough cuts; C-2: General purpose; C-3: Light finishing cuts; C-4 Precision boring. *Steel and Steel alloys:* C-5: Rough cuts and heavy feeds; C-6: General purpose; C-7: Light finishing cuts (heavy feeds)/C-7A: Light finishing cuts (fine feeds); C-8 Precision boring.

carbide tools: Metal-cutting tools in which the hardness and high-temperature strength of tungsten, titanium, or tantalum carbides are employed on their tips to permit cutting speeds in rock or metal up to 100 times that obtained with alloy steel tools.

Carbolite®: A registered trademark for an artificial abrasive made from silicon carbide [SiC]. Not as tough as Carborundum®, used as an abrasive for grinding materials of low tensile strength.

Carbolon®: A proprietary trade name for silicon carbide [SiC].

Carboloy®: A proprietary trade name for cemented tungsten carbide. See tungsten carbide, cemented.

carbon: A non-metallic bioelement. An extremely important alloying element in all ferrous metals because it imparts strength and hardness by the formation of carbides. Carbon exists in several allotropic forms amorphous (activated carbon, coal, coke, charcoal, lampblack, soot, etc.) and two crystalline forms diamond and graphite. The presence of carbon in steel, usually in excess of 0.60% for non-alloyed types, is essential for raising the hardenability to the levels needed for tools. Raising the carbon content by different amounts up to a maximum of about 1.3% increases the hardness slightly and the wear resistance considerably. The amount of carbon in tool steels is designed to attain certain properties (such as in the water-hardening category where higher carbon content may be chosen to improve wear resistance, although to the detriment of toughness) or, in the alloyed types of tool steels, in conformance with the other constituents to produce well-balanced metallurgical and performance properties.

Symbol: C
Physical state: Solid
Periodic Table Group: IVA
Atomic number: 6
Atomic weight: 12.0111
Valence: 4; [sub. (2), (3)]

Density (g/cc): 3.52 (diamond); 2.25 (graphite); 1.75-2.10 (amorphous)
Specific heat: 0.079 (diamond); 0.016 (graphite)
Melting point: 6,381°F/3,527°C (amorphous); 6,417°F/3,547°C (diamond)
Source/ores: limestone, coal, graphite
Oxides: CO, CO_2, C_2O_3
Crystal structure: Cubic diamond; hexagonal diamond; hexagonal graphite; rhombohedral graphite

carbonado: Low-grade industrial diamonds used in metal cutting and grinding wheels.

carbon-arc welding: A welding method using a non-consumable carbon electrode, with or without filler metal and inert gas shielding. See also arc-welding.

carbonitriding: A case hardening process in which low carbon steel is heated and held above Ac_1 (the lower transformation temperature) in a gaseous atmosphere containing hydrocarbons, carbon monoxide [CO], and ammonia to cause simultaneous absorption of both carbon and nitrogen into the surface and, by diffusion, create a concentration gradient. The process is completed by cooling at a rate that produces the desired properties in the workpiece. Temperatures of 1425-1625°F/774-885°C are used for parts to be quench hardened, while lower temperatures, 1200-1450°F/649-788°C, may be used where liquid quench is not required.

carbonization: Conversion, through destructive distillation, of an organic material into substantially pure carbon, accompanied by the escape of volatile compounds.

carbon monoxide: [CO] A colorless, odorless, tasteless gas. The product of incomplete combustion of carbon. A common air contaminant. High concentrations have caused many fatalities. A dangerous fire hazard when exposed to flame. Used in metallurgy to carburize steel at high temperature; in nickel refining.

carbon potential: A measure of the ability of an environment containing active carbon to alter or maintain, under prescribed conditions, the carbon content of the steel exposed to it. In any particular environment, the carbon level attained will depend on such factors as time, temperature, and steel composition.

carbon restoration: Replacing the carbon lost in the surface layer from previous processing by carburizing this layer to substantially the original carbon level.

carbon steels: An alloy of iron having carbon as its chief alloying element. Mild (low-carbon) steels contain 0.02-0.25% carbon; medium steels contain 0.25-0.7% carbon; and, high-carbon grades contain 0.7-1.5% carbon. Other elements normally present are manganese (usually limited to 1.65%), silicon (usually limited to

0.60%), and residual impurities of sulfur and phosphorus (each less than 0.05%). Also known by the names, "plain carbon steel," "ordinary steel," and "straight carbon steel."

Carborundum®: A registered trademark of the Carborundum Corp. for abrasives and refractories of silicon carbide [SiC], fused alumina, and other materials. It is produced artificially from coke and sand in an electric furnace at about 7,000°F/3,871°C. Properties: Mohs hardness 9.17; Density 3.06-3.20. Noncombustible, good heat dissipation, highly refractory.

carburization: The addition of carbon to molten or heated metals with a solid, liquid, or gaseous carburizing medium, resulting in the formation of different properties in the finished metal. This process is often used to produce a case-hardened surface by subsequent quenching. See also carburizing.

carburizing: A case hardening process of forcing the addition of carbon into the surface of low carbon steels to a controlled depth by packing it in a carbonaceous medium which can be solid, liquid, or gaseous and heating it in a furnace above Ac_1 (the transformation range) and holding it at that temperature. This process is often used to produce a case-hardened surface by subsequent quenching. The resulting depth of carburization, commonly referred to as case depth, depends on the carbon potential of the medium used and the time and temperature of the carburizing treatment. Carburizing temperatures range from 1550-1750°F, with the temperature and time at temperature adjusted to obtain various case depths. Steel selection, hardenability, and type of quench are determined by section size, desired core hardness, and service requirements.

Three types of carburizing are most often used: (1) *liquid carburizing* involves heating the steel in 20-50% molten barium cyanide or sodium cyanide. Liquid carburizing can produce case depths up to 0.010 in and up to 0.040 in when activated in a salt bath such as barium chloride. The case absorbs some nitrogen in addition to carbon, thus enhancing surface hardness; (2) *gas carburizing* involves heating the steel in a gas of controlled carbon content that can be closely controlled. Gas pack carburizing can produce case depths up to 0.125 in; (3) *pack carburizing*, which involves sealing both the steel and solid carbonaceous material in a gas-tight container, then heating at 1650-1750°F/899-954°C for up to 8 hours. Pack carburizing can produce case depths up to 0.050 in. The carbonaceous materials are usually wood charcoal with sodium, potassium, or barium carbonates, cyanides, etc. With any of these methods, the part may be either quenched after the carburizing cycle without reheating or air cooled followed by reheating to the austenitizing temperature prior to quenching. The case depth may be varied to suit the conditions of loading in service. However, service characteristics frequently require that only selective areas of a part have to be case hardened. Covering the areas not to be cased, with copper plating or a layer of commercial paste, allows the carbon to penetrate only the exposed areas. Another method involves carburizing the entire part, then removing the case in selected areas by machining, prior to quench hardening. To clean up waste cyanide salts from case hardening of steel, react the salts at 1,202-1,292°F/650-700°C with waste ferric hydroxide [$Fe(OH)_3$] sludges from various operations.

carriage bolt: A bolt with an oval head and a square shank portion that holds the bolt from turning, while the nut is being screwed into place. Carriage bolts have various thread patterns and are fitted with square nuts.

cartridge brass: Also known as 70-30 brass, spinning brass, spring brass, extra-quality brass. Alloys nominally containing 67-70% copper and 30-33% zinc.

case: (1) The surface layer of an iron-base alloy which has been suitably altered in composition and can be made substantially harder than the interior or core by a process of case hardening; (2) the term case is also used to designate the hardened surface layer of a piece of steel that is large enough to have a distinctly softer core or center.

case-hardening: A process consisting of one or more heat-treatments for low-carbon steels used to impart substantially harder outer surface layer, or case, while the interior, or core, remains soft and tough. This is often accomplished by heating the steel in an absence of air while packed in a carbonaceous medium in available form (usually wood charcoal with sodium, potassium, or barium carbonates, cyanides, etc.), cooling it to black heat, reheating to a high temperature, and quenching. Typical case-hardening processes are carburizing, cyaniding, carbonitriding, and nitriding.

cassiterite: A natural tin dioxide [SnO_2] and the principal ore of tin. Also known as tinstone.

cast: To form a metallic shape or object (a casting) by permitting molten metal or other material to solidify in a mold.

cast alloy: A general term applied to various cutting tool materials made from nonferrous metals in a cobalt base. They are capable of cutting speeds 50-70% higher than high-speed steels (HSS) and can tolerate cutting temperatures up to about 1,400°F/760°C. These alloys may contain various amounts and combinations of elements including 35-55% cobalt, 25-35% chromium, 10-20% tungsten, 0-5% nickel, 1.5-3% carbon, and vary small amounts of other elements. Small amounts of iron may be present in the form of an impurity.

casting: (1) The process, or art, of making castings by pouring molten metal into a mold and letting it harden. Casting methods include centrifugal casting, cored casting, coreless casting, die casting, investment casting, lost wax casting, metal mold casting, near-net-shape casting, plaster and ceramic mold casting, sand-mold casting, semipermanent mold casting, slush molding; (2) a finished or semi-finished solid object produced by this method of solidification of molten metal or other material in a mold.

cast iron: A general term used to describe a wide range of ferrous alloys; a saturated solution of carbon in iron. The amount of carbon can vary from 1.7% to about 6%, depending upon the amount of silicon (usually about 1% to 3%), manganese, phosphorus and sulfur present in the solution. One desirable composition widely used in industry contains 3.25-3.50% carbon, 0.40-0.70% (0.50% preferred) manganese, 1.90-2.20% (2.00% preferred) silicon, 0.60-1.00% phosphorus, and not more than 0.10% sulfur. The four basic types of cast iron are white cast iron, gray cast iron, malleable iron, and ductile iron. In addition to these basic types, there are specific forms of cast iron called alloy cast iron, chilled cast iron, and compacted graphite cast iron. Cast iron weighs about 0.26 lb/in³; tensile strength 15,000 - 30,000 lb/in² according to grade. The strength begins to decrease at about 500°F/260°C. Cast iron is also used to describe remelted pig iron. Cast iron is usually machined dry.

castle nut: Sometimes called *castellated,* a nut having slots across the outer end made to receive a cotter pin inserted through the bolt to prevent loosening, slipping off, or turning.

Castner process: An apparatus used to produce sodium metal from fused sodium hydroxide [NaOH] using an electrolytic diaphragm cell. The cell contains heavy iron anodes surrounding the cathode in the bath and a porous iron gauze diaphragm or membrane located midway between the electrodes. The sodium hydroxide [NaOH] is fed to the anode side of the membrane, and flows through the diaphragm to the steel cathode, where the sodium metal is collected and hydrogen and oxygen are liberated and may be collected.

catalyst: A substance that initiates or affects the rate of a chemical reaction, but that is neither changed nor consumed in the reaction.

cathode: The negative electrode of an electrolytic cell, to which positively charged ions travel when an electric current is passed through the cell. In electroplating, the metal that is being coated is the cathode. In a battery, the cathode is the positive electrode (usually graphite) at which reduction processes such as deposition of electroplated metal or evolution of hydrogen operate.

cathodic protection: Reduction in corrosion rate or prevention of corrosion of a metallic surface by making its potential more negative and thermodynamically stable. Accomplished by use of sacrificial anodes of zinc, magnesium, or electrodeposited coatings in contact with the metals, or by using impressed currents from inert carbon, lead, platinum or other anodic materials.

cation: A positively charged ion of an electrolyte which tends to move toward and collect at the cathode (negative pole) when subjected to electric potential.

caulking tool: A cold-deformation tool that is similar to a chisel in shape, but with a flat end. The caulking tool is used to burr down the edges of plates that have been riveted in order to mechanically seal them and make them steam and water tight. See also riveting, fullering.

caustic: Any strongly alkaline material that has a corrosive or irritating effect on organic materials, including living tissue. In chemistry it usually refers to sodium hydroxide [NaOH] (caustic soda), potassium hydroxide [KOH] (caustic potash), calcium hydroxide [CaOH] (caustic lime), barium hydroxide [BaOH] (caustic baryta), or other materials with similar corrosive properties.

caustic embrittlement: The intergranular cracking of steel stressed beyond its yield point, caused by exposure to caustic solutions at temperatures above 155°F/68°C. The cohesion between the ferrite grains is broken down by the hydroxide ions.

cavitation: Metal surface erosive damage marked by mechanical pitting, occurring from the formation of vapor bubbles within a liquid subjected to tension from rapid and intense pressure changes or vibration. The local rupture effect is due to the shock waves created by collapse of the bubbles or "holes" in the liquid. The pressures exerted by cavitation may be in the range of 30,000 psi.

cavitation damage: The removal of metal caused by the formation and collapse of bubbles, "holes," or cavities in turbulent fluids. See cavitation.

cazin: A solder containing 82% cadmium and 18% zinc.

CBFC: Abbreviation for Copper and Brass Fabricators Council.

CBN: Abbreviation for cubic boron nitride (coated carbide).

C-clamps: A general-purpose clamp having a cast or drop-forged frame in the shape of a "C" and a screw having a "V" type thread, but larger sizes are usually made with square threads.

Cd: Symbol for the element cadmium.

CD: Abbreviation for continuous dressing.

CDA: Abbreviation for the Copper Development Association.

Ce: Symbol for the element cerium.

cementation: (1) In general, a process for introducing elements into the outer surface layer of a metal objects by means of high-temperature diffusion that results in the formation of an inter-metallic alloy layer at the interface of the coating and basis metals; (2) a process used to coat steel or iron with another metal such as aluminum, chromium, copper, or zinc, by immersing the basis metal into a powder of a second metal and heating to a temperature below the melting point of either the basis or coating metal; (3) the name of a process for converting wrought iron bars to steel by packing the bars in powdered charcoal, sealing with clay, and heating to 1380-1650°F/749-899°C for several days to gradually diffuse sufficient carbon into the metal to create steel. The amount of carbon and the depth of penetration is dependent on the temperature, the time of cementation, and the packing material used. See also sherardizing, calorizing.

cemented carbides: Powder metallurgy products containing more than 90% hard carbide powders (tungsten carbide [WC], titanium carbide [TiC], tantalum carbide [TaC], molybdenum carbide [Mo_2C], and vanadium carbide [VC]) that are compressed and subsequently sintered in a binding metal, usually cobalt, iron, or nickel. Also known as cementite (iron carbide [Fe_3C]), sintered carbides, and hard metals. Used principally as tips for metal-cutting tools or as throw-away inserts that can withstand temperatures up to 1,700°F/927°C, permitting cutting speeds up to 2 to 4 times that achieved with tools of high speed steels (HSS).

cemented carbide composites: See cemented carbides.

cemented tungsten carbide: See tungsten carbide, cemented.

cementite: A compound of iron and carbon, known chemically as iron carbide [Fe_3C], that is formed in the manufacture of pig iron and ordinary steels of more than 0.85% carbon. It is characterized by an orthorhombic crystal structure, hardness, and brittleness. When it occurs as a phase in steel, the chemical composition will be altered by the presence of manganese and other carbide-forming elements. Prolonged heating may cause the cementite (iron carbide [Fe_3C]) to decompose, forming graphitic carbon. The name comes from steel made by the cementation process, which contains a high percentage of this carbide.

center distance, gear: The distance between the parallel axes of spur gears and parallel helical gears, or between the crossed axes of crossed helical gears and worm gears. Also, it is the distance between the centers of the pitch circles.

center drill: A combination tool for drilling and countersinking a shallow hole in one operation, used in predrilling operations such as accurately locating and guiding the drill bit to a hole center. The center drill has a cutting edge angle of 60°. Used as well to make bearing surfaces for lathe centers that are properly shaped. Also known as combination drill and countersink, center drills are available in several sizes.

center gage: A small gage used for testing angles of lathe centers and threading tool points, as an aid in grinding, and setting screw cutting tools at the correct angle relative to the work. The center gage is made of steel, usually tempered, about 1/16 in. thick, 2 in. long and about 3/4 in. wide. The American center gage has a 60° point at one end and a 60°V-notch at the other end. The 60° angle is the standard angle for lathe centers and threading tools used in the U.S. standard and V threads. The metric gage has a 55° angle for the Whitworth and English standard. The edges have smaller 60°or 55° V-notches of different depths, and the sides are usually graduated in to aid in finding the number of threads to the inch of a screw. The American gages are graduated to show thousands of an inch and the metric gages are graduated to read in millimeters and half millimeters.

centerless grinding: A form of cylindrical grinding that does not involve the use of chucks or center holes. Centerless grinding involves the use of a machine that passes the workpiece between a grinding wheel revolving at high speed and an opposed slowly moving regulating wheel that applies pressure against the workpiece. The workpiece is supported on a work rest blade (often called a slide) in the grinding throat. The regulating wheel imparts a uniform rotation to the work, giving it the same peripheral speed, which is adjustable; it also provides a strong grip, and is a more reliable workpiece control. In addition, the work-rest comprises guides carrying the job to the wheels and removing it again when the operation is finished. The centerless grinder produces a more accurately cylindrical surface.

There are three general methods of centerless grinding, namely through-feed, in-feed, and end-feed methods. The *through-feed method* is applied to straight cylindrical parts. The work is given an axial movement by the regulating wheel and passes between the grinding and regulating wheels from one side to the other. The rate of feed depends upon the diameter and speed of the regulating wheel and its inclination, which is adjustable. It may be necessary to pass the work between the wheels more than once; the number of passes depends upon such factors as the amount of stock to be removed, the roundness and straightness of the unground work, and the limits of accuracy required. The work rest fixture also contains adjustable guides on either side of the wheels that directs the work to and from the wheels in a straight line.

When parts have shoulders, heads, or some part larger than the ground diameter, the *in-feed method* usually is employed. This method is similar to "plungecut" form grinding on a center type of grinder. The length or sections to be ground in any one operation are limited by the width of the wheel. As there is no axial feeding movement, the regulating wheel is set with its axis approximately parallel to that of the grinding wheel, there being a slight inclination to keep the work tight against the end stop.

The *end-feed method* is applied only to taper work. The grinding wheel, regulating wheel, and the work rest blade are set in a fixed relation to each other and the work is fed in from the front mechanically or manually to a fixed end stop. Either the grinding or regulating wheel, or both, are dressed to the proper taper.

center line average (CLA): See roughness average.

center line gage: A tool comprised of a pair of conical probes having 60°points that are attach to both the fixed and movable jaw of digimatic, dial, and vernier calipers. Used for measuring centerline distances of workpiece holes.

center of gravity: The center of gravity of a body, volume, area, or line is that point at which if the body, volume, area, or line were suspended it would be perfectly balanced in all positions. For symmetrical bodies of uniform material it is at the geometric center. The center of gravity of a uniform round rod, for example, is at the center of its diameter halfway along its length; the center of gravity of a sphere is at the center of the sphere. For solids, areas, and arcs that are not symmetrical, the determination of the center of gravity may be made experimentally or may be calculated by the use of formulas that can be found in reference books such as *Machinery's Handbook.*

center distance, gear: The distance between the parallel axes of spur gears and parallel helical gears, or between the crossed axes of crossed helical gears and worm gears. Also, it is the distance between the centers of the pitch circles.

center punch: A hand punch with a good, solid shank to withstand hammer blows and a taper toward the point to allow the mark to be seen clearly. The shank is knurled to provide good finger grip; the top end is slightly chamfered to prevent the edge from becoming burred from constant hammer blows. Used for marking the center of a point or position, usually for starting a drill. The point of the center punch should be a sharper angle than the point of the drill, to insure the drill starting true. The point of a center punch is usually ground to a conical shape of 90°.

Centigrade (C): A unit of temperature in which the interval between the freezing point of water and the boiling point is divided into 100 units, or degrees, with 0° representing the freezing point and 100° the boiling point. To convert a temperature given in degrees Centigrade to the corresponding degrees Fahrenheit by multiplying it by 9/5 (or 1.8) and adding 32 to the product, being careful to respect the signs if the centigrade temperature was negative. The Centigrade scale was conceived by the Swedish scientist A. Celsius (1701-1744); his name is generally and officially applied to the scale, even though Centigrade is more meaningful.

central plane, gear: In a worm gear, this is the plane perpendicular to the gear axis and contains the common perpendicular of the gear and worm axes. In the usual case with the axes at right angles, it contains the worm axis.

centrifugal casting: A casting made by pouring metal into a rotating or revolving mold, used for the production of hollow cylindrical castings of consistent and controllable wall thickness, thereby avoiding the necessity for central cores.

Ceralumin®: A registered trademark for a line of aluminum alloys nominally containing 1-3% copper, 1-2% silicon, 0.3-1.5% iron, 0.1 - 1.0% magnesium, 0.05 - 0.3% niobium.

ceramels: A material having the combined properties of metals and ceramics made from imbedding refractory oxides or carbides in a metallic base.

ceramics: (1) The art and science of making objects from earthy raw materials such as clay, especially when the process requires the application of heat as in firing, baking, or burning metallic colors into them; (2) a cutting tool material (substrate) made from aluminum oxide [Al_2O_3], zirconium oxide [ZrO_2], beryllium oxide [BeO], or silicon nitride [Si_3N_4]. Ceramics are capable of speeds 2 to 4 times faster than cemented carbide, but have less toughness and thermal shock resistance. They are capable of cutting hardened steels without the use of cutting fluids.

ceramic-tipped tools: Metal cutting tools made with ceramic inserts made from fused, sintered, or cemented metallic oxides.

cerargyrite: [AgCl] Hornsilver. A native isomer of silver chloride and chief silver ore.

cerium: Rare earth metallic element. Highly reactive; a strong reducing agent; burns in air like magnesium, but is more intensive; decomposes in water. Used in cerium-iron alloys; pyrophoric alloys; production of spheroidal graphite cast-iron; jet engine parts. Improves the high-temperature strength and ductility of magnesium and aluminum alloys.

Symbol: Ce   **Density (g/cc):** 6.92; 8.24 (á)
Physical state: Gray solid   **Specific heat:** 0.0448
Periodic Table Group: IIIB   **Melting point:** 1,468°F/798°C
Atomic number: 58   **Boiling point:** 6,199°F/3,426°C
Atomic weight: 140.12   **Source/ores:** cerite, monazite, bastnasite
Valence: 3,4

cermets: A term derived from the combined first syllables of *cer*ramic and *met*al. A semi-synthetic cutting tool material made by powder metallurgy techniques in which a ceramic is chemically or mechanically bonded to metals such as titanium carbide [TiC], aluminum oxide [Al_2O_3], titanium boride [TiB_2], tantalum carbide [TaC], uranium dioxide [UO_2], zirconium carbide [ZrC], etc. This process combines the high-temperature strength and hardness of the ceramics with toughness, and mechanical/thermal shock resistance of the metal component. Cermets can operate at close to 1,832°F/1,000°C, and for short periods at 3992°F/2200°C. Molybdenum disilicide [$MoSi_2$], molybdenum aluminide, and nickel aluminide are cermets that are capable of being coated on materials by vapor deposition and by flame-spraying.

cerous: Containing trivalent [cerium(III) or cerium(3+)] cerium.

cesiated: Cesium coated.

cesium: A rare element of the alkali metal family. A silver-like soft solid at room temperature; liquid above melting point. Highly reactive: reacts with water, forming explosive hydrogen, and explosively with oxygen. Contact with oxidizers may cause fire and explosions.

Symbol: Cs   **Density (g/cc):** 1.87
Physical state: See above   **Specific heat:** 0.0482
Periodic Table Group: IA   **Melting point:** 83°F/28.5°C
Atomic number: 55   **Boiling point:** 1,242°F/679°C
Atomic weight: 132.91   **Source/ores:** Pollucite
Valence: 1   **Mohs hardness:** 0.2

CGA: Abbreviation for the Compressed Gas Association.

cgs: Abbreviation for centimeter-gram-second (SI) units.

CGS-II®: Registered Trademark of the General Electric Company for its diamond superabrasive designed for grinding steel and cemented carbide composites.

chalcocite: Also known as copper glance. A native cuprous sulfide (Cu_2S); an ore of copper. Mohs hardness: 2.5-3.

chalcopyrite: Also known as copper pyrites, yellow copper. A native copper-iron sulfide ($CuFeS_2$) and important ore of copper. A metallic, yellow ore. Mohs hardness 3.5-4.

chamfer: (1) *noun*: A beveled surface on a workpiece used to eliminate an otherwise sharp edge that can become damaged; (2) *verb*: The act of cutting or "softening" the edge of a workpiece at an angle of less than 90°; (3) in thread design, the conical surface at the starting end of a thread; (4) in tapping, the tapering of the threads at the front end of each land of a tap by cutting away and relieving the crest of the first few teeth to distribute the cutting action over several teeth.

chamfer tap: A general name for taps containing threads at the front end of the land that are cut away and relieved (there are no threads) at the crest of the first few teeth. The purpose of this modification is to allow easier starting by distributing the cutting action over several teeth as the threads are being cut. When the tapering amounts to 7 to 10 threads, the tap is called a *taper* tap; 3 to 4 threads, a *plug* tap. A *bottoming* tap may have 1 to 2 tapered threads or have full threads to the bottom of the hole.

chamfer relief: A tap term describing the gradual decrease in land height from cutting edge to heel on the chamfered portion, providing clearance for the cutting action as the tap advances.

charcoal: Also known as activated carbon. A highly porous and amorphous form of carbon obtained from the incomplete combustion, destructive distillation, or calcination of organic material such as wood, animal bones, or other carbonaceous material. It has many uses including chromium electroplating.

charge: (1) A term used as a noun to describe the materials, load or burden, usually metals or ores, that are placed into a furnace for melting; (2) a verb used to describe the act of placing materials, etc., into a furnace for melting.

Charpy impact test: An impact-fracture test for metals used to determine the toughness and shock resistance of a standard specimen to the generation of a crack. The specimen usually contains a V-notch and is supported at both ends in a horizontal position, as a simple beam. The test is conducted by allowing a falling pendulum to strike the specimen with a single blow behind the notch causing the specimen to break. Then energy absorbed in fracturing the specimen, as determined by the height rise of the pendulum, is a measure of impact strength or notch toughness of the metal.

chatter: A problem affecting the accuracy and finish of metals, caused by rapid vibration of the tool away from the work or *vice versa* and identified by the little ridges, grooves, or lines (chatter marks) appearing at consistent intervals on the surfaces of the workpiece. The spacing of the chatter marks depends upon the frequency of vibration. Chatter is often self-sustaining and can be prevented or reduced by correcting the problem, which can be caused by excessive speed, too light a feed or rate of table travel, loose spindle, poor choice of cutting material, unbalanced workpiece, etc.

chatter marks: See chatter.

chelate: A type of molecular structure in which a heterocyclic ring is formed by unshared electrons of neighboring atoms. An organic compound in which atoms of nonmetals form rings by bonding to a central (metal) atom as part of each ring molecules. The nonmetal atoms are called *ligands*.

chelating agent: A substance added to a system; it combines with metal ions to form ring molecules, thereby precluding the normal ionic effects of the metals present. Chelating agents are added to electroplating baths to minimize the effects of certain metallic ions.

chemical: Any element, chemical compound, or mixture of elements and/or compounds.

chemical attack: Gradual degradation of the surface of a metal in a liquid or gaseous environment, chiefly oxidation or selective attack such as pitting and cracking. Temperature of the surface layer can be a significant factor in chemical attack.

chemical cutting fluids: Aqueous solutions of various chemical compounds and additives such as chlorine, sulfur, and phosphorus that form a solid lubricating film between the tool and workpiece surface. They generally do not contain petroleum products. Plain fluids containing no wetting agents or lubricant additives are used mainly for surface grinding operations while those with wetting agents and chemical lubricants are used for a wide range of machining operations. See also additives, cutting oils.

chemically active metals: Generally refers (but not restricted) to chemicals such as sodium, potassium, beryllium, calcium, powdered aluminum, zinc, and magnesium. These metals can cause violent reactions with certain other substances and materials.

chemical milling (CHM): The process of producing metal parts by using controlled chemical etching in corrosive acid, alkaline pickling, or etching baths to remove surface metal. When the workpieces are immersed in the etching solution, unwanted metal is uniformly dissolved and removed from all exposed surfaces. Used for making components where exact tolerances are required.

chemical polishing: A process used for polishing the surface of a metal in a appropriate solution, such as hexafluorophosphoric acid (HPF_6), phosphorofluoridic acid (FH_2O_3P), difluorophosphoric acid (HPO_2F_2), etc., without the use of mechanical abrasion or electric current.

Chemoy®: A proprietary process involving the immersion of steel articles in an alkaline oxidizing solution that imparts a black, rust-resistant, oxide film.

chi: The Greek letter x.

chill cast: Metal produced in molds having metal plates or "chills" inserted in the face of a sand mold. The chills are used to promote relatively rapid solidification of the molten metal; this rapid cooling results in the formation of cementite (iron carbide [Fe_3C]) and white cast iron. This process produces a finer grain and a more uniform distribution of impurities than sand casting.

chilled cast iron: Gray iron castings having wear-resisting surfaces of white cast iron. These surfaces are designated by the term *chilled cast iron* since they are produced in chill cast molds.

China silver: A nonferrous alloy nominally containing 50-70% copper, 10-30% zinc, 10-20% nickel.

Chinese art metal: A nonferrous alloy nominally containing 64-79% copper, 15-20% zinc, 10-15% lead, and 1% tin.

Chinese bronze: Copper alloys nominally containing 72.5-80% copper and 15-22% tin. Also a general term for alloys nominally containing 72-75% copper, 15-19% lead, 10-14% zinc, and 1-5% tin.

Chinese white copper: An alloy nominally containing 40% copper, 31-32% nickel, 25% zinc, and 2-3% iron.

chipbreaker: A tool feature such as a groove that prevents a continuous chip from growing to such a length that it ruins the work, or becomes a nuisance or safety hazard.

chip driver: See spiral point.

chip load: See ipt. (inch per tooth).

chips: Small pieces of material removed from a workpiece by cutting tools, or by abrasion.

chisel: See cold chisel, cape chisel, diamond chisel, round chisel.

chisel edge: A drill term used to describe the edge at the ends of the web that connects the cutting lips.

chisel edge angle: A drill term used to describe the angle included between the chisel edge and the cutting lip as viewed from the end of the drill.

chisel steel: Although this term is often used to describe a straight carbon steel containing 0.85-0.95% carbon, it is also used for low alloy tungsten-chromium steel and alloy steels nominally containing 97.6-98.7-% iron, 1-2% chromium, and 0.3-0.4% carbon, or 96.7% iron, 3% nickel, and 0.4% carbon.

Chlorimet®: A proprietary product of the Duriron Co. for a range of nickel-base cast alloys, containing typically nickel, molybdenum, chromium, balance mainly iron and very small amounts of carbon.

Chlorimet®2: A proprietary product of the Duriron Co. containing 64.9% nickel, 32% molybdenum, 3% iron (max), 1% silicon, and 0.10% carbon.

Chlorimet®3: A proprietary product of the Duriron Co. containing 60.93% nickel, 18% molybdenum, 18% chromium, 3% iron (max), 1% silicon, and 0.07% carbon.

chlorination: The process of bubbling chlorine gas through the melt or ore-roast of metals, such as magnesium, used for degassing and removing entrapped oxides.

3-chlorocoumarin: [$C_9H_5O_2Cl$] A light yellow, crystalline solid used in tin plating.

chloroplatinic acid: [$H_2PtC_{16}\cdot6H_2O$] A red brown crystalline solid used in electroplating.

chordal addendum, gear: The height from the top of the tooth to the chord subtending the circular-thickness arc.

chordal hook: A tap term used to describe the angle between the chord passing through the root and crest of a thread form at the cutting face, and a radial line through the crest at the cutting edge.

chordal thickness, gear: Length of the chord subtended by the circular thickness arc (the dimension obtained when a gear-tooth caliper is used to measure the thickness at the pitch circle).

chromadizing (chromodizing, chromatizing): Application of a metal-conditioning solution of chromic acid to the surface of aluminum or aluminum alloys, used to improve adhesion of corrosion-resistant coatings such as paints and resins.

Chroman®: A proprietary trade name for a line of nickel alloys nominally containing approximately 80/20% nickel-chromium annoy, with 1% manganese, and some may contain up to 20% iron.

chromate treatment: The treatment of metal in a solution of hexavalent chromium to produce a conversion coating of tri- and hexavalent-chromium compounds on the metal's surface, used to increase corrosion resistance and improve adhesion of subsequent paint or resin coatings.

Chromax®: A proprietary trade name for a heat and oxidation-resistant alloy of low carbon content, containing approximately 50% iron, 35% nickel, 15% chromium.

Chromax® bronze: A proprietary trade name for a corrosion-resistant, hard, high-strength bearing alloy nominally containing 67% copper, 15% nickel, 12% zinc, 3% chromium, and 3% aluminum.

Chromel® C: An alloy nominally containing 57-62% nickel, 22-28% iron, 14-18% chromium, 0.8-1.6% silicon, 0-1% manganese, and 0-0.2 carbon. It is used principally for making electric resistance alloys and offers good resistance to seawater and wet, sulfurous environments.

chrome-molybdenum steel: A light alloy of iron having a carbon content of 0.25-0.35%, and containing chromium (0.35-1.10%), molybdenum (0.08-0.35%), manganese (0.4-0.6%).

chrome plating: Electrodeposition of a thin layer of chromium on an object by passing an electric current through an aqueous solution or plating bath containing dissolved salts of chromium. The cathode is the material being plated, which can be metal or plastic. The anode is chromium that forms salts which dissociate into positively charged metal ions that are deposited on (coating) the cathode, which can be of any shape.

chrome steel: A family of alloys nominally containing 0.20-1.60% chromium and 0.2-2% carbon, although the chromium content may be as high as 25% in specialized heat-resistant and wear-resistant steels, and in stainless steels. Ball bearing steels often contain 1-1.5% chromium, 1% carbon, and 0.5% manganese or 12% chromium and 0.4-2.0% carbon.

chromet: A bearing alloy nominally containing 89-90% aluminum and 10-11% silicon.

chrome-vanadium steel: An alloy nominally containing 98.7-99.49% iron, 0.50 - 1.10% chromium, and 0.01 - 0.20% vanadium.

chromic acid: [CrO_3] Dark-purplish-red or brown crystalline solid used for chrome plating, anodizing. process engraving

chromic chloride: [$CrCl_3$] A red-violet crystalline solid used for chromium plating including vapor plating.

chromic sulfate:[$O_{12}S_3 \cdot 2$ Cr] Violet or red powder used in chrome electroplating; as corrosion inhibitors; component of metal alloys such as stainless steel.

chromite: Chrome iron ore, chrome ironstone [$FeCr_2O_4$] The principal ore of chromium and a natural oxide of ferrous iron and chromium (containing about 69% of chromic oxides), possibly with some aluminum and magnesium present. A highly heat-resistant and nonconductive material used as a furnace lining refractory. Iron-black to brownish-black in color. Hardness: 5.5.

chromium: CAS number: 7440-47-3. An alloying element used in steel and cast irons, in copper, nickel, aluminum, and cobalt; used in chrome plating, powder coating. Chromium improves hardenability; and, together with high carbon, provides both wear resistance and toughness, a combination valuable in certain tool applications. However, high chromium raises the hardening temperature of the tool steel, and thus can make it prone to hardening deformations. A high percentage of chromium also affects the grindability of the tool steel. Carcinogenic. Chromium ores classified as "metallurgical" must contain a minimum of 48% Cr_2O_3 and have chromium-iron ratio of 3:1.

Symbol: Cr	**Density (g/cc):** 7.190
Physical state: Gray solid	**Specific heat:** 0.0793
Periodic Table Group: VIB	**Melting point:** 3,407°F/1,875°C
Atomic number: 24	**Boiling point:** 4,842°F/2,672°C
Atomic weight: 51.9961	**Source/ores:** chromite
Valence: 2,3,6	**Oxides:** CrO, CrO_2, $Cr_2 O_3$, CrO_3
	Crystal structure: b.c.c.
	Brinell hardness: 70-90 (annealed)

chromium bromide: One of several compounds of chromium and boron having high melting points, are very hard and corrosion resistant, and may be suitable for use in jet and rocket engines, cermets, refractories.

chromium-bronze: Copper-tin, or copper-zinc alloys with a chromium content of up to 5%.

chromium-copper: An age-hardening alloy nominally containing 0.5 -1.0% chromium, 0.015 maximum lead, 0.04 maximum phosphorus, 0.15 maximum other chemicals, balance copper. Has good tensile strength and resistance to oxidation. Also a casting alloy containing 1% chromium.

chromium-manganese: An alloy containing 70% manganese and 30% chromium, used in copper to increase elasticity and for making hard steels.

chromium-molybdenum: An alloy containing a 50-50 mixture of chromium and molybdenum, used for making hard steels.

chromium-nickel: An alloy containing 90% nickel, 10% chromium;or, a 50-50 mixture of chromium and nickel, used in making hard steels.

chromium-nickel steel: See stainless steel.

chromium steel: See steel and stainless steel.

chromium-vanadium steel: Ferrous alloys nominally containing 1-1.5% chromium, 0.3-0.4% carbon, and 0.15-0.25% vanadium. Chromium vanadium spring wire contains 0.45-0.55% carbon, 0.60-0.90% manganese, 0.12-0.30% silicon, 0.80-1.10% chromium, and 0.15-0.25% vanadium.

chromizing: A general name for various methods used to form a protective surface layer on metals by diffusing them with chromium. Depending on material to be treated, this process can be accomplished in sealed retorts at elevated temperature of 2278-2462°F/1250 -1350°C, by ion exchange through heating in chromium chloride vapor at 1470-1650°F/800-900°C, or by electroplating with chromium followed by a diffusion treatment.

chromous formate: [$Cr(HCOO)_2$] A reddish crystalline solid used in electroplating.

Chrom-X®: A proprietary trade name for high-carbon ferrochromium steel alloy.

chuck: A device that is mounted on a machine tool spindle, used to hold a rotating cutting tool or workpiece. There are many different types of chucks that can be actuated manually, hydraulically, or pneumatically. The universal chuck has three or four jaws which are controlled to move together, making it easy to center round stock with a fair degree off accuracy. The independent chuck is made to hold work between four jaws, which are adjusted independently and allow irregular shaped pieces, as well as regular shapes, to be positioned for lathe operations. The Jacobs chuck is mounted on a taper shank to fit in either the tailstock or headstock of the lathe. It is used primarily in the tailstock for center drilling and boring.

chuck key: A device used for adjusting chuck jaws.

CIM: Abbreviation for computer-integrated manufacturing.

cinnabar: [HgS] Natural mercuric sulfide, the most important ore of mercury. A solid that is usually red, scarlet, or reddish-brown.

circular pitch: (1) In gear design, the distance between similar flanks of adjacent teeth, measured along the pitch circle. In helical gearing this is called the transverse pitch; (2) length of the arc of the pitch circle between the centers or other corresponding points of adjacent teeth. Normal circular pitch is the circular pitch in the normal plane; (3) involute spline (p): is the distance along the pitch circle between corresponding points of adjacent spline teeth.

circular saw: Any sawing machine using a circular saw blade. Also called a cold saw, used for cutting structural materials.

circular thickness, gear: The length of arc between the two sides of a gear tooth, on the pitch circle unless otherwise specified. Normal circular thickness is the circular thickness in the normal plane.

CLA: Abbreviation for center line average.

cladding: (1) A plating method consisting of an outer layer, usually mechanically applied to a substrate metal, used to prevent corrosion; (2) the process in which two or three metals are bonded together by various methods in which each metal diffuses sufficiently into the other to form a permanent intermetallic bond and an alloy. The base metal is usually carbon or low-alloy steel but can be nonferrous metals as well. The cladding can be from 5 to 25% of the total thickness.

clamps: See c-clamps, cam-actuated clamps, clamp straps, eccentric clamps, edge clamps, floating clamps, high-rise clamps, hinged clamps, hydraulic clamps, latch-action clamps, screw clamps, spiral clamps, strap clamps, swing clamps, toggle clamps, u-clamp, up-thrust clamps.

Clark's patent alloy: A copper alloy nominally containing 74.5% copper, 14.5% nickel, 7.2% zinc, 1.9% tin, and 1.9% cobalt.

class of thread: An alphanumerical designation to indicate the standard grade of tolerance and allowance specified for a thread.

clearance: (1) Generally, space allowed to prevent interference. However, the term *clearance* should not be used in specifications without indicating clearly just what it means; (2) in gear design, the amount by which the dedendum in a given gear exceeds the addendum of its mating gear. It is also the radial distance between the top of a tooth and the bottom of the mating tooth space.

clearance diameter: A drill term used to describe the diameter over the cutaway portion of the drill lands.

clearance fit: (1) A fit having limits of size so specified that a clearance always results when mating parts are assembled; (2) the relationship between assembled parts when clearance occurs under all tolerance conditions; (3) in thread design, a fit having limits of size so prescribed that a clearance always results when mating parts are assembled at their maximum material condition.

cleavage: The separation, split, or fracture of metals and minerals that takes place in crystal units along definite lines called crystallographic or cleavage planes.

These cleavage planes are closely related to the structure or crystalline form of the metal or mineral.

cleavage plane: See cleavage.

climb milling: A form of milling in which the rotating teeth of the milling cutter "climbs" or comes down on the work in the same direction as the feed at the point of contact. Differs from the standard milling operation in that, instead of the job passing under the cutter against the rotation of the teeth, it is fed in the same direction as the path the teeth take. The teeth cut downward rather than upward. Advantages of this method over standard milling include the following: play between feed-screw and nut is eliminated; increased tool life since chips pile up behind the cutter; improved surface finish since chips are less likely to be carried by the tooth; easier chip ejection since chips fall behind the cutter; and, decreased power requirements since a higher rake angle can be used on the cutting tool. Climb milling is especially important when machining cobalt, titanium and nickel based alloys. The cutting machine should be equipped with a backlash eliminator attachment. See also milling.

clinometer: A device used to measure the angle of a slope.

close annealing: See box annealing.

closed dies: Dies for forging shapes to closely predetermined tolerances, used so that the deformed material is enclosed within the die.

cloudburst treatment: A shot peening technique using a large amount of shot that is dumped on a workpiece.

clutch: A mechanical device used to link an engine and transmission. The clutch disconnects the motor from the transmission, allowing gears to be changed. When the clutch is re-engaged the engine and transmission resume contact causing both to turn together at the same new speed.

CNC: Abbreviation for computer numerical control.

Co: Symbol for the element cobalt.

CO_2 process: A process used to harden sand molds by using a solution of sodium silicate to moisten the sand and subsequently blowing carbon dioxide gas through the mold to react with the silicate. The reaction causes the sand grains to fuse into a solid, porous mass.

CO_2 welding: An arc welding technique using carbon dioxide as the gas shield. When welding, steel carbon dioxide is used to replace argon or helium in TIG welding. See also TIG.

coarse cut file: The coarsest of all American pattern file and rasp cuts.

coarseness of cut: A series of terms used to describe the relative number of teeth per unit length, the coarsest having the least number of file teeth per unit length; the smoothest, the most. American pattern files and rasps have four degrees of coarseness: coarse, bastard, second and smooth. These degrees of coarseness are only comparable when files of the same length are compared, as the number or teeth per inch of length decreases as the length of the file increases. The number of teeth per inch varies considerably for different sizes and shapes and for files of different makes. Curved tooth files have three degrees of coarseness: standard, fine and smooth. Swiss pattern files usually have seven degrees of coarseness: 00, 0, 1, 2, 3, 4, 6 (from coarsest to finest).

coarse pitch cutter: A cutter having approximately the same number of teeth as its diameter size in inches.

coated abrasives: An abrasive product that is affixed to a flexible backing such as paper, fiber, or textiles, or a combination of these materials, after the particles have been coated with an adhesive. Common examples are sand- or emery-paper. In the metal industry, aluminum oxide and silicon carbide are used extensively.

coated carbide: Coated carbides have a thin layer of very hard material deposited on their surface. The coatings most often used are titanium nitride [TiN], titanium carbide [TiC], aluminum oxide [Al_2O_3], or a mixture of these compounds. This material can be deposited by either physical or chemical vapor deposition. Coated carbides have advantages over uncoated carbides of the same grade by permitting a significant increase in tool life and cutting speed, and by adding crater and abrasion resistance to high-productivity machining operations. See CVD and PVD.

cobalt: CAS number: 7440-48-4. An alloying element in superalloys, electrical resistance alloys, high-temperature tool steels, spring, and bearing alloys. Cobalt resists softening at high temperatures, increases wear resistance, and increases hot hardness of steels; used in applications where that property is needed, such as cobalt-chromium high-speed tool steels, cemented carbides, cermets, and jet engines. Substantial addition of cobalt raises the critical quenching temperature of steel with a tendency to increase the decarburization of the surface, and reduces toughness. An animal carcinogen and an oxidizing agent.

Symbol: Co
Physical state: Gray metal
Periodic Table Group: VIII
Atomic number: 27
Atomic weight: 58.933
Valence: 2,3

Density (g/cc): 8.6
Specific heat: 0.0827
Melting point: 2,723°F/1,495°C
Boiling point: 5,198°F/2,870°C
Source/ores: cobaltite, linnaeite, smaltite, chloanthite
Oxides: CoO, Co_2O_3, Co_3O_4
Brinell hardness: 124 (cast); 300 (electrodeposited)
Crystal structure: f.c.c. (α); h.c.p. (ϵ)

cobalt alloy, base: An alloy containing 31-47% cobalt, 20-24% chromium, 20-24% nickel, 13-16% tungsten, 0-3% iron, 0-1.2% manganese, 0.2-0.5% silicon, 0-0.2% carbon, 0-0.2% lanthanum.

cobalt alloy, base (ASTM A567-2): An alloy containing 48-58% cobalt, 24-26% chromium, 9.5-12% nickel, 7-8% tungsten, 2% iron, 0-1% manganese, 0-1% silicon, 0.4-0.6% carbon.

cobalt ammonium sulfate: [$CoSO_4 \cdot (NH_4)_2SO_4 \cdot 6H_2O$] Ruby red crystalline solid used in cobalt plating.

cobalt brass: A copper alloy nominally containing 45-61% copper, 17-25% zinc, and 22-30% cobalt.

cobalt-chromium-molybdenum steels: A ferrous alloy containing 3.05% molybdenum, 2.15% chromium, 1.33% cobalt, and 0.65% carbon.

cobalt glance: See cobaltite.

cobalt-gold alloy: Cobalt alloys nominally containing 20-60% gold, used in magnetic films.

cobaltite: [$CoAsS$] An important cobalt ore, also known as cobalt glance. Silver-white to gray mineral, metallic luster. Contains 35.5% cobalt, density 6-6.3, Mohs hardness 5.5.

cobaltous chloride: [CoC_{l2}] A pale blue crystalline solid used in electroplating, magnesium refining, and as a solid lubricant.

cobaltous sulfate: [$CoSO_4$] Red to lavender or dark bluish crystalline solid or red powder used in plating baths for cobalt and nickel.

Cobaltron®: A proprietary steel alloy nominally containing 10-11% chromium, 2.25% cobalt, 1.5% carbon, 1.25% molybdenum, and 0.25% tungsten. See also cobalt steels.

cobalt steels: High speed tool steels containing 5-12% cobalt and varying amounts of chromium, molybdenum, tungsten, and vanadium in addition to iron and a small amount of carbon.

Cobrate®: A registered trademark of the Aqualon Co. for corrosion inhibitors used primarily for copper and copper alloys.

Cobratec®: A registered trademark of the PMC Specialties Group for corrosion, staining, and tarnish inhibitors for copper and brass. Used for cutting oils, lubricating oils, and antifreezes.

cohesion: The attractive force holding together the molecules of a substance.

coil breaks: Wrinkles, creases, or ridges that emerge across the surface of a metal sheet perpendicular to the direction of coiling that can occur when the sheet has been coiled hot and uncoiled cold.

coinage bronze: An alloy consisting of 95.5% copper, 3-4% tin, and 1-1.5% zinc. Used for some "copper" coins. The U.S. one-cent piece is 95% copper and 5% zinc.

coinage nickel: A cupronickel alloy consisting of 75% copper and 25% nickel. U.S. coins in the denomination of 10, 25, and 50 cents, and $1.00 (Susan B.Anthony) are made from a sandwich of coinage nickel cladding over copper.

coinage silver: Also known as coin silver. Alloys consisting of silver, copper, nickel, and possibly zinc. U.S. silver coins once were an alloy of 90% silver and 10% copper. Pre-1920 English silver coins were an alloy of 92.5% silver and 7.5% copper. The latter alloy is known today as sterling silver.

coining: A closed-die squeezing operation usually performed cold on metal or powdered metal in which all surfaces of the work are confined or restrained within the die.

coin silver: See coinage silver.

coke: The carbonaceous residue (70-80%) of bituminous coal, coal tar, pitch, and petroleum after the volatile constituents have been distilled off. Used chiefly for reduction of iron ore in blast furnaces and as a source of gases used for synthesizing a wide range of compounds.

coke oven gas: By-product gas from smelting iron ore obtained by the passage of hot air over the coke in blast furnaces. It is rich in hydrogen (more than 50%), methane, nitrogen, carbon monoxide [CO], and heavier hydrocarbons.

Colclad®: A proprietary trade name for a range of clad steels products containing a corrosion-resistant cladding of nickel or Monel, or stainless steel and construction steel bonded together.

cold anodizing: A method for increasing the natural oxide coating on light alloys by dipping them in a cold aqueous solution of 50% nitric acid [HNO_3] for 1-2 minutes.

cold chisel: A hand tool made in several shapes and are identified by the shape of their cutting edges (the angle can vary from 50° to 75°, with 60° for most general work). They are used for chipping flat surfaces, for cutting cold metal, removal of rivet heads, and cutting nuts and bolts which are rusted fast. The chisel steel is generally octagonal in shape and is high enough in carbon content to hold and edge and low enough in carbon content to withstand a forging heat. Available in various sizes from 1/4 in to 1 in diameter, and 4 in. to 8 in. length.

cold flow: The permanent deformation of a material occurring from prolonged compression or extension at or near room temperature. In metals this is known as creep.

cold galvanizing: A process used to protect ferrous metals by coating with zinc-rich paints. These paints are made from finely powdered zinc and a binder suspended in a solvent such a chlorinated rubber or polystyrene. When the solvent evaporates, a coating or film remains containing a high percentage (93-95%) of sacrificial zinc. See also galvanized iron.

cold short: A name given to the metal condition of brittleness existing in some metals at temperatures below the recrystalization temperature. Cold short metal cannot be worked under the hammer or by rolling, or be bent when cold without cracking at the edges. Such a metal may be worked or bent when at great heat, but not at any temperature that is lower than about that assigned to dull red.

cold shut: A casting defect in the surface of a metal resulting from low casting temperature, two streams of liquid meeting and failing to unite; or, a portion of the surface that is separated, in part, from the main body of metal by entrapped oxides formed when the casting was poured.

cold treatment: exposing to suitable sub-zero temperatures for the purpose of obtaining desired conditions or properties, such as dimensional or micro-structural stability. When the treatment involves the transformation of retained austenite, it is usually followed by a tempering treatment.

cold-welding: Solid-phase welding using high pressure such as hammering or rolling at temperatures at or near room temperature.

cold work: The plastic deformation produced when a metal is subjected to an external force such as drawing, rolling, hammering, or bending at a temperature well below its recrystallization temperature. This causes permanent strain-hardening that may be removed by annealing. See also annealing.

collar gage: See ring gage.

collet: (1) A cone-shaped split-sleeve bushing with a hole for holding circular or rodlike tools and workpieces (i.e. drill, reamer, or tap) by their outside diameter during grinding and machining. The split-sleeve allows the hole to be reduced in size. A collet generally provides greater gripping force and precision than a chuck utilizing jaws; (2) a small, self-centering chuck used on a lathe.

Colmonoy®: A proprietary trade name for a line of nickel-, ferrous-, or other- base alloys, all containing chromium and boron, and having self-fluxing properties.

Colmonoy® No. 6: A proprietary trade name for a corrosion-resistant nickel-based alloy nominally containing approximately 75% nickel and chromium bromide.

colorizing: A cementation process for coating the outer portion of steel with aluminum by heating the steel with aluminum powder, producing a thin, tightly adherent coating. See also cementation.

columbite: An ore of tantalum.

columbium: The original name for the element niobium (official name established in 1947); still used, chiefly by metallurgists.

columnar structure: In casting, a coarse structure of long parallel columns of grains, having the long axis at right angles to the mold face.

Combarloy®: The proprietary trade name for a high-conductivity copper-silver alloy.

combination depth and angle gage: See depth gage.

combination dies: A die used for multiple castings, having two or more different and distinct cavities.

combination drill and countersink: See center drill, step drill.

combination drill and reamer: See step drill.

combination squares: A tool containing square, spirit level, and protractor all mounted on an adjustable blade that can be made to slide along the head and be clamped at any desired place. The sliding blade contains a concave center groove which travels a guide in the head of the square and can be pulled out and used simply as a rule. The spirit level in the head can be used as a simple level and can be used to square a piece with a surface and at the same time ascertain whether one or the other is level or plumb. Also, the head of the combination square may be removed from the blade and replaced with an auxiliary *center head,* an instrument used to find the center of cylindrical pieces such as a shaft. See also try square.

combined carbon: The carbon content in steel or cast iron that is present, other than free carbon.

combustible liquid: A liquid material having a flash point above 141°F/60.5°C, and that burns relatively slowly.

Comet metal: Alloys nominally containing 67% iron, 30% nickel, 2.2% chromium, and small amounts of copper and manganese.

commercial bronze: Wrought copper alloys containing 90% copper and 10% zinc.

commercial bronze, leaded: Alloys containing 89% copper, 9.25% zinc, and 1.75% lead.

comminuation: The process of fragmenting a material into fine particles or powder by grinding, pounding, rasping, rubbing, cutting, pulverizing, etc. Used in metallurgy as a synonym for pulverization or trituration.

commutator: A device used to interrupt or reverse an electric current.

compacts: The name used to describe small objects, usually briquets, of metal or alloy, made by the compression of metal or alloy powder in a die or powder metallurgy pill press.

comparator: (1) A device used to calibrate instruments; (2) a machine designed to measure screw threads and similar parts by projecting an enlarged shadow on a chart to compare the screw with the desired standard.

Compax®: A registered Trademark of the General Electric Company for its polycrystalline diamond (PCD) superabrasive products including tool blanks.

complementary angles: Any two angles whose sum is 90°, whether they are adjacent or not, are called *complementary.*

component: An ingredient or constituent part.

composites: A general name applied to a mixture of any combination of structural materials that are designed to blend the desirable physical properties, making the final product superior to those of either component. The definition includes cermets.

compound, chemical: A substance consisting of two or more elements that have united chemically.

compound die: Any die designed to perform multiple operations simultaneously on a workpiece during a single stroke of the press, such as blanking and piercing (punching).

compound rest: A lathe fixture that slides and is attached to the carriage cross-slide; used for holding the toolpost.

compound slide: An essential part of the lathe that supports the tool post and swivels the cutting tool on the horizontal plane by adjusting its base, which is graduated 90°each way from the center.

compressed gas: Any material or mixture having in the container a pressure exceeding 40 psi at 70°F/21°C, or a pressure exceeding 104 psi at 130°F/54°C, regardless of the pressure at 70°F/21°C; or any liquid flammable material having a vapor pressure exceeding 40 psi absolute pressure at 100°F/38°C as determined by the American National Standard Method of Tests for Vapor Pressure of Petroleum Products (Reid Method) Z11.44-1973 (ASTM-American Society for Testing Materials D 323-72).

Compressed Gas Association (CGA): 1235 Jefferson Davis Hwy, Arlington, VA 22202-4100; Telephone: 703/412-0900. FAX: 703/412-0128. Web: http//:www.cgnet.com A nonprofit technical association devoted to the diffusion of knowledge of gases, including data and utilization.

compression: To press or push the particles of a member closer together.

compressive strength: Resistance to that force tending to crush a material.

computer-aided design (CAD): Engineering design using a computer graphics system to develop mechanical, electrical/electronic, and architectural designs.

computer numerical control (CNC): An automated system used to control machine tool operations by sending programmed instructions to a microcomputer that, in turn, operates working parts and electric motors for each machine.

concave: A depressed or hollow surface. The reciprocal of convex.

concentration: A process for enriching the mineral content of an ore by the physical separation and removal of impurities, waste material, or *gangue*.

conchoidal: A mineralogy term used to describe a type of surface formed by fracturing a hard solid by impact.

conchordial marks: A term used to describe the curved, roughly parallel, marks found on hard surfaces exhibiting fractures from fatigue in metal, or impact on a hard solid. These marks are suggestive of spider web design, or those found on bivalves (sea shells such as conch) from which the term is derived.

conditioning heat treatment: A preliminary heat treatment used to prepare a material for a desired reaction to a subsequent heat treatment. For the term to be meaningful, the treatment must be specified.

conductivity: The property of a substance that describes its ability to transfer heat, electricity, light, or sound. In metals, thermal and electrical conductivity are usually related because they are both due to the flow of electrons, the atomic nuclei remaining stationary. Conductivity is the reciprocal of resistivity.

conductivity, electrical: See conductivity.

conductivity, thermal: See conductivity.

cone: (1) An inner ring of a tapered roller bearing. *See also* cup; (2) the conical part of a gas flame as it exits the orifice of the tip.

con-eccentric relief: See thread relief.

Constantan: Generic name for a group of electrical resistance alloys nominally containing from 40-55% nickel, 45-60% copper, and minor amounts of iron and manganese.

Constantin: Generic name for a group of electrical resistance alloys nominally containing 54-54.3% copper, 44-46% nickel, 1.3% manganese, 0.4% iron.

contact cement: An adhesive that feels dry to the touch and that adheres to itself instantly on contact. Both surfaces to be joined (adherends) are coated with contact adhesive and allowed to dry for a specified period of time, properly aligned, and pressed together, instantly forming a permanent bond.

contact corrosion: Corrosive action between dissimilar metals that are in contact and in the presence of an electrolyte. For example, when aluminum is in contact with steel, a galvanic reaction is created, and an electric current will flow in the presence of atmospheric moisture, the electrolyte.

contact ratio, gear: The ratio of the arc of action to the circular pitch. It is sometimes thought of as the average number of teeth in contact. For involute gears, the contact ratio is obtained most directly as the ratio of the length of action to the base pitch.

contact stress, gear: The maximum compressive stress within the contact area between mating gear tooth profiles. It is also known as hertz stress.

container: Defined by OSHA as any bag, barrel, bottle, box, can, cylinder, drum, reaction vessel, storage tank, or the like that contains a hazardous chemical. Pipes and piping systems and engines, fuel tanks or other operating systems in a vehicle, are not considered containers.

continuous casting: A casting process in which an ingot, bar, billet, tube, or other shape is continuously solidified while it is being poured, so that its length is not restricted by dimension of the mold.

continuous mill: A rolling mill containing a battery of synchronized rolls that are set up in succession to process metal that undergoes continuous and successive reductions in size as it passes at increasing speed through the mill.

contouring CNC system: A numerical control system designed to direct the tool or workpiece to move along any curved path or at any angle.

controlled cooling: A term used to describe any controlled cooling process. Used after welding or hot-working to achieve a desired microstructure, minimizing the risk of cracking, hardening, or internal damage from thermally-induced stresses, hydrogen content, phase changes, etc.

conventional milling: Also known as up-milling. Milling in which the workpiece is rotating in the opposite direction of the table feed at the point of contact. Chips are cut starting with minimal thickness at initial engagement of cutter teeth with the work, and increase to a maximum thickness at the end of engagement. Conventional milling is recommended for milling castings or forgings with very rough surfaces due to sand or scale and should be used in all applications where the machine has backlash. See also milling.

conventional stress: Force divided by initial cross-sectional area.

conversion coatings: Surface coating that provide corrosion resistance all by themselves, and also increase the corrosion resistance and adhesion of subsequent coatings and paint films. They consist of a compound of the metal, formed by chemical or electrochemical treatment of the metal; e.g., oxide and phosphate coatings on steel, chromate coatings on aluminum, cadmium, magnesium, zinc.

converter: (1) A device used to change energy, electronic signals, or matter to another form; (2) a furnace (such as a Bessemer converter) in which a blast of air (that may be oxygen enriched) flows through or across a bath of molten metal or *matte* in order to oxidize the metal of impurities.

convex: A domed or rounded exterior services. The reciprocal of concave.

cope: In sand casting, the name given to the upper part of a foundry molding box or flask. See also drag.

copper: CAS number: 7440-50-8. An alloying element for brass, bronze, etc. A tough, reddish-brown, ductile and malleable metal. Next to that of silver, one of the best conductors of heat and electricity. It is the only metal which occurs native, abundantly in large masses. The strength of copper decreases rapidly with rise of temperature above 400°F/204°C. Between 800°F/427°C-900°F/482°C its strength is reduced about half that at ordinary room temperatures. There are about 300 copper-base alloys available, most of them are brasses and bronzes.

Symbol: Cu
Physical state: Red metal
Periodic Table Group: IB
Atomic number: 29
Atomic weight: 63.546
Valence: 1,2

Density (g/cc): 8.96
Specific heat: 0.0788
Melting point: 1,983°F/1,084°C
Boiling point: 4,653°F/ 2,567°C
Source/ores: Azurite, chalcopyrite, chalcocite, covellite, cuprite, malachite (azurmalachite), native copper, copper oxide, copper glance
Oxides: CuO, Cu_2O, CuO_2
Crystal structure: f.c.c.

copper amalgam: A hard alloy nominally containing approximately 74 -76% mercury and the balance copper.

copperas: [$FeSO_4$] A native ferrous sulfate. White copperas is zinc sulfate. Blue copperas is copper (cupric) sulfate. Green copperas is a ferrous sulfate that is a byproduct form the pickling of steel and other industrial processes. Yellow copperas, is copiapite, a sulfate of iron [$Fe_4S_5O_{18} \cdot H_2O$].

Copper and Brass Fabricators Council (CBFC): 1050 17th Street NW, Suite 440, Washington, DC 20036. Telephone: 202/ 883-8575. FAX: 202/331-8267.

copper cyanide: [$Cu(CN)_2$] A green powder used for electroplating copper on iron.

copper, deoxidized: Copper metal that is specially treated with phosphorus to eliminate porosity and free up all or part of the 0.05% oxygen that is normally present as cuprous oxide [Cu_2O]. The treatment produces copper that is more ductile than ordinary copper without the use of residual metallic or metalloidal deoxidizers.

copper, electrolytic: Refined by electrolysis, this is the purest form of commercially available copper.

copper glance: See chalcocite.

copper glycinate: [$(NH_2CH_2COO)_2Cu$] A blue crystalline solid used in electroplating.

copper-hardened rolled zinc: A zinc alloy containing 1% copper.

copper lactate: [$Cu(C_3H_5O_3)_2 \cdot 2 H_2O$] A greenish-blue crystalline solid or powder used as a fungicide in copper plating baths.

copper-lead: A heavy-duty bearing metal consisting of 60 -75% copper, up to 5% silver, 4% manganese, 4% tin, 2% antimony, 2% nickel, and the balance lead.

copper matte: A copper sulfide product containing approximately 50 -70% copper, obtained from copper smelting. Oxides or metals may also be present.

copper nitrate: [CuN_2O_6] A blue-green crystalline solid used in electroplating; a nitrating agent; putting a special finish on iron.

copper oxide: [CuO] black; [Cu₂O] red. A brownish-black, red, or yellow crystalline solid or powder used in electroplating.

copper phenolsulfonate: A green crystalline solid used in electroplating.

copper pyrites: See chalcopyrite.

copper scale: A coating of cupric and cuprous oxides formed on heated copper.

copper solder: An alloy containing about 28.5% lead and 71.5% tin.

copper steel: Corrosion-resistant ferrous alloys containing up to 1% copper, usually less.

copper yellow: See chalcopyrite.

Coppralyte®: A registered trademark of the DuPont Corporation for a group of products for electroplating copper.

core: (1) The interior central portion of an iron-base or ferrous alloy that after case hardening is substantially softer than the surface layer or *case*. See also case hardening; (2) foundry core, that part of a foundry mold used to cast internal cavities; (3) the central part, as in the core of a reactor, core of the earth.

core binder: Resinous material of various types (i.e., urea-formaldehyde resin, coal tar pitch, casein, etc.), used to improve the strength of foundry molding cores.

core blower: A device used for making foundry cores. Air under high pressure is used to blow and compress a core-sand mixture into a vented core box for subsequent casting.

core drill: A multi-fluted twist drill used to enlarge holes.

coring: A variable composition between the surface and core of a material and occurring across separate crystals of a casting, with the highest purity material at the center.

corner joint: A joint formed between two components located approximately at 90° to each other.

coronel: A nickel alloy nominally containing approximately 30% molybdenum, 6% iron, and the balance nickel.

corrosion: The gradual electrochemical or chemical change to metals or alloys caused by reaction with the environment. This reaction is accelerated by the presence of moist acids or alkaline environments, and corrosion products often taking the form of metallic oxides or sulfides. Corrosion can be combated by various means including the use of chemical treatments that form corrosion-resistant oxide films on the metal's surface, corrosion inhibitors, film-forming rust-preventives, and protective coatings such as paints, plastics, galvanizing, sherardizing; plating or cladding with other metals such as chromium, cadmium, copper, nickel, zinc. Corrosion can also be prevented or inhibited through modification of the composition of metals by incorporating noncorrosive metals, and creating alloys such as stainless steel.

corrosion embrittlement: The severe loss of ductility resulting from corrosion. May be due to the breakdown of the protective oxide layer and deep encroachment of the corrosive agent into the intergranular structure of a metal. May not be visually apparent.

corrosion fatigue: Metal damage or failure due to the effect of repeated or fluctuating fatigue stresses in a corrosive environment. Corrosion fatigue is characterized by shorter life than would be encountered by the material had the stresses or the corrosive environment been applied independently or successively.

corrosion fretting: Metal destruction caused by vibration or the oscillating movement of close-fitting metal surfaces.

Corrosion Inhibitor CS®: A proprietary product of the Calgon Corp. for a corrosion inhibitor for all kinds of metal and alloy surfaces in closed heating and cooling systems.

corrosion resistance: A common term used to describe the ability of metals to withstand the damaging effects of chemical action, including localized attack and stress corrosion cracking. See also corrosion.

Corrosist®: A proprietary trade name for a chlorine-resistant nickel alloy.

corrosive: The ability to dissolve or destroy. A corrosive material is any liquid or solid with pH ranges of 2-6 or 12-14, and that causes visible destruction or irreversible alteration of living tissue, or a liquid that has a severe corrosion rate or oxidation rate on metals A liquid is considered to have a severe corrosion rate if its corrosion rate exceeds 0.250 inch per year on steel at a test temperature of 130°F/54°C.

corrosive material: Any agent, solid, liquid, or gaseous, that attacks metals or that burns, irritates, or destructively attacks organic tissues by chemical action. Some widely used chemicals that have corrosive properties are as follows: Acetic anhydride, bromine, chlorine, fluorine, glacial acetic acid, hydrochloric acid, hydrofluoric acid [FH], nitric acid [HNO_3], potassium hydroxide [KOH], sodium hydroxide [NaOH], sulfuric acid [H_2SO_4].

corundum: A natural or artificial abrasive containing aluminum oxide [Al_2O_3], and variable amounts of impurities. Very hard, used as an abrasive in polishing operations. Mohs hardness 17. Also known as emery, which is an impure corundum containing iron oxides. Artificial corundum in a pure, crystalline, granular form is called Alundum® or Aloxite®.

costellizing: A chemical process for phosphatizing by immersing a metal to be treated in a dilute phosphoric acid saturated with ferrous phosphate [$Fe_3(PO_4)_2 \cdot 8H_2O$].

cotter pin: Also known as split cotter, split pin, etc. Made from soft steel wire and shaped like a pin with a split up the center, this locking device or fastener is slipped through a hole drilled crosswise through a nut and bolt or behind a castle nut in a stud, shaft, or similar member. Its ends are bent and spread apart to retain it in place, keeping the nut from loosening, slipping, turning, or otherwise unscrewing.

counterbore: A cutting tool having a small end called a *pilot* used to guide the drill bit into the hole and keep it centered. The pilot should be oiled prior to counterboring. See also counterboring.

counterboring: A drilling process of enlarging previously drilled holes. This process can be used to enlarge one end of a hole to provide countersinking or a flush seat for a screw (such as a fillister-head cap screw) or nut, or to obtain better surface finish and tolerances when machining cast, forged, pressed, or extruded materials. Counterboring is called spot-facing if the depth is shallow. See also spot-facing.

countersink: A boring tool used to enlarge the entrance of a hole with a conical depression or beveled edge to receive the head of a screw, usually a flat headed screw that can sit flush with the surface of a workpiece. These bits have different cutting edges depending on the screws being used. An 82° countersink is used for flat head screws. Countersinks may be used in a hand drill, bit brace, or drill press. See also center drill.

countersinking: The process of using a countersink. The size of the countersink hole depends upon the head of the screw.

covered electrode: A rod of filler-metal (electrode) consisting of a metal core wire having a relatively thick coating used in arc welding. The coating material provides shielding for the molten metal from the atmosphere, improves the properties of the weld metal, and stabilizes the arc through ionization. The covering is usually mineral or metal powders mixed with cellulose or other binder.

Cowles process: A direct method for manufacturing aluminum alloys from aluminum ores by reacting with carbon in an electric furnace containing the alloying metal.

Cox process®: A proprietary electrodeposition and dipping process used for the removal of rust and the addition of a corrosion-resistant coating on steels.

Cr: Symbol for the element chromium.

crack arrester: A process or design feature used to halt or interfere with the propagation of brittle cracks, usually in large welded steel structures. The arrester element may be a weld made with special electrodes, riveted joints, or the deliberate addition of non-metallic inclusions in the steel.

crater wear: A condition resembling a smooth, regular depression produced behind the cutting edge of a cutting tool or on the "rake face" or top of an insert. This is caused by chips sliding across these surfaces.

creep: A three-stage, continuous deformation, elongation, or nonrecoverable strain of a metal, or metallic part, under constant stress at constant temperature over time. The elongation may be considerably less than that required to cause rupture or breaking. Primary creep stage is the strain occurring at a transient or diminishing rate. Secondary creep stage is the strain occurring at a minimum and almost "steady state" rate. Tertiary creep stage is the strain occurring at an accelerating rate. See also cold flow.

creep ductility: The elongation of a creep specimen at failure (the breaking or rupture point). See also creep.

creep limit: (1) The maximum stress that will cause a given amount of strain in a given material at a constant temperature; (2) the maximum nominal stress that can be withstood over time by a specified material under constant load, at a constant temperature without observable creep. May be used synonymously with creep strength.

creep strength: The constant nominal stress that will cause a specified creep rate (unit of creep in a given time) at constant temperature. May be used synonymously with creep limit.

crest, thread: That (top) surface of a thread which joins the flanks of the thread and is farthest from the cylinder or cone from which the thread projects. The crest of an external thread is at its major diameter; the crest of an internal thread is at its minor diameter.

crest truncation: In thread design, the radial distance between the sharp crest (crest apex) and the cylinder or cone that would bound the crest.

cresyl phenyl phosphate: A chemical mixture of cresyl and phenyl phosphates used in extreme-pressure lubricants, hydraulic fluids. Combustible.

crevice erosion: A type of localized corrosion of a metal inside or outside a crevice, cavity, or pocket, caused by the concentration effects of dissolved salts, electrolytic composition differential, aeration, gases including oxygen. For example, stainless steels depend on the available oxygen in the environment for corrosion-resistance, but can be susceptible to crevice erosion in areas of relatively low oxygen concentration.

crimping: A forming operation consisting of small corrugations or serrations used to set down, lock, or otherwise close a seam in sheet metal.

critical: (1). Pertaining a point of delicate balance; an abrupt change; a turning point (beyond which safety may be jeopardized); (2) a relatively precise condition that follows a planned sequence, is reproducible, and approaches a point of delicate balance.

critical cooling rate: The minimum continuous cooling rate that is just sufficient to suppress an undesired transformation; for steel, the slowest rate at which it can be cooled to prevent the transformation of austenite.

critical humidity: The point at which the atmospheric relative humidity will increase the corrosion rate of a specific metal.

critical mass: The minimum amount of fissionable material required to sustain a self-sustaining chain reaction in a nuclear reactor.

critical point: (1) The temperature or pressure at which a change in crystal structure, phase, or physical properties occurs. Same as transformation temperature; (2) in an equilibrium diagram, that specific value of composition, temperature and pressure, or combinations thereof, at which the phases of a heterogeneous systems are in equilibrium. In other words, the critical point is that condition when two phases are about to become a single phase.

critical range or critical temperature range: Synonymous with transformation range, which is preferred.

critical point: In heat treatment of steel, the temperature at which some definite change takes place in the steel's physical properties. Critical points are important because steel must be heated to a temperature above the upper critical point and then quenched. When critical points for a steel is known, the furnace heat can be regulated using a pyrometer.

critical speed: A rotation speed of shafts and other bodies at which vibrations due to unbalanced forces reach a maximum. Above the speed, the vibrations decrease, but often attain other maxima at higher speeds, called second, third, and higher critical speeds.

critical strain: Plastic deformation in many metals and alloys that is just sufficient enough to cause the rapid growth of large grains during heating (or annealing) without involving phase transformation.

critical temperature: The point at which phase changes occur.

crocus: A fine, soft, red iron oxide material used as an abrasive, for polishing antimony.

crooksite: A mineral containing about 17% titanium, selenium, copper and silver.

crop: The defective end portions of an ingot that are removed as scrap prior to rolling.

cross feed: (1) In lathe work, movement of the cutting tool across the end of the workpiece; (2) in milling or surface grinding, movement of the table toward or away from the column.

cross-peen hammer: A common hammer used by machinists. The head is made of tool steel having one face that is flat, and the other a wedge-shaped cross-peen positioned with the edge perpendicular to the handle. The flat face is used for driving center punches, chisels, and for various other general purposes. The wedge-shaped end is used for purposes such as hand swaging or peening. The straight-peen hammer is similar with the edge of the wedge-shaped end positioned parallel to the handle.

cross rolling: The process of correcting directionality effect in metal sheet or plate by subsequently rolling at about 90° to the principal direction of the rolling operation.

cross-slide: A fixture attached to the lathe carriage, that can be moved in and out, used for holding the compound rest.

crown: A profile or contour on a sheet, plate, or roll where the thickness or diameter increases from edge to center.

CRT: Abbreviation for cathode ray tube.

crucible: (1) A conically-shaped vessel or pot having a rounded base and made of a refractory material, used for melting, fusing, or incinerating metals or other substances. Some types are equipped with a cover; (2) a steel industry term used to describe the firebox or hearth of a blast furnace for collecting the molten metal.

crucible furnace: An electric furnace made by imbedding high resistance wire in a refractory material.

crucible steel: A high quality "pot" steel, made by the crucible process.

cryogenic liquid: Defined by the United States Department of Transportation as a refrigerated liquefied gas having a boiling point colder than −130°F/−90°C at one atmosphere, absolute.

cryogenic treatment: A low-temperature treatment [ranging from −120 to −300°F/−84 − 184°C)] for improving the toughness of hardened tool steel by continuing the transformation of retained austenite into martensite, which is more desirable.

crystal: The fundamental unit of the solid phase or substance. The building blocks of a homogeneous solid having highly ordered characteristic shapes and cleavage planes due to the arrangement of an internal structure of atoms, ions, or molecules which form a definite, repeatable, symmetrical pattern in three-dimensions. The arrangement of the structure is called a *lattice*.

crystal face: The repeated characteristic surface of a crystal or a plane parallel to it.

crystalline fracture: A fracture producing bright facets, the plane faces, and sharp edges of a crystalline structure. See also granular fracture.

crystallization: The phenomenon of crystal formation by change from the molten, liquid, or gaseous state to a solid state of ordered, characteristic shape from nucleation and accretion. The most common example of crystallization from nature is the freezing of water into ice. It is a process used in industry for purifying materials by evaporation and solidification.

crystallogram: A photographic record containing an x-ray diffraction pattern of a crystal.

crystallography: The science and study of crystals and their structures. When applied to metals, this science is called metallography.

crystal structure types: The basic lattices of crystals which are categorized according to their symmetry are cubic, tetragonal, monoclinic, orthorhombic, hexagonal, prismatic, rhombohedral, triclinic, etc. They may be body-centered, face-centered, or base-centered. The cubic form has the highest symmetry. The form of a crystal is called its "habit."

Crystolon®: A manufactured abrasive produced from silicon carbide, coke, sawdust, sand, and salt and heated in an electric furnace at temperatures from 3308°F/1820°C to 4082°F/2250°C. Used for grinding and polishing brass, cast and chilled iron, etc.

Cs: Symbol for the element cesium.

CSA: Abbreviation for the Canadian Standards Association.

Cu: Symbol for the element copper.

cubic boron nitride (CBN): An ultra-hard cutting tool material (substrate) made from a white powder having a graphitelike, hexagonal plate structure and a metallic or ceramic binder. Compressed at 10^6 pounds per square inch, its hardness is just below that of diamond. Primarily used to cut machine hardened steels, this material is available as solid inserts, as tips brazed to carbide insert carriers, or grinding wheels.

Cubond®: A registered trademark of SCM Corp. for a copper brazing paste. Consists of metallic copper powder on cuprous oxide pigments of high purity in organic or petroleum vehicles that impart satisfactory suspension properties.

cup: (1) Any shallow cylindrical part or shell closed at one end; (2) an outer ring of a tapered roller bearing. *See also* cone.

cupal: Copper-plated aluminum.

Cupaloy®: A registered trade name for a corrosion-resistant alloy of copper, silver, and chromium, having high electrical conductivity.

cupellation process: A process for the separation of silver, gold, or other nonoxidizing noble metals from unrefined base metals. A metallic mixture of the base metal and lead are placed in a cupel, which is a shallow, porous crucible, and roasted in a blast of air in a muffle furnace or kiln. The base-metal oxides or impurities are volatilized or absorbed in the cupel, leaving the pure noble metal "button" to be decanted.

cupola: A cylindrical vertical furnace, similar to a blast furnace, used in a foundry for melting iron or other metals for casting.

cupola malleable iron: A malleable iron that exhibits good fluidity and will produce sound castings. It is used in the making of pipe fittings, valves, and similar parts and possesses the useful property of being well suited to galvanizing. The American National Standard Specifications for Cupola Malleable Iron (ANSI/ASTM 197-79) specifies a minimum tensile strength of 40,000 lbs/in^2 ; a minimum yield strength of 30,000 psi; and a minimum elongation in 2 in. of 5%.

cupping: An operation that produces a cup-shaped part.

cupralum: Copper with a lead cladding used for inert anodes.

cupric: Copper(II) or Copper(2+). Form of the word copper used in naming divalent copper compounds which are more common and more stable than cuprous compounds.

cuprite: Native copper oxide [Cu_2O] containing a high percentage (about 90%) of copper. Mohs hardness 3.5.

cupro-nickel: Cupro-nickel 30% containing 30% nickel. Cupro-nickel 10% contains 10% nickel and 1.3% iron.

Cupronickel®: A registered trademark for an alloy of copper and nickel that is strongly corrosion-resistant, especially to seawater. Contain 88.35% copper, 10% nickel, 1.25% iron, and 0.40% manganese. Used in coinage; in condenser plates and evaporator tubes for heat exchanges.

cuprotungsten: An alloy of copper and tungsten.

cuprotungstite: [$CuWo_4$] A native ore of tungsten.

cuprous: Copper(I) or Copper(1+). Form of the word copper used in naming monovalent copper compounds which are less stable than cupric compounds.

cuprous cyanide: [$Cu_2(CN)_2$] A cream-colored, amorphous powder used in electroplating.

cuprous potassium cyanide: [$KCu(CN)_2$] A crystalline solid used in electroplating.

cuproxide: [CuO] A native copper oxide ore.

cure: An adhesive term for the chemical reaction that caused a physical property changes, such as hardening of an epoxy during the bonding process. A *post cure* involves additional cure such as the application of heat or ultraviolet light, following initial cure.

curing: (1) Conversion of a raw material to finished and useful product, generally by the application of heat and/or chemicals, over a period of time, that induce physicochemical changes; (2) setting or hardening.

curling: Forming an edge of circular cross section along a sheet or at the end of a shell or tube.

current density: In an electroplating, the electric current per unit area of the object or surface being plated, expressed in amperes per square centimeter or amperes per square decimeter.

curved cut: File teeth which are made in curved contour across the file blank. The curved tooth file has teeth that are in the form of parallel arcs extending across the face of the file, the middle portion of each arc being closest to the point of the file. The teeth are usually single cut and are relatively coarse. They may be formed by steel displacement but are more commonly formed by milling.

Cusiloy®: A registered trademark for silicon-bronze alloys used for springs and wire. Contains 1-3% silicon, 1-2% sn, 0.5 -1% iron and the balance copper.

cut, file: A term used to describe file teeth with respect to their coarseness or their character (single, double, rasp, curved, special, etc.).

cutoff: A lathe operation for parting off a workpiece that is mounted in a lathe chuck. Also used to describe parting off using a cutoff wheel.

cutoff tool: Also called a *parting tool*. A lathe tool or blade made from relatively thin steel, used for parting off stock, cutting grooves, and cutting to a shoulder. The sides of a cutoff tool have top-to-bottom clearance tapers and should never be ground. However the front and top can be ground as follows: end relief angle, approximately 4-5° is adequate; an angle of 10-15° ground across the end of the cutoff tool will provide a better cut than a square end; a back rake of 5° is optional.

cutoff wheel: Thin composition or plastic disks impregnated with abrasives such as alumina, carborundum, or diamond. Normally used to cut or slot hardened steel, ceramics, and other materials, they are rotated at very high speeds (6,000-20,000 rpm) and are fed edgewise into the workpiece. Also known as slitting wheel. An abrasive cutoff saw is similar to a cutoff wheel. It uses abrasive disks instead of a metal blade. See also abrasive cutoff wheel.

cutting clearance: A term used to indicate the space between the punch and die on each side.

cutting fluids: Natural or synthetic liquids used to cool the cutting tool and workpiece, wash away chips, lubricate the bearing formed between the chip and lip of the cutting tool. Used to enable the cutting tool to produce a good finish, extend the cutting tool's life, improve dimensional accuracy, and protect the finished product from rust and corrosion. Cutting fluids are used on cutting tools such as drills, taps, dies, reamers, milling cutters, lathe cutting tools, and power saws. See also additives, cutting oils.

cutting speed: The linear or circumferential or peripheral distance traveled in one minute by a point on the tool or the workpiece in the principal direction of cutting; expressed in feet or meters per minute.

CVD: Abbreviation for chemical vapor deposition (coated carbide).

CWB: Abbreviation for the Canadian Welding Bureau.

CWS: Abbreviation for the Canadian Welding Society, Inc.

CWDI: Abbreviation for the Canadian Welding Development Institute (Institute Canadian de Development de La Soudure).

cyanide hardening: A process of case hardening an iron-base alloy, usually low-carbon steel, by holding above Ac_1 in contact with a molten sodium cyanide salt-bath. When the alloys are treated in this way they simultaneously absorb carbon and nitrogen from the fused salt. Cyaniding is usually followed by quenching in water or oil, or other suitable heat-treatment, to produce a thin case of high hardness. Also known as cyaniding. To clean up waste cyanide salts from case hardening of steel, react the salts at 1,202-1,292°F/650-700°C with waste ferric hydroxide [$Fe(OH)_3$] sludges from various operations.

cyaniding: See cyanide hardening.

cyanoacrylate adhesives: Also known as instant adhesives, Superglue®, or Eastman 910®. One-part adhesives that cure within seconds and form full-strength bonds within 24-hours, used for bonding plastics, metals, rubbers, leather, also strain relief of wires.

cycle annealing: An annealing process employing a predetermined and closely controlled time-temperature cycle to produce specific properties of microstructure.

cylindrical grinding: See also centerless grinders.

cycloid: The curve formed by the path of a point on a circle as it rolls along a straight line. When this circle rolls along the outer side of another circle, the curve is called an epicycloid; when it rolls along the inner side of another circle it is called a hypocycloid. These curves are used in defining the former American standard composite tooth form.

D: Abbreviation for diamond.

Δ, δ: See delta.

damping: Decreasing, hindering, or slowing-down the time of motion vibration or oscillations.

damping capacity: The ability of a material to dissipate the energy of mechanical strain.

D'Arcet metal: An alloy used in fire sprinklers and fusible links and plugs, nominally composed of 50% bismuth, 25% lead, and 25% tin.

dasymeter: An instrument used to measure the heat loss of a furnace by analysis of waste products and gases.

datum: A theoretically exact point, axis, or plane derived from the true geometric counterpart of a specified datum feature. A datum is the origin or point of reference from which the location or geometric characteristics of features of a part are established.

datum feature: A specified feature of a part that is used to establish a datum.

datum reference frame: Sufficient datum features, those most important to the design of the part, are chosen to position the part in relation to a set of three mutually perpendicular planes, jointly called a datum reference frame. This reference frame exists only in theory and not on the part. Therefore, it is necessary to establish a method for simulating the theoretical reference frame from the actual features of the part. This simulation is accomplished by positioning the part on appropriate datum features to adequately relate the part to the reference frame and to restrict motion of the part in relation to it. These reference frame planes are simulated in a mutually perpendicular relationship to provide direction as well as the origin for related dimensions and measurements. Thus, when the part is positioned on the datum reference frame (by physical contact between each datum feature and its counterpart in the associated processing equipment), dimensions related to the datum reference frame by a feature control frame or note are thereby mutually perpendicular. This theoretical reference frame constitutes the three-plane dimensioning system used for datum referencing. Often, a single reference frame will suffice. At other times additional datum reference frames may be necessary where physical separation or the functional relationship of features require that datum reference frames be applied at specific locations on the part. Each feature control frame must contain the datum feature references that are applicable.

datum target: A specified point, line, or area on a part used to establish a datum.

datum target line: A datum target line is indicated by the symbol "X" on an edge view of the surface, a phantom line on the direct view, or both. Where the length of the datum target line must be controlled, its length and location are dimensioned.

datum target point: A datum target point is indicated by the symbol "X," which is dimensionally located on a direct view of the surface. Where there is no direct view, the point location is dimensioned on two adjacent views.

dead: Referring to a molten alloy such as steel where there is no evolution of gas.

dead center: See center.

dead smooth file: Also known as a Swiss-type file or Swiss pattern file. A fine cut file with 100 or more teeth per inch, used for finishing work and available in grades varying from # 00 to # 6.

dead soft steel: Low-carbon steel that is fully annealed with less than 0.15% carbon.

deca-: See deka-.

decarburization: The loss or removal of carbon from the surface of iron or steel alloy during hot-working as the result of heating (into the austenitic region) in an oxidizing or other decarburizing medium that reacts with the carbon.

decarburize: Decarbonize

deci-: A prefix denoting a multiple of 10^{-1} or 1/10. For example, a decigram is one tenth of a gram; a deciliter is one tenth of a liter = 100 cc = 3.38 fl. oz (U.S).

decomposition: Breakdown of a material or substance into parts, elements or simpler compounds that may be caused by heat, chemical reaction, electrolysis, decay, biodegradation, or other process.

decontamination: The removal of hazardous substances from employees and their equipment to the extent necessary to preclude the occurrence of foreseeable adverse health effects.

dedendum: (1) *gear*: The distance from the pitch circle to the root circle. The depth of tooth space below the pitch circle or the radial dimension between the pitch circle and the bottom of the tooth space; that circle which passes through the theoretical bottom of a gear tooth; (2) *thread*: the distance from the pitch diameter to the root of the thread.

deep drawing: The drawing of deeply recessed, cup-shaped parts from sheet or strip material when the depth of the recess equals or exceeds the minimum part width.

deep drawing zinc: Commercial rolled zinc nominally containing 99.4-99.88% zinc, 0.06-0.3% lead, and 0.06-0.3% cadmium.

deflection: The deviation of a body from a straight line or plane when a force is applied to it.

deformation: (1) The altering of the shape, flow, or elasticity of a material without rupture; disfigurement, as the elongation of a test piece under tension test.

degassification: The elimination of dissolved gases from metals. Usually performed by adding deoxidizing agents to molten metal or by using *ultrasonics;* or, on solid metal by heating in air, by using vacuum, and/or prior to coating or plating.

degreaser: A chemical solvent used to remove oils and fats, useful for surface preparation, pre-painting, and pre-bonding. Some commonly used industrial degreasers used or for metals are carbon tetrachloride, cyclohexanone, dichloroethylene, dichloropentane, ethylene dichloride, propylene dichloride, *sym*-tetrachloroethane, 1,1,1-trichloroethane, and trichloroethylene. Nearly all degreasers are chemicals of health and environmental concern.

degree: (1) A division or interval marked on a scale, generally a difference in temperature (as °F) or direction (as °angle); (2) condition in terms of some unit or in relation to some standard.

deka-: A U.S. metric prefix denoting a multiple of 10 times. Sometimes spelled deca- (SI standard).

Delhi hard: A ferrous alloy containing 79.4-81.3% iron, 16.5-18% chromium, 1.1% carbon, 0.75-1% silicon, 0.35-0.5% manganese.

deliquescent: Refers to substance which absorbs moisture from the air to the point of becoming liquid.

delta (Δ, δ): The fourth letter of the Greek alphabet and symbol for deviation or total elongation (deflection).

delta iron: one of the four solid phases of pure iron which occurs between the temperatures of 2,782-2,554°F/1,528-1,234°C.

delta metal: Alloys nominally consisting of 55-60%copper, 38.2% zinc, 1.8% iron, and possibly small amounts of nickel, lead, manganese. Similar in composition to Tobin metal.

demagnetizing: Removal of magnetic properties.

dendrite: A tree- or fern-like branching pattern of crystals grown in cast metals during slow cooling through the solidification range.

dendritic segregation: Amorphous distribution of alloying elements through the branching patterns or dendrites. See also dendrite.

denitrification: The removal of nitrogen or nitrates.

density: A measure of the mass per unit volume of a substance, usually expressed in lb/in^3 or g/cm^3 (g/cc) at a temperature of 73.4°F/23°C. Density information is used mainly to calculate the amount of material required to make a part of a given volume, the volume being calculated from drawing dimensions. For example, lead has much greater density than aluminum.

dental amalgam: Also known as silver filling. Alloys nominally containing 33% silver, 52% mercury, 12.5% tin, 2% copper, and 0.5% zinc.

deoxidized copper: See copper, deoxidized.

deoxidizer: An agent used to remove oxygen from molten metals.

deposition rate: A metal surfacing term describing the amount of material that forms or adheres to a base metal per unit of time.

depth gage: A measuring device or tool which contains a narrow rule or rod with a sliding head (also know as sliding stock or base) set at right angles to the rule and with a means of clamping the slide to lock the reading. Used for determining the depth of holes, recesses, shoulders, slots, keyways, grooves, etc. The body of the tool is placed across the hole while the graduated rule or rod is lowered into the hole or recess to be measured. Direct reading may be obtained, or the gage can be fixed to a particular measurement and used for testing depth. Another depth gage is the combination depth and angle gage which can contain and optional hook mounted at the end of the rule and be used to measure from a recess or undercut. Refinements of the basic depth gage includes the vernier depth gage, and micrometer depth gage. See also hook rule.

depth of cut (DOC): The distance a cutting tool is advanced into the revolving workpiece, measured at right angle to the piece.

depth of engagement, involute spline: The radial distance from the minor circle of the internal spline to the major circle of the external spline, minus corner clearance and/or chamfer depth.

desiccant: A drying agent. A substance that removes moisture from a substance or system.

design size: The design size is the basic size with allowance applied, from which the limits of size are derived by the application of tolerances. Where there is no allowance, the design size is the same as the basic size.

desulfurization: The removal of sulfur.

detergent: A cleaning agent. Any cleansing agent that reduces the surface tension of water and prevents the precipitation of soluble sludge. A detergent concentrates where oil and water interface, and act as an emulsifying agent.

detonation: A decomposition reaction which is extremely rapid and self-propagating until it reaches explosive violence.

detonator: Primer; a substance used to set off or ignite an explosive material.

Devarda's alloy: Also known as Devarda's metal. An alloy containing 50% copper, 45% aluminum, and 5% zinc.

deviation: Departure, error or variation from a standard. In ISO usage, the algebraic difference between a size and the corresponding basic size. Upper deviation is the algebraic difference between the maximum limit of size (actual, maximum, minimum) and the corresponding basic size. Lower deviation is the algebraic difference between the minimum limit of size and the corresponding basic size. Fundamental deviation is that one of the two deviations closest to the basic size. The term deviation does not necessarily indicate an error.

dezincification: A form of attack on some brasses caused by a dissolution of both the cooper and zinc components. Although this form of corrosion is marked by a deep copper-red surface color, red brasses commonly used for plumbing are resistant to dezincification.

Diakon®: A registered trademark for a methyl methacrylate plastic.

dial indicator: An instrument containing a graduated dial, used by inspectors, toolmakers and machinists in setup and inspection work. Available with inch or metric graduation dials that show the amount of error in size or alignment of a part. Dial indicators can be used on snap gages, depth gages, etc. Depending on the degree of accuracy required, the dials may be graduated in thousandths of an inch (0.001 in.) up to 50 millionths of an inch (0.0005 in);or, in hundredths of a millimeter (0.01 mm) to two-thousandths of a millimeter (0.002 mm).

diametral pitch: (1) *gear:* The ratio of the number of teeth to the number of inches of pitch diameter-equals number of gear teeth to each inch of pitch diameter. Normal diametral pitch is the diametral pitch as calculated in the normal plane and is equal to the diametral pitch divided by the cosine of the helix angle; (2) *involute spline (P):* The number of spline teeth per inch of pitch diameter. The diametral pitch determines the circular pitch and the basic space width or tooth thickness. In conjunction with the number of teeth, it also determines the pitch diameter. See also pitch.

diamond: A naturally occurring or synthetically-made high-purity crystalline form of carbon, the hardest known mineral. Mohs harness = 10. Industrial diamonds as well as those made synthetically are used as abrasives, for metal cutting, and as bearings in delicate mechanisms. Metal cutting diamond material can be natural or synthetic, and may be single, clustered, or sintered polycrystalline. They last from 10 to 450 times as long as cemented or coated carbide tools.

diamond chisel: Also called a diamond-pointed chisel. A cold chisel with a narrow blade, similar to a cape chisel, but with a square section at the end that is ground away to a single bevel forming a cutting edge at one corner, shaped like a diamond. Used for chipping through plates, cutting sharp-bottomed grooves, V-shaped oil grooves, and square, sharp corners such as those found in slots. Sometimes call a lozenge.

diamond dresser: A hand tool containing one or more fixed diamond chips. Used for dressing grinding wheels after they have been roughed out with less expensive forms of cutters. See also grinding wheel dresser.

diamond pyramid hardness test: See Knoop hardness test and Vickers harness test.

diaspore: A natural hydrous aluminum oxide ($Al_2O_3 \cdot H_2O$) used as an abrasive. Mohs hardness 6.5-7.

dibenzyl disulfide: A substance used in extreme-pressure lubricating oils and greases; silicone oils.

die: (1) A tool used to cut outside (external) threads; (2) a tool or device used to produce a specific shape or design in a material such as a metal either by stamping, die casting, extrusion, or forging, or by compacting powdered metal; (3) a mold used in die casting; (4) a device through which alloys are drawn to make wire, filaments, etc.

die, blanking: A die for cutting or punching out flat blanks on a press.

die casting: A casting process for producing a precise shape by forcing molten metal or alloy into a hardened steel dies or molds under pressure using hydraulics.

die clearance: A term used to refer to the angular clearance provided below the cutting edge so that the parts will fall easily through the die. The term clearance as here used means the space on one side only; hence, for round dies, clearance equals die radius minus punch radius.

die, curling: A special form of die used to bend the edge of the workpiece in the shape of circular beads along a straight or curved axis.

die holder: A plate or block used to mount the die block.

die, inverted: A die in which the conventional positions of the male and female members are reversed.

dielectric: A term used to describe an insulator or non-conductor of electricity.

dielectric fluid: electrical insulating material.

die, multiple: A die used to produce two or more identical parts in a single press stroke.

die pad: A movable plate or pad in a female die, usually for part ejection by mechanical means, springs, or fluid cushions.

die, perforating: A die in which multiple holes are punched either simultaneously or progressively in a single press stroke.

die, plain: The simplest form of die. Used to cut out a given shape (in the shape of the punch) from sheet metal in a single stroke of the press.

die, self centering: A die containing a small conical point used to guide the punch into stock that was previously center-punched at the point of punching.

diethylamine:[$C_4H_{11}N$] Colorless liquid that smells like ammonia. Used in electroplating. A fire hazard and highly toxic. Wear protection and consult MSDS prior to using this substance.

diethylene glycol:[$(HOC_2H_4)_2O$] A clear, syrupy liquid used in mold release agents.

differential heating: Any heating process by which the temperature is made to vary non-uniformly throughout the object being heated so that on cooling, different portions may have such different physical properties as may be desired.

diffraction: The bending of a ray of light or beam of radiation at the edge of an object or interference pattern.

diglycol laurate: A light, straw-colored, oily liquid used as an antifoaming agent, and in emulsions for lubrication, cutting, and spraying oils.

dihydroxy diphenylsulfone: A white crystalline solid used in electroplating.

dilauryl phosphite: A clear combustible liquid used to make organophosphorus compounds for extreme-pressure lubricants.

dimension: A numerical value expressed in appropriate units of measure (such as diameter, length, angle, or center distance) and indicated on drawings along with lines, symbols, and notes to define the geometrical characteristic of an object.

dimethylamine:[$(CH_3)_2NH$] A flammable gas or highly alkaline liquid used in electroplating.

dimpling: The process of indenting sheet metal, so as to permit the head of a rivet or a bolt to be countersunk or fastened flush with the surface of the sheet.

DIN: German Institute for Normalization (DIN) standards are developed by a non-profit organization of approximately 130 standards committees with representatives from all technical areas.

dip brazing: See brazing.

dipstick: A metal stick containing marking and inserted into in a reservoir, used to measure high- and low-levels of fluids such as oils and transmission fluids.

direct chill casting: A semicontinuous casting process for making billets by pouring the molten metal into a water-cooled mold of limited length. The base of the mold is a movable platform that is gradually lowered during the solidification of the metal prompted by contact with the cooling water.

direct quenching: Quenching carburized parts directly from the carburizing operation.

DIS: Abbreviation for Ductile Iron Society.

discrimination: Also known as resolution. The smallest increment into which a gage is divided.

Discaloy®: A registered trademark of the Westinghouse Corporation for an austenitic iron-base alloys nominally containing 55% iron, 25% nickel. 13% chromium, 3% molybdenum, 2% titanium, 0.7% manganese, 0.7% silicon, 0.5% aluminum, 0.05% carbon. Used primarily for making gas-turbine disks in jet engines.

dishing: Forming a concave surface of large radius in a workpiece.

dissociation: A specific kind of decomposition resulting from the breaking apart of a molecule.

dividers: Steel instruments that are similar in construction to calipers, but containing legs of equal length that are often round and that terminate in sharp points ground to give an included angle on the point of about 25°. Dividers are used for measuring distances between points, for transferring distances when taken from a scale such as a graduated rule, and for scribing circles and arcs. Although firm joint dividers are available, they are hard to set accurately; consequently, the most commonly used dividers contain a curved spring joint connecting the legs and an adjusting screw and nut that allows the measuring points to come together evenly. Many steel rules contain deeply cut graduations that facilitate setting dividers directly and accurately from them.

dividing head: A lathe or milling machine fixture, also known as an indexing head or index plate, used to accurately divide the circumference of a workpiece or part into many equal spaces, for milling splined shafts, the flutes in reamers, gear and sprocket teeth, etc. To accomplish this, the dividing head has several concentric circles of carefully spaced holes, each circle having a different number of holes and a crank arm that is fitted on the dividing spindle and provided with a peg for a stop which can be pushed into any hole. The peg is made to be moved along the arm so that any circle of holes can be used.

dog: (1) A general name used to describe any projecting piece which strikes and moves some other part; (2) a device, such as a lathe dog, used for clamping a workpiece so that it can be revolved by faceplate of a lathe. The bent tail dog is used for driving round, square, hexagonal or other regular work and the clamp dog for rectangular work. Some better grades of dogs contain safety screws, and are balanced to eliminate vibrations.

dolomite: A native calcium magnesium carbonate $[CaMg(Co_3)_2]$, the most prevalent ore of magnesium.

Dore silver: Silver containing a small amount of gold.

double angle point drill: Describes a drill point devised by first grinding a larger included angle, and then a smaller included angle on the corner. The length of the corner angle should be 1/3 the original cutting lip length. Initially developed for drilling at high speeds, especially for cast irons and other very abrasive materials. Their main advantage is preventing burning and chipping at the outer periphery of the cutting lip or cutting edge and can be used to reduce breakthrough burr in soft materials.

double annealing: Heating a steel casting well above the transformation range (Ac_3) to promote homogeneity and to coalesce sulfides that may be present, cooling it rapidly just below the transformation range (Ac_1) to minimize the formation of coarse ferrite grains, and reheating immediately to just above the transformation range (Ac_3), and slow cooling to soften the metal and refine the grain size.

double cut, file: The double cut file has a multiplicity of small pointed teeth inclining toward the point of the file arranged in two series of diagonal rows that cross each other. For general work, the angle of the first series of rows is from 40 to 45° and of the second from 70 to 80°. For double cut finishing files, the first series has an angle of about 30° and the second, from 80 to 87°. The second, or upcut, is almost always deeper than the first or overcut. Double cut files are usually employed, under heavier pressure, for fast metal removal and where a rougher finish is permissible. See also single cut.

double seaming: The process of joining metal edges of sheet stock with each edge being flanged, curled, and crimped.

double tempering: A treatment in which quench hardened steel is given two complete tempering cycles at substantially the same temperature for the purpose of ensuring completion of the tempering reaction and promoting stability of the resulting microstructure.

dowel pin: Straight or plain tapered pins that are fitted into reamed holes to position two mating parts. When a dowel pin is positioned in a mechanism that makes it impossible to be removed by driving out, the dowel pin is sometimes threaded and screwed into place.

Dowmetal®: A registered trademark of the Dow Corporation for a series of alloys nominally containing principally 3 -12% aluminum and more than 85% magnesium.

downfeed: See infeed.

downloading: The process of transferring digitalized data from a large central computer to another, usually smaller, remote system or device such as a CNC machine tool. The opposite of upload.

drag: A name given to the lower part of a foundry molding box. See also cope.

dragons blood: A deep-red natural resinous material derived from fruits of the rattan palm found in Indonesia and Borneo. Used to protect zinc plates from etching acid.

draw: (1) To gradually reduce the diameter of a metal wire or rod by pulling it through dies of successively diminishing size; (2) a foundry casting term used to describe that area subject to gross contraction or the formation of cavities, caused by insufficient feeding.

drawing: A term used to describe various processes. (1) reducing the cross section of wire or tubing by pulling it through a series of dies of successively diminishing apertures; (2) a production of seamless hollow vessels from flat metal blanks; forming recessed parts by forcing the plastic flow of metal in dies; (3) used to describe a forging process (drawing down) used to reduce the cross section of bar stock; (4) the term "drawing the temper" (for reducing hardness in quenched steels) has become synonymous with tempering (the preferred term)

drawing the temper: See drawing.

dressing: The operation of truing, cleaning, and sharpening the edged of grinding wheels, used to improve their overall performance. See grinding wheel dresser, diamond dresser.

drift: (1) A kind of hammer-driven hand punch of differing sizes, shapes, and uses. Some drifts have a specific use, such as for removing chucks and taper-shank drills from the drilling machine spindle, while others are a simple form of broach used for forming a shaped hole. Still other drifts are used to force holes into alignment or to cut through the thin fin left between holes when a large hole is or section is cut out of a block by the method of drilling a line of small holes round the shape to be cut out. The drill drift is used to remove drills from sleeves and sockets and are available in different sizes in the form of a wedge having a rounded edge on the tapered side and a flat working edge. In removing a drill, the round edge is placed against the sleeve and flat edge against the top of the drill, and care should be taken to insure that the point or tip does not hit the table of the press or the work. Drill drifts are manufactured as "plain" or "safety," the latter having a spring-loaded handle; (2) the uncertain motion of an indicating pointer on an instrument gage.

drift punch: Sometimes called an aligning punch. A long, tapered hardened steel punch used for aligning bolt and rivet holes in two or more workpieces, and to drive out pins and rivets.

drill: Rotary end-cutting tools for originating or enlarging holes. Drills have one or more cutting lips and one or more straight or helical flutes for the passage of chips and admission of cutting fluids to the cutting edges when the drill is working in a hole. Drills are classified by the material from which they are made, method of manufacture, length, shape, number and type of helix or flute, shank, point characteristics, and size series. They are also classified as *conventional* and *special purpose*. Two-fluted twist drills are standard tools for cutting holes in solid metal. Drill sizes are manufactured in four diameter ranges: Number (sometimes called wire gage) series, letter series, fractional series, and metric series. See also drill types, letter drills, number drills, fractional drills, and metric drills.

drill diameter: The diameter over the margins of a drill measured at the point.

drill drift: See drift.

drill gage: (1) A flat plate drilled with graduated, different-sized holes and each marked with correct drill size and number; (2) an L-shaped piece of thin steel plate, with graduated arm marking an angle of 121° with the body on one side of the gage. The angle between the center line of the drill and its cutting edges should be 59° for the all-purpose drill. The drill gage is used to test this angle, and the length of the cutting edge, during the operation of drill grinding.

drilling: A machining operation for making holes in a metal workpiece with a rotating tool called a twist drill. Drilling is performed on drill presses, lathes, vertical milling machines, jig boring machines, and other machines. Reaming and tapping are also classified as drilling operations. See also tapping and reaming.

drill press: A machine tool used for small, light work and using smaller drills. This small press consists of an adjustable table for holding the work and a frame in which are mounted a vertical revolving spindle to carry the drill which is connected to a handle, and motor or other mechanism to drive the drill at varying rates of speed. This is the simplest drill press, it can be mounted on a floor stand or used on a bench, in which case it may be called a bench drill. See also gang drill, radial drill, multiple-spindle drill press, turret drill press.

drill size gage: A flat metal plate perforated with various sized holes to be used in determining the size of twist drills.

drill types: Drills may be classified based on the type of shank, number of flutes or hand of cut. *Straight shank drills*: Those having cylindrical shanks which may be the same or different diameter than the body of the drill. The shank may be with or without driving flats, tang, grooves, or threads. *Taper shank drills*: Those having conical shanks suitable for direct fitting into tapered holes in machine spindles, driving sleeves, or sockets. Tapered shanks generally have a driving tang. *Two-flute drills*: The conventional type of drill used for originating holes; *three-flute drills* (core drills): drill commonly used for enlarging and finishing drilled, cast or punched holes. They will not produce original holes; *four-flute drills* (core drills): used interchangeably with three-flute drills. They are of similar construction except for the number of flutes; *right-hand cut*: When viewed from the cutting point, the counterclockwise rotation of a drill in order to cut; *left-hand cut*: When viewed from the cutting point, the clockwise rotation of a drill in order to cut.

drill vise: A vise used on a drill press to hold work being drilled.

drop forging: A forging made under a drop hammer falling under its own weight.

drop hammer: A forging hammer that uses gravity to produce its force. See also hammer forging.

dross: Slag, scum.; the impurities floating on the surface of molten metal during smelting. The slag may be high in silicates.

dry copper: A brittle copper containing excessive red copper oxide [cuprous oxide (Cu_2O)].

dryseal: A pipe thread connection for both internal and external application, used for liquid- or gas-pressure joints where the use of pipe compound or a sealer is objectionable. See also American Standard Dryseal Pipe Thread.

dtg: Abbreviation for the term, "difficult to grind."

ductile: Easily drawn out; flexible, pliable.

ductile cast iron: Nodular iron, spheroidal graphite cast iron. A ductile form of cast iron that can be extended by pulling. See also ductile.

ductile crack: A slow crack propagating through a material that is accompanied by external plastic deformation and that continually absorbs significant energy from outside the body.

Ductile Iron Society (DIS): 28938 Lorain road, Suite 202, North Olmsted OH. Telephone: 440/734-8040. FAX: 440/734-8182. WEB: http://www.ductile.org

ductility: The ability of a solid material to be stretched, pulled, or rolled into shape without rupture or destroying the integrity of the material.

Duponol®: A rgistered trademark of DuPont for a series of surface-active agents based on lauryl sulfate. These have detergent, emulsifying, dispersing, and wetting properties and are used in electroplating.

Duracool®: A registered trademark of the Nalco Chemical Corporation for a line of metalworking coolants and corrosion inhibitors.

Duralumin®: A proprietary trade name for a series of alloys nominally containing 93-95% aluminum, 3.5-5.5% copper, 0.5% magnesium, 0.25-1.2% manganese, 0.3-1.2% iron, and small amounts of silicon. Duralumin can be heat treated until it has the strength of steel.

Duranic: An alloy nominally containing 92.5-96.5% aluminum, 2-5% nickel, and 1.5-2.5% manganese.

Duranickel: An alloy containing 93.85% nickel, 4.5% aluminum, 0.5% silicon, 0.4% iron, 0.4% titanium, 0.3% manganese, 0.05% copper.

Duranickel®301: A registered trademark of Inco Alloy International for an alloy containing 94% nickel and 4.5% aluminum.

Durco®D-10: A proprietary product of Duriron Co. for an alloy nominally containing 65.5% nickel, 23% chromium, 5% iron, 3.5% copper, 2% molybdenum, 1% manganese.

Durichlor® 51: A proprietary product of Duriron Co. for a high-silicon, acid and chlorine resistant alloy nominally containing 79.45% iron, 14.5% silicon, 4.5% chromium, 0.9% carbon, 0.65% manganese.

Durimet® 20: A registered trademark of Duriron Corporation for an austenitic stainless steel, containing nominally 29% nickel, 20% chromium, 3% copper, 2% molybdenum, 1% silicon, 0.07% (max) carbon. Machinable, weldable, and resistant to sulfuric acid [H_2SO_4].

Duriron®: A proprietary trade name of Duriron Co. for high-silicon alloy nominally containing 77.87-78.88% iron, 14.5-15.5% silicon, 4.5% chromium, 0.9% carbon, 0.65-0.66% manganese, and 0.57% phosphorus.

Durite®: A registered trademark for a phenol-formaldehyde plastic.

Duronze®: A proprietary trade name for a copper alloy nominally containing silicon.

Dutch gold: A metallic foil used for gilding, made from copper, or a brass alloy containing 15-20% zinc.

Dy: Symbol for the element dysprosium.

dye-penetrant test: A non-destructive testing method using low viscosity, low surface tension dyes to identify small flaws, cracks, and porosity in metal components.

dyn: Abbreviation for dyne.

dynamic balancing: Also known as running balance. Used to describe the adding or reducing counterbalancing weights to secure the perfect running balance so that rotating parts such as shafts and pulley will run without vibration.

dyne: The c.g.s. unit of force. The force which acting upon a unit mass of one gram for one second produces a velocity of one centimeter per second (cm/sec.).

dysprosia: See dysprosium oxide.

dysprosium: A rare earth metallic element of the lanthanide group. A noncorroding, hard, alloying metal; used in magnets.

Symbol: Dy	**Density (g/cc):** 8.550
Physical state: Silvery metal	**Melting point:** 2,565°F/1,407°C
Periodic Table Group: IIIB	**Boiling point:** 4,644°F/2,562°C
Atomic number: 66	**Source/ores:** Monazite, bastnasite, gadolinite
Atomic weight: 162.50	**Oxides:** Dy_2O_3
Valence: 3	**Crystal structure:** Orthorhombic

dysprosium oxide: [Dy_2O_3] A white powder that is used with nickel in cermets.

dystectic mixture: Alloys nominally containing such proportions of constituents as to yield the maximum melting point, so that if the proportions are altered, the melting point is lowered.

E

E: Symbol for Young's modulus.

earing: The formation of waviness or scallops (ears) around the top edge of a deep-drawn part caused by differences in the directionality of the sheet metal used.

earth, alkaline: See alkaline earth.

earth, rare: See rare earth.

Easy-out: See Ezy-out.

EBC: Abbreviation for electron beam cutting.

EBM: Abbreviation for electronic beam machining.

eccentric: A machine component that converts rotary motion to straight-line motion.

eccentricity: The deviation of the centers of two circles from each other.

eccentric relief: See Thread Relief.

ECS: Abbreviation for Electrochemical Society.

eddy currents: The electric currents set up by alternating currents in a conductor by a changing magnetic flux, near or in an electric circuit.

eddy-current testing: A testing method used in nondestructive testing in which eddy-current flow is induced in a conductor referred to as the test object. Changes in the flow caused by variations in the object are reflected by changing magnetic flux into a nearby coil or coils for subsequent analysis using suitable instrumentation and techniques.

economics, cutting tool: A concept prescribing that cutting tool material or grade should ideally be selected to yield the highest productivity by providing the most economical metal removal rate based on the lowest cost, while providing precise and consistent tool life.

edge, file: Surface joining faces of a file. May have teeth or be smooth.

edging rolls: Rolls used during the rolling process of metal sheet and plates to control edge shape and/or width.

Edison Welding Institute (EWI): 1250 Arthur E. Adams Drive, Columbus OH 43221. Telephone: 614/688-5000. FAX: 614/688-5001. Web: http://www.ewi.org

EDM: Abbreviation for electrical discharge machining.

effective clearance, involute spline (c$_v$): The effective space width of the internal spline minus the effective tooth thickness of the mating external spline.

effective face width, gear: That portion of the face width that actually comes into contact with mating teeth, as occasionally one member of a pair of gears may have a greater face width than the other.

effective number of teeth: Describes the effective or useful number of teeth which includes the complete tooth. The distinction is due to the fact that some types of cutters, such as a staggered tooth slotting cutter or a step mill, the effective number of teeth will be less than the total number of teeth.

effective space width, involute spline (S$_v$): An internal spline relationship equal to the circular tooth thickness on the pitch circle of an imaginary perfect external spline that would fit the internal spline without looseness or interference (considering engagement of the entire axial length of the spline). The minimum effective space width of the internal spline is always basic. Fit variations may be obtained by adjusting the tooth thickness of the external spline.

effective thread: A thread design term used to describe the effective (or useful) thread which includes the complete thread, and those portions of the incomplete thread which are fully formed at the root, but not at the crest (in taper pipe threads it includes the so-called black crest threads); thus excluding the vanish thread.

effective tooth thickness, involute spline (t_v): An external spline relationship equal to the circular space width on the pitch circle of an imaginary perfect internal spline that would fit the external spline without looseness or interference, considering engagement of the entire axial length of the spline.

effective variation, involute spline: The accumulated effect of the spline variations on the fit with the mating part.

efficiency, gear: The actual torque ratio of a gear set divided by its gear ratio.

EIA: Abbreviation for Electronic Industries Association. Also, one of the binary coded decimal systems used to write N/C programs.

ejector: A spring-actuated ring, collar, or disk used to remove work or blanks which usually have a tendency to remain in the die of a punch press.

elastic bond grinding wheels: A type of bonded abrasive wheel made from natural or synthetic abrasive materials mixed with reinforced rubber, shellac, plastic, or synthetic resins. These wheels are strong and can be made very thin. Used for slotting; cutting off stock such as pipes, tubing, and wire; and for grinding hard-to-hold parts Some are called resinoid grinding wheels. See also bonded abrasive wheels.

elastic coefficient: See Young's modulus.

elasticity: The ability of a material to assume its original form, volume, size, and shape after a force causing deformation or distortion is removed.

elastic limit: The maximum stress or force to which a material may be subjected, and yet completely spring back or return to its original shape and dimensions with no strain remaining, upon completed release of the stress.

electrical mean line: A term related to the measurement of surface texture, the centerline established by the selected cutoff and its associated circuitry in an electronic roughness average measuring instrument.

electric arc welding: See arc welding.

electric furnace: Used for special types of steel, as well as for high-temperature (range up to 5,432°F/3,000°C) reactions such as the manufacture of synthetic abrasives, diamonds, silicon, silicon carbide [SiC], etc.

electrician's solder: See solder and/or eutectic solder.

electric steel: (1) Steel made in an electric furnace; (2) low-carbon steel alloys used in electrical machines such as alternators and transformers.

electrobandsaw: A specialized production bandsaw with an electrical current running through the saw blade, used to cut thin metals and products made from them: radiator cores, honeycomb structural materials, stainless steel, thin-wall tubing, etc. See also bandsaw.

electrochemical corrosion: Degradation of metals or alloys associated with the passage of a direct electrical current. The current is produced by chemical action and accelerated by the presence of acids or bases.

electrochemical machining: Finishing an article by making it an electrical anode. The cathode is shaped as a "mirror image" of the required form when material is transported to the cathode from the anode by reverse electroplating.

154 Electrochemical Society, Inc. (ECS)

Electrochemical Society, Inc. (ECS): 65 South Main Street, Pennington NJ 08534-2839. Telephone: 609/737-1902. FAX: 609/737-2743. E-mail: acct@electrochem.org . Web: http://www.electrochem.org An international, nonprofit, scientific organization concerned with the broad range of phenomena related to electrochemical and solid-state science and technology.

electrodeposition: Any electrochemical process in which an element or mixture of elements, usually a metal or alloy, is deposited or precipitated on an electrode. An important advantage of electrodeposition is its ability to provide exact thickness control on irregular shapes with fine and complex cavities.

electroforming: The production by electrodeposition of metal in finished or semifinished form, as the production of sheets, tubes, or patterns by electrolysis. A pattern of the form to be reproduced is made of plastic, plaster, or wax, and the mold surface is coated with a conductive release agent such as graphite. Consequently, a suitable metal is electrodeposited on the mold surface, which is a negative of the object being produced.

electroless coating: A protective, rust inhibiting, metal coating which is deposited on another solid metal in a solution or bath by chemical reduction, without application of an electric current. The advantage of this process is that uniform coatings can be formed in tubes, deep holes, etc.

electrolysis: The process of changing or decomposing in a chemical compound (electrolyte) by the passage of an electrical current either in its natural form, or in solution, or in a molten form. There are variations of this process including electroplating, electrodeposition, electrocoating, and electroforming.

electrolyte: (1) A chemical substance, mixture, or solution that conducts an electric current; (2) any substance whose solution has the property of conducting the electric current. All soluble acids, bases, and salts are electrolytes. In a battery, for example the electrolyte is a mixture of sulfuric acid and water.

electrolytic: Pertaining to electrolysis and electrolytes; decomposition by an electric current.

electrolytic cell: An electrochemical device in which electrolysis takes place, consisting of electrodes (the anode and cathode) that are connected together, and bridged with an electrolyte, in which electrolysis occurs when an electric current is passed through the electrodes and electrolyte.

electrolytic cleaning: A process for cleaning and de-rusting metal surfaces in which the object to be cleaned is made the cathode in an appropriate electrolyte.

electrolytic copper: See copper, electrolytic.

electromagnetic: Pertaining to electricity and magnetism.

electrometallurgy: Application of many of the techniques and principles of electrochemistry to the production of certain metals including aluminum and titanium.

electron: The fundamental carrier of electrical current; a negatively charged sub-atomic particle having a mass that is 9.1095×10^{-28}g. (1/1837 of the mass of hydrogen atoms). See also atom.

electron beam machining (EBM): A limited-production machining process using thermal energy from an invisible, highly-focused, pulsating stream of high-speed electrons, traveling at more than half the speed of light, which bombards the workpiece, causing the outer surface layer to vaporize so that small amounts of material are removed with each operation. Electron beam machining is most efficient when performed in a vacuum, and uses the same equipment as electron beam welding, but at lower power output. Used for drilling tiny holes and narrow slots, especially in very hard or difficult-to-machine material.

electron beam microprobe analyzer: An instrument used for chemical microanalysis of small samples of material in an area that is usually less than a micrometer in diameter. An electron beam is focused on the area of interest and bombards it with x-rays that are dispersed and analyzed in a crystal spectrometer to provide a qualitative and quantitative evaluation of chemical composition.

electron beam welding (EBW): A welding process using an invisible beam of electrons produced by a high voltage generator; the electrons are accelerated from a heated filament and focused into a narrow beam by a magnetic field, generating very high temperatures. The focused beam bombards the base metal, causing it to melt and fuse together. Electronic beam welding is most efficient when performed in a vacuum, and uses the same equipment as electron beam machining, but at higher power output. It is used for complicated weldments of tool steels.

electropainting: The electrolytic deposition of a thin layer of paint on a metal surface. In this process the metal surface has been made the anode.

electrophoresis: The migration of particles, usually suspended or colloidal macro-molecular particles, through a fluid under the action of an electric field.

electrophoretic plating: The process of depositing a surface film of colloids by means of electrophoresis.

electroplating: The process of depositing a smooth, homogeneous film or coating of metal or mixture of metals upon the conductive metallic or plastic surface of an object (cathode) by direct current electrolysis. See also electrodeposition.

electropolishing: A process of electrolytic dissolution (actually the reverse of electroplating) for polishing and cleaning metal surfaces. The object to be polished is made the anode in an electrolytic circuit, the cathode usually being carbon. The electrolytes used are polishing acids: hydrofluoric, nitric, phosphoric, and sulfuric.

electrorefining: Any method for the refining and purification of metals by means of electrodeposition.

electrostatic separation: Separation of components, usually of fine powder, by acceleration in an electrostatic field, charged with moving electrodes of opposite polarity and high voltage.

electrowinning: A term used in metallurgy to describe a reduction method used for the recovery of metals from their ores or other impurities by electrodeposition on a cathode made of the same metal being recovered.

element(s): The simplest form of a pure substance that cannot be further decomposed (broken down into simpler components) by chemical means.

elongation: In testing, the extension of the gage length of a tensile specimen, divided by the original gage length, measured after fracture of the specimen, usually expressed as a percentage of the original gage length.

elution: The process of extraction one solid from another.

elutriation: The process of washing or separating powder particles by suspending them in a solution, allowing them to precipitate, and then decanting them.

embossing: A process of forming a pattern, shallow indentations, or raised designs on the surface of metal objects.

embrittlement: A term used to describe hardening and loss of ductility in metals such as steel or ferrous alloys upon exposure to a specific environment, often resulting in material failure. In metals, the primary causes include exposure to hydrogen, corrosion, overheating, and exposure to chemicals. Thermal shock can cause embrittlement in the walls pf pressurized-water vessels such as reactors, and can result in rupture. See also corrosion embrittlement, caustic embrittlement, hydrogen embrittlement, burning.

emery: A natural, impure abrasive containing 35-70% aluminum oxide [Al_2O_3], and variable amounts of iron oxide.

emf: Abbreviation for electromotive force.

emulsifying agent: Emulsifier. A substance used to stabilize an emulsion by reducing the surface tension or holding droplets in a film, thereby protecting them.

emulsifiable mineral oil: Also called water-soluble oil or soluble oil. A type of mineral oil containing an emulsifying agent and other chemicals that allow it to mix with water, used as a cutting fluid. Usually milky-white or cloudy. Always mixed by adding the oil to the water (not the other way around). A 1:20 solution of oil added to water can be used for average severity machining operations. A 1:40 solution of oil added to water can be used for average grinding operations. *Note:* Do not use water-based cutting fluids on magnesium. Water reacts with hot magnesium releasing explosive hydrogen gas.

emulsion: A liquid that is a microscopically heterogeneous mixture of two normally immiscible fluids, in which a minute amount of one fluid such as water is finely suspended in another fluid such as oil.

end milling cutter: Also called end mills. A milling cutter with cutting edges on both its circumference (face) and end (periphery). The teeth on the circumference may be straight or helical. End mills can be solid with teeth and shank in one piece, or the shankless shell-type for mounting on an arbor. Solid end mills typically have an integral straight or taper shank for mounting. The straight shank contains a flat surface for securing it in the mill end holder with a set screw.

endothermic: Absorbing heat. An endothermic reaction is one in which heat is absorbed.

endurance limit: See fatigue limit.

English bearing metal: A metal alloy nominally containing approximately 53% tin, 33% lead, 11% antimony, and 3% copper.

English metal: An alloy used in jewelry making containing approximately 85% tin, 8% antimony, 2% copper, 2% nickel, 2% tungsten, 1% bismuth.

English white bearing metal: A bearing alloy nominally containing 77% tin, 15% antimony, 8% copper.

engraver's acid: A synonym for nitric acid [HNO_3].

enthalpy: A thermodynamic concept representing the total heat content of a substance, defined by the equation $H = E + pV$ where H is the enthalpy, E is the energy, p the pressure, and V the volume of a system.

entry angle: See angle of cutter entry.

environment: The surroundings or physical, mechanical, or chemical conditions in which a material exists.

EPA: The Environmental Protection Agency, a federal agency of the U.S. government.

EP additive: Abbreviation for extreme-pressure additive.

epoxy resin: A group of synthetic thermosetting resins. They adhere to smooth surfaces and have high resistance to corrosion, weathering, chemicals, and electricity. Among the many uses are coatings, strong-bonding adhesives for composites and metals, glass, and ceramics; casting metal-forming tools and dies; encapsulation of electrical parts; cements and mortars; nonskid road surfacing; rigid foams, etc.

epsilon (E, ϵ): The fifth letter of the Greek alphabet, used to denote symbol for electromotive force (ϵ).

equilateral triangle: A triangle having all three sides of equal length. Each of the three angles in an equilateral triangle equals 60°.

equilibrium: A state of balance among contending forces.

equilibrium diagram: A phase diagram showing the limits of composition, temperature, and structure in which the various phases or constituents of an alloy are in equilibrium or stable.

Er: Symbol for the element erbium.

Era metal: A steel alloy nominally containing 72% iron, 20-21% chromium, and 6-7% nickel.

erbium: A rare earth metallic element of the lanthanide group. An alloying element used with titanium; in magnetic alloys, nuclear controls, and special alloys.

Symbol: Er	**Density (g/cc):** 9.066
Physical state: Silvery metal	**Melting point:** 2,771°F/1,522°C
Periodic Table Group: IIIB	**Boiling point:** 5,185°F/2,863°C
Atomic number: 68	**Source/ores:** Monazite, bastnasite
Atomic weight: 167.26	**Oxides:** Er_2O_3
Valence: 3	**Crystal structure:** h.c.p. (α); b.c.c. (β)

Erichsen test: A cupping test for measuring the ductility of sheet metal and its suitability for pressing or deep drawing. A specimen section of sheet metal is lightly clamped except at the center, and subsequently deformed by a 20-mm (diameter) conical, spherical-end plunger until fracture occurs. The depth in millimeters of the cup-shaped depression at fracture is a measure of the ductility.

erosion: The deterioration or wearing away of a surface of a material by the abrasive action of moving fluids.

erosion corrosion: The combined effects of erosion and corrosion on a metal surface.

error of observation: The difference between the true value and the observed value.

esu: Abbreviation for electrostatic units.

eta (H, η): The seventh letter of the Greek alphabet, used to denote the symbol for efficiency (η); coefficient of viscosity (η).

etchant: A solution used for chemical etching of metals.

etch figure: The characteristic pattern corresponding to crystallographic planes within the metallic material produced when a polished surface is etched by a suitable reagent.

etg: Abbreviation for the term, "easy to grind."

ethylenediaminetetraacetic acid: Also known as EDTA, $[C_{10}H_{16}N_2O_8]$. A colorless crystalline solid used in electroplating.

ethyl sulfide: $[(C_2H_5)_2S]$ A colorless, oily liquid used in electroplating. Reacts violently with water, steam, and acids.

eutectic: (1) An alloy of metals in proportions such that it has the lowest possible melting point; (2) the lowest melting point of an alloy that is obtainable by varying the proportion of the components.

eutectic alloy: Alloys containing such proportions of constituents as to yield the minimum melting point, so that if the proportions are altered, the melting point is increased. Eutectic alloys have definite and minimum melting points in contrast to other combinations of the same metals.

eutectic mixture: A mixture of substances having the lowest possible constant melting point.

eutectic point: Melting point of eutectic mixture. The lowest temperature at which the eutectic mixture can exist in liquid phase.

eutectic solder: A solder, often used for electrical work, consisting of 60-63% tin, 37% lead with a melting point of 361°F/183°C.

eutectoid steel: Steel with 0.80% to 0.85% carbon. Steel with less carbon is called hypoeutectoid steel; when it contains more, it is called hypereutectoid steel.

eutropic series: Series of substances in which the physical properties and crystal structure show regularity in their variation.

EWI: Abbreviation for Edison Welding Institute.

exfoliation: A type of corrosion that progresses approximately parallel to the outer surface of the metal, causing separation of thin layers of the metal to be elevated in the form of thin leaves.

expansion reamer: See adjustable reamer.

explosive: Any chemical compound, mixture, or device that produces a sudden, almost instantaneous release of pressure, gas, and heat when subjected to sudden shock, ignition source, pressure, or high temperature.

explosive limits: Defined by the National Fire Protection Association as the boundary-line mixture of vapor or gas with air, which, if ignited, will just propagate the flame. Known as the lower and upper explosive limits, they are usually expressed in terms of percentage by volume of gas or vapor in air. Same as flammable limits. See also lower explosive limit and upper explosive limit.

exposure testing: Analysis of the degradation of a metals and alloys by exposing samples to adverse environments. Exposure-testing techniques can include combinations of high temperature, strong sunlight, salt air, and moisture; or, extremely low temperatures; or, burial in acid soils or seawater.

external gear: A gear with teeth on the outer cylindrical surface.

external grinding: See cylindrical grinding.

external spline: A spline formed on the outer surface of a cylinder.

external thread: A thread design term used to describe a thread on a cylindrical or conical external surface.

extractive metallurgy: That branch of metallurgical science devoted to the study and practice of the technology of mining and processing of ores and their metals.

extreme-pressure additives: Lubricating oil and grease additives used to prevent metal-on-metal contact. Viscous at medium temperatures, they become somewhat fluid on heating by friction. Some additives react with the metal forming a protective coating. However, some EP additives are not suited for high speed use. This category includes cutting fluid additives that impart high film strength. Although many compounds are proprietary, they mainly contain sulfur, chlorine, and occasionally phosphorus. Also, lead salts saponified with fats, lead naphthenate, lead oleate, lead stearate, and lead soap are often used.

extreme-pressure emulsifiable oils: Oils that are mixed with water by agitation, with the addition of an emulsifiing agent to prevent droplets coalescing.

extreme-pressure lubricants: See extreme-pressure additive.

extrusion: Forcing a material such as metal or plastic through a die aperture or orifice.

extrusion die: A die in which a punch forces metal or other material to assume the contour and cross-sectional area of the die apertures or orifice through which the metal is forced.

eyeleting: The process of creating a lip around a hole in sheet metal, used to improve the fatigue strength and stiffness.

Ezy-out: A tool of varying size with sharp left-hand spirals used to remove broken bolts, screws, etc., from holes. To use an Ezy-out, a hole is first drilled in the broken bolt or screw, the size being slightly smaller than the minor diameter of the thread. An Ezy-out of the proper size is then inserted and revolved counterclockwise, causing the left-hand spirals to grip the broken part which is backed out.

F

F: (1) Symbol for the element fluorine; (2) symbol for degree Fahrenheit, a unit for measuring temperature, as °F.

Φ, φ, φ: The Greek letter *phi.*

face-centered cubic: A *unit* or *cell* of space lattice (crystalline) structure having 14 atoms with one located at each of the eight corners of the cubic unit and one at the center of each of the six faces of the cube. The atoms composing the corners of one lattice of the unit cell lie at the centers of the faces of another lattice of a cube. Metals having this atomic space lattice structure include aluminum, copper, gold, lead, nickel, palladium, platinum, rhodium, and silver.

face, file: The widest cutting surface or surfaces that are used for filing.

face grooving: See trepanning.

face mill: A special form of large end mill cutter of the *inserted-tooth type,* used to produce flat surfaces at right angles to the axis of the cutter. Face mills are made in sizes 6 in (152.4 mm) or over and have inserted teeth that cut on the periphery and face. The rotating face machines the surface of the work, while the teeth on the periphery do most of the cutting and remove most of the stock.

face of tooth, gear: That surface of the tooth which is between the pitch circle and the top of the tooth.

faceplate: The disk or plate that screws on the nose of a machine spindle or lathe and drives or carries work to be turned or bored. Sometimes the table of a vertical boring machine is called a faceplate.

facet: A flat, usually highly reflective, surface of a crystal.

facing: The process of making a cut across the end of a workpiece. When facing on the lathe, the cutting tool should be set exactly in line with the workpiece center, and the cutting edge of the cutting tool should be set at an angle of 80° to the face.

FAO: Drawing abbreviation used to designate a part that must be finished all over.

farad: The name of the SI electromagnetic unit of capacitance. It is defined as that capacity which is charged to a difference of potential of one volt by one coulomb of electricity. The microfarad (10^{-6}) is the unit in common use.

faraday constant: The quantity of electricity that can liberate (deposit or dissolve) one electrochemical equivalent in grams of a metal during electrolysis. One faraday = approximately 96,500coulombs.

Faraday's laws: (1) The quantity of an electrolyte decomposed is proportional to the strength of the current passing through the solution; (2) the masses of any substance deposited or dissolved by a given quantity of electricity are proportional to their chemical equivalent weights.

Farmer drill: A kind of straight-fluted drill bit, used for soft metals, brass, copper, and thin steel.

fatigue: See metal fatigue.

fatigue life: The number of stress cycles that can be endured for a stated test condition, prior to failure.

fatigue limit: The maximum stress range below which a material can presumably endure an infinite, or predetermined limiting number of stress cycles without fracture. If the stress is not completely reversed, the value of the mean stress, the minimum stress, or the stress ratio should be stated.

Fe: Symbol for the element iron. From *ferrum,* the Latin name for iron.

feather: Sometimes called a spline. A sliding key, nearly always permanently fixed to the sliding piece, used to prevent a pulley, key, or other part from turning on the shaft but allowing it to move lengthwise as in the feed shaft used on most lathes and other tools.

feature: The general term applied to a physical portion of a part, such as a surface, hole, pin, tab, or slot.

feature of size: one cylindrical or spherical surface, or a set of two opposed parallel surfaces associated with a size dimension.

fee: A device used to move, deliver, or dispense stock to a die.

feed: Also called feed rate. The rate at which work is advanced relative to the position of the cutting tool. An important factor in determining the rate of metal removal and overall machine efficiency. Quality of finish, effects on the machine, and safety are important considerations in selecting the proper feed rate. Usually expressed in inches per minute (ipm).

feeler gage: Also known as a thickness gage or gapper. It consists of a holder containing a series of thin stainless steel blades or "leaves" or wires of varying thicknesses, somewhat like a pocket knife. The blades or wires are marked in thousands of an inch. These leaf type may be used singly, or together, thus enabling the user to make up any desired dimension within the limits of the tool. Used to measure the distance or small gaps, or to feel slight variations between two surfaces.

feldspar: General name for a group of igneous, crystalline mineral consisting chiefly of alumina, with varying amounts of silicates of barium, calcium, potassium, and sodium. Commercially, this term often refers to potassium feldspars such as potassium aluminosilicate. Used as an abrasive and bond for abrasive wheels; in cermets.

ferrite: (1) An allotrope of iron, having the body-centered cubic (alpha) form.

Commonly occurs in steels, cast iron, and pig iron. Alpha (α-), beta (β-), and delta (δ- or Δ-) iron are the common varieties of ferrite; (2) a metallurgical term for carbon that combines with iron to form cementite (iron carbide [Fe_3C]).

ferro-: A prefix indicating content of metallic iron, as in ferroalloys or ferroaluminum.

ferroaluminum: Ferroalloys nominally containing 80% iron and 20% aluminum.

ferroboron: Alloys containing 10-25% boron, added to special steels as a hardening agent and deoxidizer.

ferrobronze: An alloy nominally containing 91.2% copper, 8% iron, 0.8% chromium.

Ferrocarbo®: A proprietary product of Carborundum Corp for silicon carbide [SiC] used as a deoxidizer and graphitizer in the production of gray iron or steel. Machinability and strength of the iron or steel are increased without loss of hardness.

ferrocarbon titanium: A ferroalloy used to dexoidize molten steel.

ferrochromium: Also known as chrome ferroalloy, carbon ferrochromium. An alloy containing principally iron, chromium, with carbon and silicon. Available in several grades, generally containing 50-70% chromium, 22-42% iron, 1-10% carbon, 1-7% silicon, and 0.01-0.02 sulfur.

ferromagnesium: A ferro alloy of magnesium and iron.

ferromanganese: A tough steel alloy containing approximately 48-50% manganese. Also reported to be typically 80% manganese, 6% carbon, balance iron. Used in steelmaking.

ferromolybdenum: A steel alloy containing from 55 to 75% molybdenum.

ferronickel: A ferro alloy nominally containing 25% nickel and 0.8% carbon.

ferronickel valve steel: A ferro alloy nominally containing 32% nickel and 0.2% carbon.

ferroniobium: An alloy of iron and niobium. Used for making stainless steels and alloys for welding rods.

ferrophosphorus: An alloy of iron and phosphorus used in the steel industry for controlling and adjusting the phosphorus content of special steels. Available in two grades, 18% or 25% phosphorus. Used to increase fluidity in steel casting and for preventing thin sheets of steel from sticking together when rolled and annealed in bundles.

ferrosilicon: A ferro alloy containing 2% silicon and 0.4% carbon.

ferrotitanium: An alloy used to add titanium to steel and for the introduction of titanium into stainless steels. Typical analysis 40% titanium, 5-20% silicon, 0-10% aluminum, balance iron and carbon

ferrotungsten: An alloy of iron and tungsten used as a means of adding tungsten to steel. Engineering steels typically contain 70 - 80% tungsten, up to 0.6% carbon, balance iron. Tool alloy typically contains 5% tungsten, 0.5% carbon, balance iron. High speed ferrotungsten typically contains 18% tungsten, 6% chromium, 0.3% vanadium, 0.7% carbon, balance iron.

ferrous-: A prefix indicating compounds containing divalent iron indicated by the notation, iron(II) or iron(2+).

ferrovanadium: An iron-vanadium alloy used to add vanadium to steel. Vanadium is used in engineering steels to the extent of 0.1-0.25% and in high-speed steels to the extent of 1-2.5% or higher.

ferrozirconium: An alloy for steels of varying silicon contents. Steels of high silicon content typically contains 12-15% zirconium, 39-43% silicon, 40-45% iron. Steels of low silicon content typically contain 35-40% zirconium, 47-52% silicon, 8-12% iron.

FIA: Abbreviation for Forging Industries Association.

fiber stress: Local stress acting along the fibers of a material or through a small area usually parallel to a neutral axis.

filament: A fine, threadlike, continuous fiber usually made by extrusion from drawn material such as metals, glass, boron, metal carbide, and silicon carbide [SiC].

file: A hand tool used for cutting and smoothing. A file consists of a blade or body with a tang that fits into a wooden handle. The blade is hardened and tempered, and teeth of a suitable kind are cut into the blade's faces and edges, although the edges of some files are smooth and uncut (termed "safe"). Files are classified according to their shape or cross-section and according to the pitch or spacing of their teeth and the nature of the cut. The cross-section may be quadrangular, circular, triangular, or some special shape. The outline or contour may be tapered or blunt. In the former, the point is more or less reduced in width and thickness by a gradually narrowing section that extends for one-half to two-thirds of the length. In the latter the cross-section remains uniform from tang to point. The length generally varies from 4 to 18 in., and the length extends from the point to heel, and does not include the tang. Special shapes are available for specific jobs.

file test (hardness): A test of resistance to deformation above which a material cannot be cut with ordinary files, determined by using a set of files of graduated hardness.

filler: (1) In brazing, soldering, and welding, the metal or alloy deposited in the joint between metals being joined, thereby effecting successful union; (2) in plastics, inert mineral powders such as asbestos, barytes, calcium carbonate (whiting), carbon black, mica, silicates, soft clays, or slate flour. These inert minerals are used to color, extend and dilute, or to modify physical properties such as stiffness and hardness. Also fillers may be used to reduce the overall cost of basic materials.

fillet: (1) A female or internal radius; (2) involute spline: The concave portion of the tooth profile that joins the sides to the bottom of the space.

fillet curve, gear: The concave portion of the tooth profile where it joins the bottom of the tooth space. The approximate radius of this curve is called the fillet radius.

fillet gage: See radius and fillet gage.

fillet root splines, involute spline: Splines in which a single fillet in the general form of an arc joins the sides of adjacent teeth. Fillet root splines are recommended for heavy loads because the larger fillets provided reduce the stress concentrations. The curvature along any generated fillet varies and cannot be specified by a radius of any given value.

fillet stress, gear: The maximum tensile stress in the gear tooth fillet.

film: A general term used to describe a membrane or covering layer such as lubricants or oxides on the surfaces of metals. For example, oxide films are formed automatically on the surface of aluminum protect it from corrosion.

FIM: An abbreviation for full indicator movement. See runout.

fin: The thin, excess edge left by the parting of a mold or die.

fine silver: The purity expressed in parts per 1000, usually 999 minimum.

fish eyes: See flakes.

fit: (1) A general term used to signify the range of tightness that may result from the application of a specific combination of allowances and tolerances in the design of mating parts; (2) the relationship resulting from the designed difference, before assembly, between the sizes of two mating parts which are to be assembled.

fits, description: The classes of fits are arranged in three general groups: running and sliding fits, locational fits, and force fits.

Running and Sliding Fits (RC): Running and sliding fits, for which limits of clearance are intended to provide a similar running performance, with suitable lubrication allowance, throughout the range of sizes. The clearances for the first two classes, used chiefly as slide fits, increase more slowly with the diameter than for the other classes, so that accurate location is maintained even at the expense of free relative motion.

These fits may be described as follows:

RC1 Close sliding fits are intended for the accurate location of parts that must assemble without perceptible play.

RC2 Sliding fits are intended for accurate location, but with greater maximum clearance than class *RC1*. Parts made to this fit move and turn easily but are not intended to run freely and, in the larger sizes, may seize with small temperature changes.

RC3 Precision running fits are about the closest fits that can be expected to run freely, and are intended for precision work at slow speeds and light journal pressures, but are not suitable where appreciable temperature differences are likely to be encountered.

RC4 Close running fits are intended chiefly for running fits on accurate machinery with moderate surface speeds and journal pressures, where accurate location and minimum play are desired.

RC5 and *RC6* Medium running fits are intended for higher running speeds, or heavy journal pressures, or both.

RC7 Free running fits are intended for use where accuracy is not essential, or where large temperature variations are likely to be encountered, or under both these conditions.

RC8 and *RC9* Loose running fits are intended for use where wide commercial tolerances may be necessary, together with an allowance, on the external member.

Locational Fits (LC, LT, and LN): Locational fits are fits intended to determine only the location of the mating parts; they may provide rigid or accurate location, as with interference fits, or provide some freedom of location, as with clearance fits. Accordingly, they are divided into three groups: clearance fits (LC), transition fits (LT), and interference fits (LN).

These are described as follows:

LC Locational clearance fits are intended for parts that are normally stationary, but that can be freely assembled or disassembled. They range from snug fits for parts requiring accuracy of location, through the medium clearance fits for parts such as spigots, to the looser fastener fits where freedom of assembly is of prime importance.

LT Locational transition fits are a compromise between clearance and interference fits, for applications where accuracy of location is important, but either a small amount of clearance or interference is permissible.

LN Locational interference fits are used where accuracy of location is of prime importance, and for parts requiring rigidity and alignment with no special requirements for bore pressure. Such fits are not intended for parts designed to transmit frictional loads from one part to another by virtue of the tightness of fit. These conditions are covered by force fits.

Force Fits: (FN): Force or shrink fits constitute a special type of interference fit, normally characterized by maintenance of constant bore pressures throughout the range of sizes. The interference therefore varies almost directly with diameter, and the difference between its minimum and maximum value is small, to maintain the resulting pressures within reasonable limits.

These fits are described as follows:

> FN 1 Light drive fits are those requiring light assembly pressures, producing more or less permanent assemblies. They are suitable for thin sections or long fits, or in cast-iron external members.

> FN 2 Medium drive fits are suitable for ordinary steel parts, or for shrink fits on light sections. They are about the tightest fits that can be used with high-grade cast-iron external members.

> FN 3 Heavy drive fits are suitable for heavier steel parts or for shrink fits in medium sections.

> FN 4 and FN 5 Force fits are suitable for parts that can be highly stressed, or for shrink fits where the heavy pressing forces required are impractical. See also standard fits designations.

fixed renewable bushings: See renewable bushings.

fixed spline: A spline that is either shrink fitted or loosely fitted, but piloted with rings at each end to prevent rocking of the spline, which results in small axial movements that cause wear.

fixture: A device that holds the work while the cutting tools are in operation, but does not contain any special arrangements for guiding the tools. A fixture, therefore, must be securely held or fixed to the machine on which the operation is performed, thence the name. A fixture is sometimes provided with a number of gages and stops, but not with bushings or other devices for guiding and supporting the cutting tools. For comparison, see jig.

flakes: Short, discontinuous internal fissures or cracks; characteristic white, disk-like, crystalline appearance on fractured surfaces of ferrous metals, caused by the release of internal hydrogen during cooling, following hot working. Also known as fish eyes, snowflakes, and shatter cracks.

flame annealing: A process in which the surface of an iron-base alloy is softened by localized heat applied directly by a high-temperature flame.

flame cleaning: A process for cleaning metal of paint, rust, scale, dirt, grease, moisture, etc., by running a broad, flat flame over the surface from portable cylinders of compressed gas, usually butane or oxyacetylene.

flame hardening: A case hardening process involving rapid heating with a direct high-temperature gas flame, such that the surface layer of the part is heated above the transformation range, followed by cooling at a rate that causes the desired hardening to a depth from about 1/32 in to 1/4 in. Steels for flame hardening are usually in the range of 0.30-0.60% carbon, with hardenability appropriate for the case depth desired and the quenchant used. The quenchant is usually sprayed on the surface a short distance behind the heating flame. Immediate tempering is required and may be done in a conventional furnace or by a flame-tempering process, depending on part size and costs.

flammable limits: Range of gas or vapor concentrations (percent by volume) in air that will burn or explode if an ignition source is present. See also explosive limits.

flammable material: Generally used to describe any material that can be set on fire instantly when exposed to spark or flame, and that burns rapidly and continuously. More specifically, it can means "catches on fire and burns rapidly." Both the National Fire Protection Association (NFPA) and the U.S. Department of Transportation (DOT) define a flammable liquid as one having a flash point not more than 141°F/60.5°C.

flank: A thread design term used to describe either surface connecting the crest with the root. The flank surface intersection with an axial plane is theoretically a straight line.

flank angle: A thread design term used to describe the angles between the individual flanks and the perpendicular to the axis of the thread, measured in an axial plane. A flank angle of a symmetrical thread is commonly termed the half-angle of thread.

flank diametral displacement: A thread design term. In a boundary profile defined system, flank diametral displacement is twice the radial distance between the straight thread flank segments of the maximum and minimum boundary profiles. The value of flank diametral displacement is equal to pitch diameter tolerance in a pitch line reference thread system.

flank of tooth, gear: That surface that is between the pitch circle and the bottom land. The flank includes the fillet.

flaring: The process of forming an outward flange on a tubular part.

flash: A term used in forging, casting, or butt welding to describe the thin "fin" of excess metal extruded or forced out along the joint lines between mating surfaces of molds, dies, etc.

flashback: A phenomenon characterized by vapor ignition and flame traveling back to the source of the vapor.

flash point: The lowest temperature at which a substance decomposes to a flammable gaseous mixture, or the vapors that in contact with spark or flame will easily ignite and burn rapidly. The lower the flash point of a liquid, the higher the risk of fire.

flash welding: A welding process used primarily for butt ends of bar and tube in which the weld is created by pressure and heat. While an electrical current is passed through the abutted members being welded, the members are drawn apart slightly, until an electrical arc is created; the arc produces heat which is slightly below the melting point, and then the members are subsequently pressed together to produce the weld.

flat file: A file that is always double cut on the faces and single cut on the edges. These files taper both in width and thickness toward the end away from the tang. One of the most commonly used files, it is used for heavy metal removal, cutting, and chipping and leaves a poor workpiece finish. See also vixen file.

flat root splines: Those splines in which fillets join the arcs of major or minor circles to the tooth sides. Flat roof splines are suitable for most applications. The fillet that joins the sides to the bottom of the tooth space, if generated, has a varying radius of curvature. Specification of this fillet is usually not required. It is controlled by the form diameter, which is the diameter at the deepest point of the desired true involute form (sometimes designated as TIF).

flats: See faceted insert.

flaw (F): A surface texture term used to describe material inclusions, scratches, holes, cracks, and other unintentional, unexpected, and unwanted deformations in the surface of a workpiece.

flexible spline: A spline that permits some rocking motion such as occurs when the shafts are not perfectly aligned. This flexing or rocking motion causes axial movement and consequently wear of the teeth. Straight-toothed flexible splines can accommodate only small angular misalignments (less than 1° before wear becomes a serious problem. For greater amounts of misalignment (up to about 5°, crowned splines are preferable to reduce wear and end-loading of the teeth.

flint: A crystalline form of quarts or silica used as an abrasive.

flotation: A process used in extractive metallurgy to separate and concentrate minerals from waste ores by agitating the ground and pulverized mixture of solids with oil, special chemicals, and frothing agents, floating them on water, and agitating the mixture with compressed air. The wet impurities (gangue) settles, and the concentrated ore is floated off.

flue: An escape outlet for gases, dusts, and liquids.

flue dust: See fly ash.

fluid friction: Viscosity, the internal resistance to flow exhibited by a fluid.

fluoboric acid: [HBF_4] Colorless liquid used to clean metal surfaces prior to welding and in electroplating baths.

fluorescent penetrant inspection: A method of non-destructive testing using an ultraviolet light and a dye penetrant or fluorescent penetrating oil that is brushed or sprayed on the part being inspected. The sample is wiped dry, dusted with an absorbent powder to draw the penetrating oil out of any discontinuation or cracks that are open to the surface, and placed under ultraviolet light that reveals flaws appearing as bright fluorescent colors.

fluorine: A nonmetallic, gaseous, halogen element. The most active of all chemical elements. Highly reactive; the most powerful oxidizing agent known. Some compounds are highly toxic. Used in aluminum production.

Symbol: F
Density (g/cc): 1.516 (liquid)
Physical state: Yellow gas
Melting point: −362°F/−219°C
Periodic Table Group: VIIA
Boiling point: −308°F/−188°C
Atomic number: 9
Source/ores: Fluorspar, fluorapatite, cryolite
Atomic weight: 18.9984
Crystal structure: monoclinic (α); cubic (β)
Valence: 1

fluorocarbon polymer: A term for thermoplastics that includes polymers of chlorotrifluoroethylene, fluorinated ethylene-propylene polymers, hexafluoropropylene, polyvinylidene fluoride, polytetrafluoroethylene, etc. Used for bonding industrial diamonds to metal for abrasive materials, grinding wheels.

fluosilicic acid: [H_2SiF_6] A colorless, fuming liquid used in electroplating.

fluorspar: Fluorite; calcium fluoride [CaF_2], a principal source for fluorine. Used in metal smelting and open hearth steel making.

flute: (1) Shop name for a groove, applied to taps, reamers, drills and other tools; (2) a drill term used to describe the helical or straight grooves cut or formed in the body of the drill to provide cutting lips, to permit removal of chips, and to allow cutting fluid to reach the cutting lips.

flutes: (1) Longitudinal recesses in cylindrical parts; (2) the helical grooves or straight, longitudinal channels cut or formed in a tool such as a tap to create cutting edges on the thread profile and to provide chip spaces and cutting fluid passages; (3) Surface imperfections found in formed or drawn parts.

fluting: (1) The longitudinal channels or spiral grooves cut or ground into a cylindrical part to improve cutting action and chip removal by providing chip spaces and cutting fluid passage.

flux: A substance, such as rosin, zinc chloride, sodium carbonate, or borax, that is applied as a cleaner on metal surfaces being joined by brazing, soldering, welding, etc. Fluxes are used to remove undesirable substances, including surface impurities including oxide, prevent oxidation, and lowering the surface tension of the solder, allowing solder to flow freely and unite more firmly with the surfaces being joined.

fly ash: The flue dust resulting from the combustion of certain fuels such as powdered or pulverized coal. Recoverable through the use of electrostatic precipitators. Used as a filler in plastics, in cement products, removal of heavy metals from industrial wastewaters, and in road fill.

flycutting: A specialized milling operation using one or more rotating single-point tool bits or cutters mounted on a fly cutter arbor that is attached to the spindle of a milling machine.

flywheel: A heavy wheel used to produce uniform rotation and steadying motion of machinery.

FMS: Abbreviation for flexible manufacturing system.

FN: Abbreviation for force or shrink Fit. See also fits, description.

foam: A heterogeneous dispersion of a gas in a liquid or solid.

foam depressant: A chemical agent used to decrease the accumulation of finely divided gas bubbles suspended in a liquid such as cutting fluids.

foamed metal: A light metal to which a metal hydride such as titanium hydride has been added, from which hydrogen is evolved, producing a blown structure having a specific gravity about the same as water.

fog: A heterogeneous mixture of finely divided liquid droplets suspended in air or gas; an aerosol.

fog quenching: Quenching in a mist.

foil: Metal in sheet of metal with no defined limit but usually less than 0.006 in. in thickness.

fools gold: Iron pyrite [FeS_2] often found in coal and containing small amounts of arsenic, cobalt, copper, gold, nickel, selenium. Brass-yellow; resembling gold. Used in the production of sulfur-based chemicals, inexpensive jewelry, and recovery of metals.

foot stock: The tail stock or tail block of a grinder, lathe, etc.

force: That which changes or tends to change the state of a body at rest, or which modifies or tends to modify the course of a body in motion. A fore always implies the existence of a simultaneous equal and opposite force called *reaction.*

force fit: See fits, description.

foreseeable emergency: Any potential occurrence such as, but not limited to, equipment failure, rupture of containers, or failure of control equipment, that could result in an uncontrolled release of a hazardous chemical into the workplace. (OSHA)

forge: An open fireplace usually having forced draft by fan or bellows for heating metals for forging, welding, etc.

forging: Plastic deformation of metals, usually heated to high temperatures, using compressive force such as hammering or pressing, with or without dies.

forging brass: Alloys containing 59% copper, 39% zinc, and 2% lead.

Forging Industries Association (FIA): 25 Prospect Avenue, W, Suite 300, Landmark Office Towers, Cleveland OH 44115. Telephone: 216/781-6260. FAX: 216/781-0102. E-mail: info@forging.org WEB: http://www.forging.org

formability: The capability of a material such as metal to withstand plastic deformation without fracture.

form circle, involute spline: The circle which defines the deepest points of involute form control of the tooth profile. This circle along with the tooth tip circle (or start of chamfer circle) determines the limits of tooth profile requiring control. It is located near the major circle on the internal spline and near the minor circle on the external spline.

form clearance, involute spline (c_f): The radial depth of involute profile beyond the depth of engagement with the mating part. It allows for looseness between mating splines and for eccentricities between the minor circle (internal), the major circle (external), and their respective pitch circles.

form cutters: See form-relieved milling cutter.

formic acid: [CH_2O_2] A colorless, fuming liquid used in electroplating.

forming: Processes of changing the shape of thin metal pieces by bending, stretching, impacting, stamping, etc., that does not remove metal.

forming, magnetic: The shaping of a workpiece using forces generated by the momentary discharge of a current pulse through a coil and the resultant interactions with the eddy currents induced in the workpiece, which is a conductor.

forming, pneumatic: The mechanical cold-shaping of a workpiece by forces generated by the impact of a ram speeded up by the release of compressed gas.

form tool: See form cutter.

form-relieved milling cutter: A series of specialized milling tools used for cutting irregular shapes, curved surfaces, rounding corners, etc. Cutters within this class include concave and convex cutters, corner rounding cutters, fluting cutters, spur-gear milling cutters, etc.

foundry sand: The sand/binder mixture used in making molds for casting metals. Contains titanium, zirconium, and other metals, and mixed with various binders that are resins, such as casein. Also known as molding sand or green sand.

fractional drills: Twist drills with diameters ranging in size from 1/64 in (0.40 mm) to 4 in (102 mm) or larger. In the smaller drills the diameter increases by 1/64th in. (0.40 mm) from 1/8 to1-3/8, and in the larger sizes by 1/32nd in (0.79 mm) from 1/8 to1-3/8, and 1/16th in (1.59 mm) from 2- 5/16 to 3-1/2.
Note: In most drill charts the number and letter drill sizes normally fall between the fractional drill sizes.

fractography: Examination of fracture surfaces, especially in metals by using direct observation, photographs of the fracture surface at low magnification (macrofractography), or optical electron microscopy to study the physical characteristics and crystallographic structure at high magnification (microfractography).

fracture: The character of the new surfaces, usually with sharp edges, produced by breaking a solid substance.

fracture test: One of the oldest tests for metal, involves breaking a specimen for the purpose of examining the fractured surface to determine structure, grain size, case depth, soundness, and presence of internal defects. This test can be performed the unaided eye or under a low-power microscope.

fracture toughness: A measure of the energy absorbed by a material prior to fracture.

frary metal: An alloy produced by electrolytic deposition of barium and calcium (1-2%) in molten lead (96-98%), used for low-pressure bearings at moderate temperatures and shrapnel bullets.

free-machining: Also called free-cutting. A term used to describe alloys offering or designed to have good machine qualities or machinability. This term relate to the metals' alloying elements or additives during manufacture that produce one or more desirable characteristics such as low cutting resistance, lower power consumption, better surface finish, longer tool life, the ability to give small broken chips during machining, etc. Among such alloying elements or additives are lead added to brass, sulfur or lead added to steel, lead and bismuth added to aluminum, and sulfur or selenium added to stainless steel.

frequency: The number of complete vibrations or light waves passing a fixed point unit of time.

fretting (fretting corrosion): A form of surface corrosion occurring between the contacting surfaces of metals that are under heavy load and subject to very slight relative motion or vibration. This mechanism results in the generation of local high temperatures, galling, the production of tiny deposits of metal, and subsequent oxidation. In the case of steel, these tiny oxidized particles are reddish-brown rust or ferric oxide [Fe_2O_3], commonly called "cocoa-powder" or "bleeding."

friction: The resistance offered to the motion of one body upon or through another, and is generally defined as, "that force which acts between two bodies at their surface of contact, so as to resist sliding on each other."

frictional heat: Heat generated from frictional forces between two bodies in contact, such as moving parts of engines.

friction sawing: A metal cutting operation for very hard ferrous materials, especially those that are difficult to machine or cut in any other way such as ½ in. armor plate. This method uses a special friction sawing bandsaw blade employing speeds up to 15,000 ft per minute that creates sufficient friction, thereby heating the metal to its softening point just ahead of the saw blade. Unlike conventional sawing, this methods is unaffected by a dull blade because more friction is created and faster cutting speeds are achieved.

friction welding: A welding process in which the weld is created by frictional heat and pressure, used to join dissimilar materials.

frit: A finely powdered glass or mineral glazing material that become a polished glaze or vitreous enamel surface on heating. The frit is applied (usually sprayed), melted, and bonded firmly onto a metal surface after firing.

fritting: Applying frit material at a temperature high enough (usually near the melting point) to cause melting and binding at points of contact on a metal surface. See also frit.

full annealing: A softening process in which iron-base alloy is heated to a suitable high temperature above the transformation range (austenitizing temperature (Ac_3) and, after being held for a proper time at this temperature, is gradually cooled to a temperature below the transformation range. The treated objects are ordinarily allowed to cool slowly in the furnace, although they may be removed from the furnace and cooled in some medium which assures a slow rate of cooling. This treatment can result in a clearer, stronger, and more uniform material by removing internal strains caused by previous operations, and eliminates distortions and imperfections. The austenitizing temperature to hypoeutectoid steel is just above Ac_3; and for hypereutectoid steel, usually between Ac_1 and Ac_m.

fuller: A hammerlike blacksmith's tool with a round nose, used for spreading or fulling the work as it is struck with a hammer. The hand fuller is held on the work and struck with a sledge. The anvil fuller fits in a hole in the anvil and acts as the lower tool in fullering work.

fullering tool: A chisellike tool, but with a flat end made equal in thickness and often beveled to about 80°. The fullering tool burrs down the edges of plates that have been riveted and have to be made steam and water tight. See also riveting.

fully killed steel: Steel that is fully deoxidized by the addition of silicon and augmented by manganese, aluminum, titanium, etc., while the mixture is maintained at melting temperature until all bubbling ceases. The steel is quiet and begins to solidify at once without any evolution of gas when poured into the ingot molds.

fume: A suspension of very fine particle of solids or liquids in a gas or air. Vapors from a volatile liquid, or the smoke-like particulate from the surface of heated metals. These are usually toxic.

fundamental deviation: See deviation and deviation, fundamental.

functional diameter: See pitch diameter.

fundamental deviation (ISO term): For standard threads, the upper or lower deviation closer to the basic size. It is the upper deviation ES for an external thread and the lower deviation EI for an internal thread.

fungicide: Any substance that destroys or inhibits the growth of fungi and their spores. Used in emulsifiable cutting fluids, especially those that are stored in large tanks, strained, and used over and over again.

furnace: Depending on function, a furnace may have various designs used to heat, fuse, or harden materials by heating them to high temperatures. An enclosed firebrick or refractory-lined chamber or structure that contains a heat source (coal, coke, gas, oil, or electricity) that can produce high temperatures above 572°F/300-°C and up to 5432°F/3000°C. Used for steel production, for smelting iron and other ores, and for manufacture of pyrolytic graphite, salt cake, silicon, silicon carbide [SiC], and synthetic diamonds.

fused: Material that has been cooled to a compact mass following heating to a molten state or sintering, such as dross or slag.

fusible: Capable of being melted or liquefied when heated.

fusible alloy: Low-melting point alloys that fuse (melt) on slight heating. An alloy having a melting point usually lower than the mean melting point of its constituents, and often in the range of 124-500°F/51-260°C. On of the best know of these alloys is Wood's metal, used in automatic sprinkler systems.

fusing point: Melting point.

G

G: Symbol for giga- (10^9), gas, or gravity.

Γ **or** γ*:* See gamma.

G-30®: A registered trademark of Haynes International, Inc. for a high-strength, nickel-based alloy with outstanding corrosion resistance.

ga.: Abbreviation for gage.

G-ratio: See grinding ratio.

gadolinium: A rare earth metallic element of the lanthanide group. Dust and powder are highly flammable. Used in titanium production; magnetic alloys.

Symbol: Gd	**Density (g/cc):** 7.90
Physical state: Silvery metal	**Melting point:** 2,394°F/1,312°C
Periodic Table Group: IIIB	**Boiling point:** 5,911°F/3,266°C
Atomic number: 64	**Source/ores:** Monazite, bastnaesite, samarskite
Atomic weight: 157.25	**Oxides:** Gd_2O_3
Valence: 3	**Crystal structure:** h.c.c. (α); b.c.c. (β)

gage: (1) The thickness or diameter of a material; (2) to measure accurately; (3) a tool, instrument, or device for establishing a particular dimension of an object, for determining whether or not a workpiece is within specified limits, for measuring the pressure or flow of a liquid, or to accurately position work in a die. Except for the surface gage which is *adjustable to a fixed range*, the great variety of gages are fixed. Gages are divided into the following types according to the purposes for which they are used: ring gage, plug gage, receiving gage, snap gage, caliper gage, limit gage, taper gage, height gage, depth gage, center gage, screw pitch gage, wire gage, sheet metal gage, surface gage, tap and drill gage, thickness or feeler gage, thread gage, etc. *See also* master gage, inspection gage, and working gage.

gage blocks, rectangular: Precision tools made of hardened, ground, and stabilized tool steel blocks with surfaces that are flat, parallel, and finished to an accuracy within 0.000008 in., used as the standard of precision measurement. These blocks are rectangular, approximately 3/8 by 1 3/8 in., and available in full sets containing 81 blocks containing four series as follows:

Composition of Gage Block Sets of 81 blocks	No. of bocks in series	Size range	Steps between blocks
first series	9	0.1001 to 0.1009 in.	0.0001 in.
second series	49	0.101 to 0.149 in.	0.001 in.
third series	19	0.050 to 0.950 in.	0.050 in.
fourth series	4	1 to 4 in.	1 in.

These blocks can be *wrung* together (forcing out all air from between them) in various combinations. Precision gage blocks sets are available in three quality levels, all measured at a temperature of 68°F/20°C: Class 3 or "B" quality (Working set) = +0.000008/–0.000004 in. (+0.00020/–0.00010 mm) accuracy; Class 2 or "A" quality (Inspection grade set) = +0.000004/–0.000002 in. (+0.00010/–0.00005 mm) accuracy; and Class 1 or "AA" quality (Laboratory grade or master set) = ±0.000002 in. (±0.00005 mm) accuracy. See also Jo blocks, angle gage blocks.

galena: A natural lead sulfide (PbS), also known as lead glance, a principal ore of lead.

galling: The friction-induced damage to the surfaces of one or both mating surfaces of metals sliding in direct contact, due to localized welding of high spots, with appreciable spalling and roughening of the surface.

galvanic corrosion: An electrochemical corrosion process occurring when two dissimilar metals are coupled together in the presence of the same electrolyte. The corrosion process generates an electrical current causing an exchange of atoms carrying electrical charges between the dissimilar metals, resulting in the dissolution of the metal with the higher electrode potential [the anode (+ pole)], while the cathode (- pole) is protected. If the electrode potential of the coupled metals is similar, there should be little or no galvanic corrosion.

galvanized iron: Coating of a ferrous metal with corrosion resistant coatings. This can be accomplished with a coating of tin by electrodeposition and subsequent immersion in a zinc bath, or by passing a ferrous metal through a bath of molten zinc without electricity. The latter method is called hot-dip galvanizing. See also sacrificial protection, cold galvanizing.

galvanizing: (1) Adding an atmospheric-protective finish to a metal by coating it with a less oxidizable metal; (2) to coat iron or steel with a thin corrosion-resistant coating of zinc. See also galvanized iron.

Galvomag®: A registered trademark of Dow Chemical Corp. For magnesium alloy composition used in anodes in cathodic protection.

gamma: The Greek letter, Γ, γ. Used as a symbol for shear strain.

gamma iron: One of the four solid phases of pure iron which occurs between the temperatures of 2,554°F/1,401°C and 1,648°F/898°C. In terms of hardness, *gamma*-iron lies between that of *alpha*- and *beta*- irons. It is nonmagnetic and crystallizes in the cubic system.

gang drill: A mass production drilling machine utilizing multiple drills that are *ganged* or made into a single machine, usually mounted on a table, that can perform several drill press operations. Each spindle that holds and turns the drill tool may run separately. Work such as hole drilling can be done at one spindle and passed along the table to the next spindle for countersinking, and then passed to the next spindle, etc.

gangue: The relatively valueless impurities or earthy material by-product that are mined with metallic ores. Gangue is separated from the economically valuable ore in the milling, reduction, and extraction processes, often as slag. Common gangue materials include calcite, feldspar, pyrite, quartz.

garnierite: [$(Ni, Mg)_6(OH)_6Si_4O_{11} \cdot H_2O$] An ore of nickel.

gas: The vapor state of matter. An air-like, formless, nonelastic fluid having the property of uniformly distributing itself throughout a space in air.

gas tungsten arc welding (GTAW): See TIG

gaseous: Describing a state of matter other than the solid or liquid states.

gauze: A loosely woven, fine wire netting.

Gd: Symbol for the element gadolinium.

Ge: Symbol for the element germanium.

gear: A term used in engineering for almost any kind of mechanism, but it refers especially to a toothed wheel. A pair of toothed wheels carried on separate shafts and having teeth meshing together form a convenient means of causing one shaft to drive the other at an exact speed ratio.

gear bronze: A copper alloy nominally containing 10-12% tin, 1-2% nickel, and 0.1-0.3% phosphorus.

gear cutter: A milling machine tool for making gears. They are available in standard sets of eight cutters for each diametral pitch (6,8,10,12, etc). The individual cutters in the standard set are numbered as follows: #1 (135 teeth to a rack); #2 (55-134 teeth); #3 (35-54 teeth); #4 (26-34 teeth); #5 (21 to 35 teeth); #6 (17-20 teeth); #7 (14-16 teeth); #8 (12-13 teeth).

gear rack: Straight rectangular or round stock with teeth cut on a flat surface. When meshed with a gear, the gear rack changes rotary motion to reciprocating motion.

geometric tolerance: The general term applied to the category of tolerances used to control form, profile, orientation, location, and runout.

geometry (insert): The physical makeup or characteristics of a tool insert.

germanium: A nonmetal (metalloid) element. Used in brazing alloys. An alloying element in beryllium and gold.

Symbol: Ge	**Density (g/cc):** 5.32
Physical state: Brittle, silver	**Melting point:** 1,719°F/ 937°C
Periodic Table Group: IVA	**Boiling point:** 5125°F/2830°C
Atomic number: 32	**Source/ores:** ergyrodite, germanite;
Atomic weight: 72.61	recovered in zinc production
Valence: 2,4	**Oxides:** GeO, GeO_2
	Mohs hardness: 6
	Crystal structure: Diamond cubic

German silver: A nonferrous copper-nickel-zinc alloy nominally containing 50% copper, 30% zinc, and 20% nickel. It is a substitute for high grade silver such as sterling in inexpensive jewelry. See nickel-silver.

germicide: An agent used to destroy microorganisms, especially disease germs.

getter: A substance, usually an active metal, that "cleans" (consumes or absorbs) the last traces of gas from a closed container or system. In metallurgy, a getter is added to a molten alloy or metal and binds with oxygen or nitrogen in the melt, causing removal of the gasses into the slag.

gettering: Maintaining a high vacuum by adding a *getter* which absorbs gas.

GeV: Symbol for 10^9 (giga-) electrovolts.

GI: Abbreviation for Gold Institute.

gib: A guide located alongside a sliding member to take up wear, or to ensure proper fit.

giga- (G): Prefix for 10^9.

gilding: Covering the surface of a material with a thin gold layer.

gilding metal: Alloys nominally containing 85-90% copper and 10-15% zinc. Used for coins, metals, tokens, base metal for gold plate, inexpensive jewelry, etc. A kind of alpha brass used in cold working.

gilding solution: Used for electroplating, these solutions generally contain aqueous solutions of gold chloride and potassium carbonate [$CO_3 \cdot 2K$].

glance: A general term used to describe hard, brittle minerals having a glassy, metallic luster, especially when fractured. See glance ore.

glance ore: Ores including argenite (silver glance), bismuthinite (bismuth glance), chalcocite (copper glance), cobaltite (cobalt glance), galena (lead glance), molybdenite (molybdenum glance), and stibnite (antimony glance). See also glance.

glass, metallic: Metal alloys containing an amorphous atomic structure formed by cooling of the molten alloy so rapidly (less than a second) that the formation of crystalline structure is prevented. This structure is more resistant to corrosion and is harder than crystalline equivalents.

glaze: (1) A smooth, glassy, hard surface coating for porous material such as ceramic, cermet, or ferrous metals. See also porcelain glaze; (2) the condition of an abrasive material when the grains have been dulled, caused by the abrasive being too hard or too fine for the work, or the cutting speed may be too fast.

glycols: Dihydric alcohols with many industrial uses including antifreeze (ethylene glycol) and making plastics and explosives.

glycolic acid: [$C_2H_4O_3$] A colorless crystalline solid used in polishing, and soldering compounds; copper pickling; electroplating.

go-devil: A cylindrical object, usually a scraper or brush, used for cleaning the interior surface of pipes.

gold: CAS number 7440-57-5. The most malleable metal known. Available in various forms including bullion, ingots, sheet; wire; tubing; sponge, powder, leaf, plate. The gold content is disclosed in karats, the number of parts of gold in 24 parts of alloy (with 24 karat being pure). Also fineness is expressed in parts per 1,000. Gold standard alloys of 22, 18, 14 are used in the US and UK. Also, 9 karat and 10 karat are legal standards in the UK and US respectively. Used in jewelry, decorative items, bullion, brazing alloys, dental alloys, plating; alloys used for electrical contacts. Often used commercially alloyed with copper, cadmium, silver, zinc and other metals which both change the color and add hardness.

Symbol: Au	**Density (g/cc):** 19.32
Physical state: Yellow met.	**Melting point:** 1,949°F/1,065°C
Periodic Table Group: IB	**Boiling point:** 5,085°F/2,807°C
Atomic number: 79	**Source/ores:** Native metal; calaverite
Atomic weight: 196.97	**Oxides:** Au_2O, Au_2O_3
Valence: 1,3	**Crystal structure:** f.c.c.

gold alloy: See gold, gold coinage, gold plating, gold silicon alloy, gold solder.

gold bronze: An α-aluminum bronze alloy powder containing 5.0-9.5% aluminum, 81.5-95.0% copper, and small amounts of iron, manganese, and/or tin. See also aluminum bronze. Also used as a synonym for copper.

gold button brass: A synonym for brass.

gold coinage: An alloy consisting of 90% gold and 10% copper.

gold cyanide: [Au(CN)$_3$·3 H$_2$O] A colorless liquid used as in electroplating.

gold filled: A thin gold alloy bonded to a base metal. Lower in quality than rolled.

gold foil: See gold leaf.

Gold Guard®: A proprietary trade name of Chemtronics, Inc. for a chemical cleaner for precious metals and electrical connectors.

gold hydroxide: [Au(OH)$_3$] A brown powder used in gold electroplating.

Gold Institute (GI): 1112 16th Street, NW, Suite 240, Washington DC 20036. Telephone 202/835-0185. FAX 202/835-0155. E-mail: info@goldinstitute.org WEB: http://www.goldinstitute.org

gold leaf: Gold, the most malleable of metals, can be pounded or rolled into very thin foil sheets of the order of 0.000003 to 0.0001 in. thick; so thin that one troy ounce at 0.0001 in. thickness may cover more than 65 sq ft. Used for gilding metal, plaster, and wood. Also known as gold foil.

gold oxide: [Au$_2$O$_3$] A brownish-black powder used in gold electroplating.

gold plating: The electrodeposition of gold on a base metal using sodium gold cyanide or a solution of gold cyanide in potassium cyanide.

gold, rolled: A thin layer of gold alloy that is mechanically applied to a ferrous base. Used in making inexpensive jewelry. The gold in the U.S. is not usually less than 10 karat, and the U.K. it is not usually less than 9 karat.

gold silicon alloy: An amorphous alloy formed in thin foils (about 10 microns) by instantly cooling a mixture of molten gold and silicon before a crystal structure can form.

gold solder: Alloys variously composed of gold, silver, copper, and sometimes zinc. A typical gold solder might be composed of about 42-43% gold, 30% silver, 20% copper, and 7-8% zinc.

gold trioxide: Auric acid [$AuCl_3$]. A dark-red crystalline solid used in gold electroplating.

gold washed: The lowest quality gold surface finish, thinner than plate, rolled, or filled.

go/no-go gage: See limit gage.

GP: See Guinier-Preston zone.

gr: Abbreviation for grain.

grade: (1) The ratio of rise of a slope to its length; (2) the sine of the angel of slope; (3) a designation given to a particular material based on quality, size, rank, etc. There are recognized grades used in industry, including commercial, technical, chemically pure (CP, a grade signifying a minimum of impurities, but not 100%), National Formulary (NF), refined, reactor, semiconductor, electronic, spectrophoto-metric, etc.

graduated: Divided into units by a series of marks or lines (graduations), as a scale for temperature, measure for liquids, etc.

grain: The appearance of a non-homogeneous surface.

grain size: (1) In metals, a measure of the average areas or volumes of grains, when the individual sizes are fairly uniform. Grain sizes may be reported in a plane section as number of grains per unit area, or per unit volume, average grain diameter, or as a grain-size number deduced from area measurements; (2) in uncoated abrasives, grain size generally refers to the size of the particles used, and these can vary from 10 to 600 according to the following rating system: Coarse: 10, 12, 14, 16, 20, 24; Medium: 30, 36, 46, 54, 60; Fine: 70, 80, 90, 100, 120, 150,

180; Very fine: 220, 240, 280, 320, 400, 500, 600; (3) Coated abrasives have a slightly different rating system that can vary from 12 to 600 as follows: Extra coarse: 12, 16, 20, 24, 30, 36; Coarse: 40, 50; Medium: 60, 80, 100; Fine: 120, 150, 180; Extra-fine: 220, 240, 320, 360, 400, 500, 600.

Gramp's solder: An aluminum solder nominally containing 60.5% tin, 36.3% zinc, 3% copper, 0.27% lead, and 0.19% antimony.

graney bronze: A copper alloy consisting of (approx.) 76% copper, 9% tin, and 15% lead.

Granodine®: A proprietary trade name of ICI Chemicals for a metal treatment solution containing zinc phosphate (tribasic), $[Zn_3(PO_4)_2]$ as an active ingredient.

granodized metal: A corrosion resistant steel, aluminum, or other metal that contains a conversion coating by treating with zinc orthophosphate, $[Zn_3(PO_4)_2]$.

graphical centerline: A term related to the measurement of surface texture, the line about which roughness is measured and is a line parallel to the general direction of the profile within the limits of the sampling length, such that the sums of the areas contained between it and those parts of the profile that lie on either side are equal.

graphite: Black lead; plumbago, mineral carbon. A native or synthetic crystalline allotropic form of carbon. Greasy feeling, relatively soft with a steel-gray to black, shiny metallic appearance. It is slippery and has a low coefficient of friction and it is a good conductor of heat and electricity. Graphite has many industrial uses including in cathodes in electrolytic cells, electrodes, brushes for electrical motors, foundry molds, paints and coatings, pyrometers, crucibles, retorts, self-lubricating bearings, solid lubricants, structural material. Used as lubricant where there is low speed and moderate heat. Graphite is a component in Lith-X®, a dry fire extinguishing agent suitable for lithium, titanium, and zirconium fires.

graphite metal: A bearing metal containing 68% lead, 17% antimony, and 15% lead.

graphitic carbon: Carbon that is practically pure and forms in a ferrous alloy during the cooling process. This form of carbon exists as tiny, shiny, black flakes distributed throughout the mass and may tend to weaken the metal, such as pig iron. Also known as graphitic temper carbon.

graphitic corrosion: Corrosion of gray cast iron caused by sulfate-reducing bacteria, which cannot tolerate oxygen and are common in the muds of swamps, ponds, lagoons, water, as well as in cooling towers, air washers, and other re-circulation water systems. May be controlled by the use of water treatment microbiocides.

graphitic temper carbon: See graphitic carbon.

graphitizing: An annealing treatment for ferrous alloys in which some or all of the combined carbon is precipitated as graphite. See also temper carbon.

gravimetric feeder: Used to control the flow of a wide range of particulate sizes of materials during processing operations; a belt conveyor incorporating a scale that continuously weighs the material passing over it.

gravimetry: Measurement and analysis by weight; used to determine specific gravity.

gray cast iron: A soft cast iron usually containing from 1.7 to 4.5% carbon and graphite (the allotropic form of carbon), and from 1 to 3 % silicon. It is tough with low tensile strength; it breaks with a coarse grained dark or grayish fracture. The excess carbon is in the form of graphite flakes and these flakes impart to the material the dark-colored fracture that gives it its name. Gray cast iron may easily be cast into any desirable form and it may also be machined readily. Gray iron castings are widely used for such applications as machine tools, automotive cylinder blocks, cast-iron pipe and fittings and agricultural implements. As the relative percentage of combined carbon and graphite decreases, the iron becomes more brittle and harder; its tensile strength increases and the fracture is fine grained and smooth. This grade of iron is called white cast iron or white iron.

gray iron: Another name for gray cast iron.

greases: Mixtures of animal fats, petroleum mineral oil, or oils thickened with one or more soaps, carbon, clays, calcium, graphite, lead, lithium, potassium, sodium, waxes, zinc oxide, and other substances. Greases range in viscosity and texture from soft liquids to fibrous or stringy to rubbery to stiff, and in color from transparent to black and are classified by the National Lubricating Grease Institute from 0 (softest) to 6 (stiffest). Grease specifications are determined by speed, load, temperature, environment, and metals in the desired application. In the case of machinery, grease selection should follow the recommendation of the manufacturer.

grease cup: A cup shaped lubricating fitting.

grease seal: A circular metal-backed rubber device that protects wheel bearings from contamination such as dust and moisture and keeping the shaft, bearing, or bushing lubrication from leaking.

Greek alphabet: See individual names for their technical or scientific application.

alpha	A	α	nu	N	ν
beta	B	β or 6	xi	Ξ	ξ
gamma Γ		γ	omicron O	o	
delta	Δ	δ	pi	Π	π
epsilon E		ϵ	rho	P	ρ
zeta	Z	ζ	sigma	Σ	σ or ς
eta	H	η	tau	T	τ
theta	Θ	θ	upsilon Υ	υ	
iota	I	ι	phi	Φ	ϕ
kappa	K	κ	chi	X	χ
lamda	Λ	λ	psi	Ψ	ψ
mu	M	μ	omega	Ω	ω

greenockite: A naturally occurring cadmium sulfide [CdS] containing approximately 78% cadmium.

grinding: A machining operation involving the surface attrition of a material from a workpiece by rubbing with an abrasive material imbedded in a powered wheel that is mounted in a grinder or grinding machine. Grinding is used to shape or improve the surface of a workpiece. Lapping and honing are also considered grinding operations.

grinding cracks: The formation of minute cracks in the surface of relatively hard materials created by excessive grinding and consequential local overheating.

grinding fluid: A liquid coolant or lubricant that is applied to an abrasive apparatus to assist in a machining operation, and wash away dislodged abrasive grains.

grinding machines: A family of machines that employs a grinding wheel for accurately producing and finishing cylindrical, conical, or plane surfaces. Excess stock is removed from the workpiece by feeding the work against the revolving wheel or by forcing the revolving wheel against the work Commonly used grinding machines include the surface grinder, centerless grinder, cutter grinder, tool grinder, internal grinder, and external grinder.

grinding wheel: A cutting tool made from natural or artificial abrasive grains of differing size and properties that are mixed with a suitable cement or bond, and compressed into a wheel. See also bonded abrasive, elastic bond grinding wheels, vitrified bond wheels.

grinding wheel dresser: A tool for cleaning, truing, dressing, and sharpening the face of a grinding wheel. Truing removes any high spots. There are various kinds of grinding wheel dressers in common use. The Huntington dresser consists of alternating pointed (star-shaped) or corrugated discs of hardened metal which freely rotate on a shaft containing a handle. The abrasive wheel dresser is similar to the Huntington dresser but uses an abrasive wheel mounted at a slight angle in a protective holder. The abrasive stick dresser uses replaceable single-point cutting element of hard abrasive materials such as boron carbide [B_4C] or silicon carbide [SiC]. The diamond dresser contains one or more industrial diamonds set in a holder or radius wheel dresser. These dressers are pressed, usually by hand, against the face of a rapidly revolving grinding wheel and moved from side to side prying and breaking off small particles of the grinding wheel.

grit blasting: See sand blasting.

ground joint: A metal joint that is finished by grinding the junction of two joined parts with lubricant and abrasives and that fits tightly without the need for gasket or packing material.

Grossman's alloy: An alloy nominally containing 87% aluminum, 8% copper, and 5% tin.

Gross solder: An alloy of aluminum, lead, phosphorus, tin, and zinc.

Guerin process: A proprietary name for a metal sheet forming method using a pliable rubber pad and a die.

Guettier metal: A copper alloy nominally containing 61% copper, 33% zinc, and 6% tin.

Guillaume® alloy: Guillaume metal. A proprietary name for a copper alloy nominally containing 65% copper and 35% bismuth.

gullet: The space between the cutter teeth that allows for the exiting of chips during a cutting operation.

gun drill: A single-lip, straight-fluted drill bit used for drilling very deep holes, usually self-guiding, containing coolant passages for oil, cutting fluid, or compressed air, through the shank and body of the drill. See also straight-flute drill.

gun drilling: The process of using a gun drill and high-pressure coolant to cool the drill point and flush out any chips.

gun metal: A series of bronzes formerly used to make cannons. An alloy of copper containing 85-88% copper, 5-10% tin, 2-5% zinc, and sometimes small amounts of antimony, nickel, lead, iron, and/or aluminum. See also admiralty gun metal.

gun taps: Taps having two, three, or four straight flutes with cutting edges ground at an angle to the centerline, designed to push chips ahead of the tap. Gun taps permit rapid thread cutting and minimal tap breakage. See also spiral point taps.

Gurley's metal or bronze: Alloys nominally containing 86.5% copper, 5.4% zinc, 5.4% tin, and 2.7% lead.

Gurney's metal or bronze: Alloys nominally containing 76% copper, 15% lead, and 9% tin.

H

H: Symbol for the element hydrogen.

hacking knife: A knife for cutting heavy sheet metal. The cutting blade has a wide back which can be struck with a hammer.

hacksaw blade: A very hard, narrow saw blade made of hardened and tempered carbon tool steel, high speed steel, or high speed tungsten or molybdenum alloy steel. The length of the blade is determined by the distance between the holes that are used to mount the blade in the hand frame or power saw. Blades having teeth of various sizes are used, depending on the kind of metal being cut. The number of teeth per inch, called the pitch, normally range from 14 to 32 for a hand hack saw frame, and the teeth have a standard *set*. The set comes from making the blade with the alternate teeth bent slightly outward. Fourteen pitch blades are normally used for mild material and large sections and 32 pitch blades are used for fine stock, tool steel, brass and copper tubing, thin sheet metal, etc. Eighteen pitch blades are commonly used on all kinds of metal. Coarse, ten-pitch blades can only be used in a power saw. Blades are generally mounted in the frame with the teeth pointed forward (away from the handle), so as to cut on the forward stroke only. The blade is held in square-sectioned blade holders and is pulled taut in the frame so that the blade is under sufficient tension to keep it from bending and breaking.

hacksaw, hand: A general work saw that consists of a handle, metal bow frame, and a narrow saw blade made of steel and hardened to cut metal. The frame can be of the fixed type made to take only one length of blade, or adjustable to accept blades from 8 to14 in. in length. *Close-quarter hacksaws* that have a pistol-grip handle are also available. See also hacksaw blade.

hacksaw, power: A machine that uses a straight-toothed, hardened serrated blade held firmly in a metal frame that reciprocates, cutting on both the forward and return stroke, while cutting fluid is applied to the blade. On the cutting stroke, pressure is applied by gravity or hydraulics, forcing the blade teeth into the workpiece, and automatically removed on the return stroke.

hafnium: A metallic element having high strength and resistance to corrosion. Formerly known as celtium (in France) or oceanum. Used as an alloying element in high temperature alloys, including refractory materials.

Symbol: Hf   **Density (g/cc):** 13.3
Physical state: Silvery metal   **Melting point:** 4,046°F/2,230°C
Periodic Table Group: IVB   **Boiling point:** 9,387°F/5,197°C
Atomic number: 72   **Source/ores:** Alvite; occurs in zirconium ores.
Atomic weight: 178.49   **Oxides:** HfO_2 (hafnia)
Valence: 4   **Crystal structure:** h.c.p. (α); cubic (β)

hairline cracks: Cracks found in low-alloy steels, caused by the presence of hydrogen and due to rapid cooling through the critical temperature during the manufacturing process. The cracks can often be minimized or avoided by the addition of titanium to the steel.

half round cape chisel: See cape chisel.

half round file: A file having one flat face and one curved or semi-circular face, and always double-cut on the flat face. The round face may be double- or single-cut. Used for curved or concave surfaces. Made in various lengths.

Hall process: Also called the Hall-Héroult process. The electrolytic recovery of aluminum from alumina, an aluminum oxide [Al_2O_3], whose most important source is bauxite. The process involves mixing the alumina with molten fused cryolite, a sodium-aluminum fluoride [Na_3AlF_6] in carbon-lined, rectangular steel box or "cell." at 1,713-1,832°F/950-1,000°C. Current is passed from carbon electrodes in the bath and during electrolysis the alumina is decomposed and is deposited on a metal pad in the bottom of the cell. At 24 to 48 hour intervals, aluminum is tapped and drawn off from the cell by a siphon and cast into ingots of 99% or higher purity, although higher purity can be achieved through further refining.

hammer: A tool for beating and striking, consisting of a heavy metal head that is set at right angle to a relatively long handle. Hammers come in many forms to fit various needs, but machinists favor the peening hammer which are classed with respect to the peen as: ball, straight, or cross. See also ball-peen hammer. cross-peen hammer, and straight-peen hammer.

hammer forging: Hand or machine forging using repeated blows to deform or shape the work. Compare with press forging. See also drop forging.

hand: A term used to describe the direction of rotation of a milling cutter. It is also used to describe the direction of the helical flutes on the milling cutter (cutters with straight teeth have no helix of the teeth) and the direction of the cut looking toward the spindle at the front end of the cutter. Consequently, in a right-hand cut, the milling cutter must rotate counterclockwise.

hand file: A thick, rectangular file that is parallel in its width. It tapers slightly in thickness from a point about one-third of its length from the base. The hand file is double-cut with one edge left uncut ("safe") and helpful when filing one surface only without touching the other. Primarily used for filing flat surfaces.

hand groover: A special tool made of tool steel, hardened and tempered, used for the sole purpose of driving down the seam of a grooved joint, such as used on metal pipe. It is available in various sizes, depending on the width of the lock desired.

hand of cutter: All solid-shank, shell, face and national standard drive milling cutters are designed for either right-hand (RH) or left-hand (LH) rotation. The hand is used to describe the direction of rotation of a milling cutter. To check the hand needed for any machine, first look at the front end of the cutter (toward the spindle nose or column) while it is mounted in the machine spindle. If the desired rotation is counter-clockwise, it is a right-hand cutter. If it is clockwise, it is a left-hand cutter. The hand also describes the helix of teeth or helical flutes on cutting tools including milling cutters, drills or reamers. the hand of helix is determined by viewing the end of the cutting tool. If the helical flutes twist to the right, the helix is right hand, and vice versa.

hand punch: A long steel tool designed for a variety of work. The tool is held in the hand and struck with a hammer while the other end is held against the work. Hand punches are usually made of tool steel and are knurled or octagonal shaped to provide good finger grip. See also center punch, pin punch, prick punch.

Hanover metal: An alloy containing 87% tin, 8% antimony, and 5% copper, used a s bearing material.

hard aluminum: An aluminum alloy nominally containing 77% aluminum, 12% magnesium, 11% copper or zinc.

hard chromium: A non-decorative, heavily electrodeposited coating of chromium for the purpose of engineering applications, such as wear resistance on sliding metal surfaces.

hardenability: A property in a ferrous alloy which determines the depth and distribution of hardness induced by the formation of martensite following quenching from the austenitic phase.

hardening: A general term for any process used to increase the resistance to breaking, bending, cutting, or grinding of metal by suitable treatment, usually involving heating and cooling. Heating at even temperature above the critical range gets the grain structure in the steel into the proper state. Cooling is the quenching of the steel in some medium such as water, brine, caustic solution, or oil in order to preserve the structure caused by heating. The most critical factors in hardening are (a) heating to the correct temperature, (b) holding at heat for a correct length of time, (c) selecting the correct quenching medium and proper method of quenching, and (d) selecting correct drawing temperature. For more specific terms, see also age hardening, case hardening, cyanide hardening, flame hardening, induction hardening, precipitation hardening, secondary hardening, and quench hardening.

hardenite: A common name for austenite and martensite of eutectoid structure.

hardie: A blacksmith's anvil tool having a sharp edge. It fits into a hole in the anvil and acts as the lower tool in cutting off operations.

hard lead: Also known as acid lead alloy or antimonial lead. A general name for a family of lead alloys nominally containing other metals, usually 4-28% antimony, used to stiffen the base metal for various uses including battery plates. See also antimonial lead.

hard metal: (1) A general name for a copper-tin alloy nominally containing 66.67% copper and 33.33% tin; (2) a general name for cemented carbide materials. See also cemented carbide.

hardness: The most well-known property of solid materials. Although difficult to define accurately, it is measured by determining the resistance to external forces such as wear, abrasion, cutting, scratching, or plastic deformation, usually by penetration, or the effect upon the rebound of a weight. The resistance of a metal to deformation of a needle-like indenter of specific size and shape under a known load. The most generally used hardness scales are: Brinell, for cast iron; Rockwell, for heat-treated steel and sheet metal, Knoop, for metals; diamond pyramid, metal; Shore scleroscope, for metals, and the Shore Durometer test for rubber and plastics. Not fitting this definition is the Mohs scale, a scratch test for minerals.

hardness testing: A nondestructive testing technique used to test the performance of metals. See entries for hardness, Brinnell hardness test, Knoop hardness test, Rockwell hardness test, Shore scleroscope hardness test, and Vickers hardness test.

hard solder: See solder.

hard soldering: See brazing.

hard tin: Alloys nominally containing 99.6% tin and 0.4% copper. Used for foil and collapsible tubes.

hard water: A nontechnical word used to describe the condition of water due to the presence of calcium or magnesium salts (more usually calcium carbonate or calcium sulfate), occurring on boiler walls and pipes in which hard water has been heated or evaporated.

Hastelloy®: A proprietary trade name of Haynes International, Inc. for a family of high-strength, nickel-based alloys having good corrosion resistance to acid attack, moderate wear resistance, excellent resistance to stress corrosion, and good high temperature properties with ease of welding. They typically contain 65% nickel, 30% molybdenum, 5% iron. Another alloy contains 59% nickel, 17% molybdenum, 14% chromium, 5% antimony, 5% iron. An acid-resistant grade contains 91% nickel and 9% silicon.

Haynes® 25: A proprietary trade name of Haynes International, Inc. for an alloy containing cobalt, chromium, nickel, and tungsten,

Haynes® 188: A proprietary trade name of Haynes International, Inc. for an alloy containing 31-47% cobalt, 20-24% chromium, 20-24% nickel, 13-16% tungsten, 0-3% iron, 0-1.2% manganese, 0.2-0.5% silicon, 0-0.2% carbon, 0-0.2% lanthanum.

Haynes® 242: A proprietary trade name of Haynes International, Inc. for a fluorine-resistant, nickel-molybdenum-chromium alloy.

Haynes® alloy: A registered trademark for a series of high-performance alloys nominally containing 45% cobalt, 26% chromium, 15% tungsten, 10% nickel, and small amounts of carbon and boron. Engineered to resist high-temperature corrosion while providing high-temperature strength, creep-resistance, and oxidation resistance.

Haystellite®: A registered trademark of the Cabot Corporation for a family of cast tungsten carbide products, principally in the form of hard-surfacing rods for protecting parts from severe abrasion.

hazardous classes: A collection of terms established the United Nations

Committee of Experts to categorize hazardous materials. The specific categories are: flammable liquids, explosives, gases, oxidizers, radioactive materials, corrosives, flammable solids, poisonous and infectious substances, and dangerous substances.

hazardous materials: Refers generally to hazardous substances, petroleum, natural gas, synthetic gas, acutely toxic chemicals, and other toxic chemicals. Substances or materials which have been determined to be capable of posing an unreasonable risk to health, safety, and property, usually by improper handling, shipping or storage.

hazardous waste: Defined by Resource Conservation and Reconstruction Act (RCRA) as any solid or combination of solid wastes, which because of its physical, chemical or infectious properties, may pose a health hazard when improperly managed. It must possess at least one of four characteristics--ignitability, corrosivity, reactivity, or toxicity; or, appears on special United States Environmental Protection Agency lists.

hazard warning: The OSHA definition means any words, pictures, symbols, or combination thereof appearing on a label or other appropriate form of warning that conveys the hazards of the chemical(s) in the container(s).

HBN: Abbreviation for hexagonal boron nitride.

He: Symbol for the element helium.

headstock: That component of the lathe that consists of the permanently fastened housing located at the left end of the bed. It contains the motor drive system and spindle that holds and turns the workpiece.

heat-affected zone: During cutting, brazing, or welding, that section of the base metal which was not melted, but has undergone microstructure and physical property changes due to the heat.

heat sink: A device or material used to dissipate or absorb unwanted heat from a process, such as manufacturing.

heat treatment: A process involving a combination of heating and cooling temperature cycles applied to a metal or alloy in the solid state to obtain desired conditions by changing their physical properties. Heating for the sole purpose of hot working is excluded from the meaning of this definition. There are three major steps in the heat treating of steel: hardening, tempering, and annealing.

heavy metal: A metal of atomic weight greater than sodium (23) or specific gravity greater than 4. Located in the lower half of the periodic table.

heavy metal soaps: Soaps formed by metals heavier than sodium such as aluminum, calcium, cobalt, lead, and zinc. These soaps are not water soluble and specific types are used in lubricating greases.

HEED: Abbreviation for high-energy electron diffraction.

heel or shoulder, file: That portion of a file that abuts the tang.

height: A surface texture term, considered to be those measurements of the profile in a direction normal to the nominal profile.

height gage: A precision instrument for measuring above a given surface.

height of thread: (1) The height (or depth) of thread is the distance, measured radially, between the major and minor cylinders or cones, respectively; (2) the distance between the crest and the base of a thread measured normal to the axis.

helical gear: Also called a spiral gear. Gears having teeth at an angle across its face.

helical milling: A milling operation in which a cylindrical workpiece is rotated around its axis and fed lengthwise, at a uniform rate, into a cutter, creating a spiral flute on a helical path such as those found on twist drill and tap blanks.

helical overlap, gear: The effective face width of a helical gear divided by the gear axial pitch. Also known as the face overlap.

helium: CAS number: 7440-59-7. A noble gas element. It has many industrial uses including welding.

Symbol: He	**Density:** 0.1785 g/L @ STP
Physical state: Gas	**Melting point:** –458°F/–272°C
Atomic number: 2	**Boiling point:** –452°F/–269°C
Atomic weight: 4.0026	**Source:** From natural gas
Valence: 0	

helix angle: (1) Describes the angle that a tool's leading (helical cutting) edge makes with the plane of its axis or center line. Refers to the cutting edge or flute that progresses uniformly around a cylindrical surface in an axial direction; (2) a drill term used to describe the angle made by the leading edge of the land with a plane containing the axis of the drill; (3) a thread design term. On a straight thread, the helix angle is the angle made by the helix of the thread and its relation to the thread axis. On a taper thread, the helix angle at a given axial position is the angle made by the conical spiral of the thread with the axis of the thread. The helix angle is the complement of the lead angle; (4) a gear term describing the angle that a helical gear tooth makes with the gear axis at the pitch circle unless otherwise specified.

hematite: The ore of iron, consisting mainly of ferric oxide oxide [Fe_2O_3], used for the manufacture of iron and steel. Certain varieties are used for rouge and for making paint pigments. Also known as raddle, red iron ore, bloodstone.

Hennig Purifier®: A registered trademark for a preparation having a soda-ash base and other materials. Used as a ladle addition to produce cleaner steel by aiding in removal of dissolved oxides and silicates and fluxing nonmetallic inclusions to slag.

hermaphrodite calipers: A two-legged steel instrument containing one divider-type leg with a sharp point (which may be adjustable) and one bent caliper leg and used to scribe center lines on a round bar or shaft, or to scribe lines parallel to the edge on a surface of a workpiece. Their size is measured by the greatest distance they can be opened between the two points.

Hertz (Hz): The approved international term that replaces cycles per second (cps).

Hertz stress: See contact stress, gear.

Hessian crucible: A large clay or refractory material crucible used in metallurgical work.

hexagon: A plane figure containing six sides and six angles.

hexagonal boron nitride (HBN): A white powder with a graphite-like, hexagonal plate structure. When compressed at 106 psi, it becomes hard as diamond. Used as a refractory, high temperature and furnace insulator, self-lubricating bushings, metalworking abrasive, in high-strength fibers.

hexagonal close-packed cube: A *unit* or *cell* of space lattice (crystalline) structure having 17 atoms; one located at each corner of the six corners of the two end faces of the hexagon unit and one at the center of each end face, and three equally spaced between the two end faces. Metals with this lattice arrangement include cadmium, cobalt, magnesium, titanium, and zinc. This arrangement allows for little plasticity and is difficult to cold work.

hex wrench: See Allen wrench.

Hf: Symbol for the element hafnium.

Hg: Symbol for the element mercury.

Higbee cut: See blunt start thread.

high carbon steels: Steels containing from about 50% to 1.50% carbon. Tool steels are high carbon steels and contain sufficient carbon to allow them to be hardened and tempered.

highest point of single tooth contact, gear (hpstc): The largest diameter on a spur gear at which a single tooth is in contact with the mating gear.

high-leaded tin bronze: See tin bronze, high leaded.

HighPermalloy® 49: A registered trademark for an alloy nominally containing 50-53% iron and 47-50% nickel, having high permeability at low field strength and high electrical resistance.

high shear cutter: A cutting tool having a negative radial rake and a high positive axial rake. High shear cutters offer freer cutting on a variety of alloys and thus can be used on lower horsepower machines.

high-speed spindles: High-performance spindles capable of running at over eight-thousand rpm and are balanced and/or balanceable.

high-speed steel (HSS): Also known as high-speed tool steel. So named because they are designed primarily for efficient removal of metal faster than ordinary steel, because high speed steel retains its hardness (does not lose temper) at high temperatures. Due to this property, such tools may operate satisfactorily at speeds which cause cutting edges to reach red heat. They can hold their harness up to a temperature of about 1,000°F/538°C. High speed steels usually have a carbon content of 0.65 to 1.50%, and contain such alloying elements as tungsten (12 to 17%), chromium (2 to 4.7%), vanadium (1 to 2.5%) and sometimes cobalt, molybdenum (8.5%), manganese (0.2%). High-speed steel is made in an electric furnace. For hardening and tempering high-speed steels, a molten salt bath (temperatures used may be as high as 2,400°F/1,316°C) containing calcium, barium, potassium, and sodium chlorides or nitrates to which sodium carbonate [$CO_3 \cdot 2Na$] and sodium cyanide may be added.

hob: (1) A worm-shaped, multi-fluted, steel rotary cutting tool containing multiple teeth (called gashes or flutes) arranged on a helical path, used in a gear-tooth-generating process for spur gears, helical gears, etc; (2) a hardened steel master die (or hob) used to make multiple mold cavities by hydraulically pressing it into soft steel or beryllium-copper blanks.

hobbing: A process for accurately generating worms, spur, or helical gears, or splines by rotating and advancing a hob cutter past a blank workpiece (revolving at the same rate used by the finished gear) in precise relation to each other. The speed of the hobbing machine does not vary speed in any part from start of the cut to the end of it, producing more highly accurate work than any other gear-cutting process. See also hob.

hold flanging: Turning up or drawing out a flange around a hole. Also known as extruding.

hole basis: The system of fits where the minimum hole size is basic. The fundamental deviation for a hole basis system is *H*.

hole saw: A metal cylinder with teeth around its circumference and a twist drill bolted to its center and which is used as a guide. This drill is used to drill large diameter holes in relatively thin material.

hollow punch: A steel impact tool containing a cylindrical cutting edge, used to punch holes in sheet metal.

hollow-setscrew wrench: See Allen wrench.

homogenizing: A heat-treatment process involving holding at a high temperature for sufficient time to eliminate or decrease chemical segregation by diffusion; used to attain a uniform structure and composition.

honing: (1) A precision grinding operation used for blunting and strengthening cutting edges, such as knife blades; (2) an extremely accurate, internal finishing process for bringing a hole to size, and correct geometry by using low-speed and low-pressure application of abrasives such as silicon carbide [SiC] or aluminum oxide [Al_2O_3].

honing machine: A machine with a rotating head containing a tool bit for holding abrasive inserts used to produce accurately finished holes.

hook rule: Also known as a sliding hook rule or adjustable hook rule. A measuring device made of tempered steel, cut with great accuracy, and containing an arm (the hook) projecting from the zero division at one end. The hook may be rigidly fixed or adjustable and allows the user to make accurate measurements from points where one cannot see if the rule is even with the measuring edge, against shallow shoulders, through hubs of pulleys, setting inside calipers.

hopped: A term used among file makers to represent a very wide skip or spacing between file teeth.

Höpfner process: An electrolytic process for recovering copper.

horsepower (hp): A unit of power or the rate of doing work that has been adopted for engineering work. One horsepower is equal to 33,000 foot-pound/min or 550 foot-pound/sec or 76.1 kilogram meter/sec. The kilowatt, used in electrical work, equals 1.34 horsepower; or 1 horsepower equals 0.7457 kilowatts. However, in the metric SI, the term horsepower is not used, and the basic unit of power is the watt. This unit, and the derived units milliwatt and kilowatt, for example, are the same as those used in electrical work. Maximum available horsepower at the machine spindle is a critical factor as it can limit metal cutting capacity. To calculate horsepower at the cutter use the formula: $Hpc=mrr/K$. To calculate horsepower at the motor use the formula: $HPm=HPc/E$.

hot-dip galvanizing: See galvanized iron, galvanizing.

hot-melt adhesives: Usually supplied in solid form, hot-melt adhesives liquify when exposed to elevated temperatures. After application, they cool quickly, solidifying and forming a bond between two mating substrates. Hot-melt adhesives have been used successfully for a wide variety of adherents and can greatly reduce both the need for clamping and the length of time for curing. In general, hot-melt adhesives permit fixturing speeds that are much faster than can be achieved with water- or solvent-based adhesives. Some drawbacks with hot-melt adhesives are their tendency to string during dispensing and relatively low-temperature resistance.

hot quenching: An imprecise term used to cover a variety of quenching procedures in which a quenching medium is maintained at a prescribed temperature above 160°F/71°C.

hot shortness: Steel or wrought iron with high sulfur content which tends to brittleness at high temperatures, in the hot forming range, resulting in the formation of minute flaws and cracking.

hot top: A thermally insulated or heated supply reservoir for holding molten metal on top of a mold, used to feed the ingot or casting to avoid having "pipe" or voids as it contracts during solidification.

hp: The abbreviation for horsepower.

HSLA: Abbreviation for high-strength, low-alloy (steels).

HSS: Abbreviation for high speed steel.

hubbing: The process of forcing a male die into sand under pressure, thereby forming a female die.

Huntington dresser: See grinding wheel dresser.

hydraulic press: A hydraulic press is composed of a large piston in an enclosed chamber; its top is attached to a platen that rests on the members of a metal frame when the press is open. A liquid (usually water or oil) is pumped into the chamber through a valve. Once it has been filled, the pressure per-square-inch (in^2) applied at the valve will be transmitted to every square inch of the piston and of the walls of the chamber as well*. Thus, for a piston having a cross-sectional area of 100 in^2, a pressure of 10 psi at the valve exerts 1,000 lb pressure on the bottom of the piston, causing it to rise and the press to close. The pressure on the object being pressed varies inversely with its area. Large hydraulic presses exerting pressures up to 15 tons are used for shaping steel products.

*Note: Operates on Pascal's principle: *pressure applied to a unit area of a confined liquid is transmitted equally in all directions throughout the liquid.*

hydrazine: [H_4N_2] An oily, clear liquid and strong reducing agent used in electroplating.

hydrochloric acid: The aqueous solution of hydrogen chloride (HCl). A strong, highly corrosive acid and one of the highest volume chemicals produced in the world. Used for meal pickling and cleaning. Also known as muriatic acid. Highly toxic if inhaled or swallowed, and corrosive to skin and eyes, causing burns and possible blindness.

hydrofluoric acid: [FH] A colorless liquid used in stainless steel pickling; making specialized metals. Cleaning metals such as brass, cast iron, copper; enameling and galvanizing iron.

hydrogen: A gaseous element. Hydrogen combines with all elements except the five noble gases. Hydrogen has many industrial uses including the production of high-purity metals and high temperature welding.

Symbol: H
Physical state: Gas
Periodic Table Group: 1A
Atomic number: 1
Atomic weight: 1.00794
Valence: 1

Melting point: $- 434°F/-259°C$
Boiling point: $- 423°F/-253°C$
Source: Methane, industrial by-product

hydrogen embrittlement: A condition of embrittlement (or low ductility) in metals resulting from the absorption of hydrogen. Also known as acid embrittlement.

hydrozincite: An ore of zinc.

Hymu® 80: A registered trademark for an alloy nominally containing 79% nickel, 17% iron, and 4% molybdenum, having high permeability at low field strength and high electrical resistance.

hyper-: A Greek prefix indicating "above," "excess."

hypereutectoid steel: A steel alloy containing more than 0.80% to 0.85% carbon, the eutectic composition.

Hypernic®: An alloy nominally containing 50-53% iron and 47-50% nickel, having high permeability at low field strength and high electrical resistance.

hypo-: A Greek prefix indicating "below," "under."

hypoeutectoid steel: A steel alloy containing less than 0.80% to 0.85% carbon, the eutectic composition.

hypoid gear: A kind of bevel gear having beveled teeth used for transmitting rotary motion at an angle. Used to connect shafts that are not parallel to each other, and whose center lines do not intersect each other. See also bevel gear.

hysteresis: A lagging behind or impedance of a given effect, when the forces acting upon the body are changing. The magnetic lag, or retentivity of the magnetic state of ferromagnetic material such as iron and its alloys in an alternating magnetic field. Hysteresis is analogous to mechanical inertia and the energy lost is analogous to that lost in mechanical friction. It presents a major problem in the design of electrical machines with iron cores such as transformers and rotating armatures.

Hz: The abbreviation for *Hertz*.

I

I: Symbol for the element iodine.

ID: Abbreviation for internal diameter.

ignition temperature: The minimum temperature required to initiate or cause self-sustained combustion independent of a heat source. See also autoignition temperature.

ILMA: Abbreviation for Independent Lubrication Manufacturers Association.

Illium®: Proprietary trade names of Stainless Foundry & Engineering, Inc for a corrosive-resistant stainless steel alloys.

IMA: Abbreviation for the International Magnesium Association.

impact resistance: The amount of force or energy required to fracture a standard-size sample in an impact test such as the Izod or Charpy test. The resistance is expressed in terms of the number of foot-pounds or meter-kilograms of force required to break the specimen with a single blow. Sometimes called impact energy or impact value.

impact test: Tests to determine the shock resistance or toughness of a metal when subjected to high rates of loading, usually in bending, tension, or torsion. The quantity measured is the energy absorbed in breaking the specimen by a single blow, as in the Charpy or Izod tests.

imperfect thread: See incomplete thread.

imperial metal: Copper-nickel alloys nominally containing 80% copper and 20% nickel.

In: Symbol for the element indium.

inch: A unit of length 1 in = 1/12 ft = 1/36 yd = 1/63,360 mi = 2.54000 cm.

included angle: See thread angle.

inclusion: A state of one substance being completely enclosed in or surrounded within the other, as suspended material in a crystal lattice, or a nonmetallic material either wholly or partly locked within the a solid metallic matrix.

Inco® chrome nickel: A proprietary trade name of Inco Alloys International, Inc. for a line of nickel alloys containing 79-82% nickel, 12-14% chromium, and 6-7% iron.

Incoloy® : Proprietary trade names of Inco Alloys International, Inc. for a line of corrosion resistant alloys containing varying amounts of nickel, chromium, iron, molybdenum, copper, silicon, and titanium.

Incoloy® alloy 800: A proprietary trade name of Inco Alloys International, Inc. for an alloy containing 48% iron, 32% nickel, and 20% chromium. Also listed as containing 39-47% iron, 19-23% chromium, 0-1.5% manganese, 30-35% nickel, 0-1% silicon, 0-0.8% copper, 0-0.6% aluminum, 0.6 titanium and 0-0.1% carbon.

Incoloy® alloy 825: A proprietary trade name of Inco Alloys International, Inc. for an alloy containing 40% nickel, 33% iron, 21% chromium, 3% molybdenum, 2% copper, 1% titanium.

Incoloy® alloy 901: A proprietary trade name of Inco Alloys International, Inc. for an alloy containing 42% nickel, 36.9% iron, 12.5% chromium, 5.7% molybdenum, 2.9% titanium.

Incoloy® alloy DS: A proprietary trade name of Inco Alloys International, Inc. for an alloy containing 42.7% iron, 37% nickel, 18% chromium, 2.3% silicon.

Inconel®: A proprietary trade name of Inco Alloys International, Inc. for heat and corrosion resistant alloys containing 76-80% nickel, 14-15% chromium, 6-9% iron.

Inconel® 600: A proprietary trade name of Inco Alloys International, Inc. for heat and corrosion resistant alloy nominally containing 77% nickel, 16% chromium, 7% iron.

Inconel® 700: A proprietary trade name of Inco Alloys International, Inc. for heat and corrosion resistant alloy nominally containing 47.45% nickel, 29% cobalt, 15% chromium, 3.3% aluminum, 3% molybdenum, 2.25% titanium.

Inconel® 718: A proprietary trade name of Inco Alloys International, Inc. for heat and corrosion resistant alloy nominally containing 53% nickel, 19% chromium, 19.2% iron, 5% niobium, 3% molybdenum, 0.8% titanium.

Inconel® X-750: A proprietary trade name of Inco Alloys International, Inc. for heat and corrosion resistant alloy nominally containing 73.7% nickel, 15% chromium, 7% iron, 2.5% titanium, 0.9% aluminum, 0.9% niobium.

incompatibility: (1) A term used to indicate whether a material or substance can be placed in contact with certain other products, substances, or materials. The direct contact of incompatible materials can cause corrosion or more dangerous reactions such as heat, fire, or explosions, and give off toxic vapors; (2) a computer term used to describe a quality of a system that is unable to use, work with, or manipulate data and programs devised for another computer system.

incomplete thread: A thread design term used to describe a threaded profile having either crests or roots or both, not fully formed, resulting from their intersection with the cylindrical or end surface of the work or the vanish cone. It may occur at either end of the thread.

Independent Lubrication Manufacturers Association (ILMA): 6515 Washington Street, Alexandria VA 22314. Telephone: 703/684-5574. FAX: 703/836-8503. WEB: http://www.ilma.org

indexable inserts: Single-point cutting tools of various types and shapes used for turning, threading, boring, and parting. These inserts, often called throw-away inserts, are comprised of a special toolholder to which the indexable insert is attached with a lock pin tightened with an Allen wrench. These inserts are made of various materials and have multiple cutting edges that can be reshaped, resharpened, and replaced.

indexing: (1) On turret-type machines, rotating the turret to make other tools available; (2) in dividing head operations, moving a workpiece angularly and sequentially so that equally spaced divisions can be machined.

indexing head: See dividing head.

indium: A metallic element of the aluminum subgroup. CAS 7440-74-6. Softer than lead. Used in low-melting soldering and brazing alloys, automobile and aircraft bearings, electroplated coatings, often with silver, increase tarnish resistance.

Symbol: In
Physical state: Silvery metal
Periodic Table Group: IIIA
Atomic number: 49
Atomic weight: 114.82
Valence: 1,3

Density (g/cc): 7.31
Specific heat: 0.0303
Melting point: 314°F/156°C
Boiling point: 3,776°F/2,080°C
Source: From zinc and lead production.
Oxides: In_2O_3
Crystal structure: Face centered tetragonal

indium trichloride: [$InCl_3$] A white to tan or yellow crystalline solid used in electroplating.

induction heating: A process of local heating by electrical induction wherein there is no direct electrical connection between the electrical supply and the metal being heated.

induction hardening: A case hardening process that is similar in many respects to flame hardening except that the heating is generated by electrical induction, whereby a high-frequency electric current sent through a coil or inductor surrounding the part being hardened. Quenching is usually accomplished with a water spray introduced at the proper time through jets in or near the inductor. In some instances, however, parts are oil-quenched by immersing them in a bath of oil after reaching hardening temperature.

industrial diamond: Low-grade diamonds (carbonado, ballas, black diamonds, bort, boart, boort) as well as those made synthetically in an electric furnace at high pressure and temperature (at 3,000°F/1,649°C, 1.3 million psi). Used in grinding operations; cutting tools, grinding wheels, specialized drill bits, wire-drawing dies, grinding sheel dressing tools, etc. See also diamond.

inert gas: See noble gas.

inert-gas shielded-arc welding: See MIG.

inert substances: elements and compounds that react very slowly or not at all.

inertia: In mechanics, the property of matter that causes it to resist any change in its motion or state of rest.

ingot iron: The malleable, commercially pure iron from the Bessemer process.

inhibitor: A chemical or substance that is added to another substance to prevent unwanted chemical change (e.g., rust, corrosion) from occurring.

inserted-tooth cutters: Milling cutters containing cutting teeth that can be replaced after long used or repeated sharpening, used widely for heavy production milling such as face milling cutters. The cutting teeth are made of hard material such as high-speed steel and cemented carbide.

inside diameter: See internal diameter.

inside calipers: A two-legged steel instrument with the hardened steel ends of each bent outward. Their size is measured by the greatest distance they can be opened between the two points. Used to measure the diameters of holes and distances between objects.

insoluble: Not dissolving in a solvent, or products that cannot be dissolved in each other.

inspection bench: A bench with a level top, usually made of cast iron, on which work can be inspected with precision instruments.

instant adhesives: See cyanoacrylate adhesives.

interchangeable: Sufficiently alike in size as to permit replacement without modification.

interference fit: (1) One having limits of size so specified that an interference always results when mating parts are assembled; (2) the relationship between assembled parts when interference occurs under all tolerance conditions.

intermediate annealing: Annealing at one or more stages during manufacture and before final thermal treatment.

intermetallic compounds: Compounds occurring at the interface between the metal surfaces of two or more metals that are fused together during the process of coating or brazing. In the case of galvanizing, the metals form bimetallic compounds; in other cases, they form alloys.

internal diameter (ID), gear: The diameter of a circle coinciding with the tops of the teeth of an internal gear.

internal gear: A gear with teeth on the inner cylindrical surface.

internal spline: A spline formed on the inner surface of a cylinder.

internal thread: A thread on a cylindrical or conical internal surface.

International Magnesium Association (IMA): 1303 Vincent Pt., Suite 1, McLean, VA 22101; Telephone: 703/442-8888. FAX/703/821-1824. WEB: www.intlmag.org

interrupted quenching: A quenching procedure in which the metal object being quenched is removed from the initial quench at a temperature substantially higher than that of the quenchant and is then subjected to a subsequent quenching medium having a different cooling rate than the first. Austempering, martempering, and isothermal quenching are three methods of interrupted quenching that have been developed to obtain greater toughness and ductility for given hardnesses and to avoid the difficulties of quench cracks, internal stresses, and warpage, frequently experienced when the conventional method of quenching steel directly and rapidly from above the transformation point to atmospheric temperature is employed.

interupted thread tap: A tap having an odd number of lands, with every other tooth along the thread helix removed.

Invar®: A registered trademark for a iron-nickel alloy nominally containing 36% nickel and 63.8% steel, and 0.2% carbon. This alloy has an extremely low coefficient of thermal expansion.

inverse annealing: A heat treatment, analogous to precipitation hardening or age hardening, applied to cast iron, usually to increase its hardness and strength.

investment casting: A casting made by melting or heating an expendable pattern (that is made from wax, plastic, or frozen mercury) out of a mold of hardened plaster or ceramic. Also known as precision casting, or lost-wax casting.

involute, gear: The curve formed by the path of a point on a straight line, called the *generatrix*, as it rolls along a convex base curve. The base curve is usually a circle. This curve is generally used as the profile of gear teeth.

involute spline: A spline having teeth with involute profiles.

ipm: Abbreviation for inches per minute. The table feed or machine feed. The ipm value refers to how far the workpiece or cutter advances linearly in one minute.

ipr: Abbreviation for inches per revolution. A feed volume reporting how far the cutter advances during one revolution.

ipt: Abbreviation for inches per tooth. Sometimes referred to as chip load, ipt is the linear distance traveled by the cutter during the engagement of one tooth. Although the milling cutter is a multi-edge tool, it is the capacity of each individual cutting edge that sets the limit of the tool, defined as: ipt = IPM/NT x rpm

Ir: Symbol for the element iridium.

Irco® Aluminum Coatings: A registered trademark of International Rustproof Co. corrosion-resistant coatings for aluminum that also increase the corrosion resistance and adhesion of paint films subsequently added.

Irco Lube®: A registered trademark of International Rustproof Co. for oil-absorptive manganese phosphate coatings for mating moving steel surfaces.

iridium: A metallic element of the platinum group. CAS 7439-88-5. The most corrosion-resistant element. Used in specialized alloys, often with platinum.

Symbol: Ir
Physical state: Silvery metal
Periodic Table Group: VIII
Atomic number: 77
Atomic weight: 192.22
Valence: 1,2,3,4,6

Density (g/cc): 22.56
Melting point: 4,370°F/2,410°C
Boiling point: 7,466°F/4,130°C
Source/ores: Osiridium, iridiosmium, nevyanskite; in platinum ores
Oxides: IrO_2, Ir_2O_3
Brinell hardness: (cast) 218
Crystal structure: f.c.c.

iridium chloride: [IrCl₄] Brownish-black solid used in electroplating solutions.

iridium steel: A carbon steel containing 74.43% iron, 16% tungsten, 4% cobalt, 3.5% chromium, 0.8% molybdenum, 0.67% vanadium, and 0.6% carbon.

iridosmine: A native alloy found in deposits of platinum. Variably contains 10-77% iridium, 17-80% osmium, 0-10% platinum, 0-17% rhenium, 0-9% ruthenium, 0-1.5% iron, less than 1% copper, and trace amounts of palladium.

iron: Pure iron (also known as ferrite) is a relatively soft metallic element of crystalline structure. CAS number: 7439-89-6. An extremely important metal; the only metal that can be tempered; used principally in steels and other alloys. Pure iron can exist in four solid phases having different physical characteristics, denoted by the Greek letters alpha (α), beta (β), gamma (γ), and delta (δ). See also alpha iron, beta iron, gamma iron, and delta iron.

Symbol: Fe (from ferrum) **Density (g/cc):** 7.874
Physical state: Soft, silvery **Melting point:** 2,795°F/1,535°C
Periodic Table Group: VIII **Boiling point:** 4,982°F/2,750°C
Atomic number: 26 **Source/ores:** Hematite, limonite, magnesite, siderite,
Atomic weight: 55.847 taconite (low grade ores)
Valence: 2,3,6 **Oxides:** FeO, Fe_2O_3, Fe_3O_4
Crystal structure: b.c.c. (α); c.c.p. (γ); b.c.c. (δ)

ironac: Alloys of iron and silicon with small amounts of carbon, manganese, and phosphorus.

iron, alibated: Iron covered with a thin coat of aluminum.

iron alloy, base (ASTM B163-800): Also known as alloy 800, Incoloy® 800, Pyromet® 800, Sanicro 31, Thermax® 4876. Alloys containing 39-47% iron, 19-23% chromium, 0-1.5% manganese, 30-35% nickel, 0-1% silicon, 0-0.8% copper, 0-0.6% aluminum, 0.6 titanium and 0-0.1% carbon.

iron alloy, base (ASTM XM10): Also known as alloy 21-6-9. Alloys containing

60-69% iron, 18-21% chromium, 8-10% manganese, 5-7% nickel, 0-1% silicon, 0.2-0.4% nitrogen, 0-0.1% carbon, 0-0.1% phosphorus.

iron alloy, base, Fe, Ni: See ferronickel.

iron brass: Brass nominally containing 1-9% iron.

iron, cast: See cast iron.

iron ore: An oxide of iron containing, ordinarily, from 35 to 65% of iron, and in addition, oxygen, phosphorus, sulfur, silica (sand), and other impurities.

iron oxide: (1) [FeO] ferrous oxide; (2) [Fe_2O_3] ferric oxide; (3) [$Fe(OH)_3$] ferric hydroxide [$Fe(OH)_3$]; (4) [$Fe(OH)_3$] ferrous hydroxide. [$Fe_2O_3 \cdot H_2O$] hydrated ferric oxide. Red iron oxide is primarily ferric oxide, also known as burnt sienna, ferric monoxide, ferric trioxide, ferrosoferric oxide, Indian red, iron saffron, red hematite, red iron oxide, red iron trioxide, red oxide, red rouge, Turkey red, Van Dyke red, Venetian red: C Black iron oxide is primarily ferrous oxide, also known as black rouge, acetylene black, bone black, carbon black. Brown iron oxide is primarily ferric hydroxide [$Fe(OH)_3$] and ferrous hydroxide, also known as iron subcarbonate, ferric hydrate, iron hydroxide, iron hydrate. Yellow iron oxide is another name for hydrated ferric oxide.

Iron and Steel Society (ISS): 186 Thorn Hill Road, Warrendale PA 15086-7528. Telephone: 724/776-1535. FAX: 724/776-0430. WEB: http://www.iss.org.

iron, stainless: See stainless iron.

ISO: International Organization for Standardization, Geneva, Switzerland, is the highest international authority, a federation of national standards institutes that has the purpose of setting standards on which all countries can agree.

iso-: From the Greek word isosceles meaning "the same as" or "equal."

ISO metric thread: A thread form that is the European standard, with thread dimensions and lead measured in millimeters instead of inches. Like the Unified National Thread form, the International Standards Organization (ISO) is comprised of a 60° included angle.

Isopropanolamine: [C_3H_9NO] A liquid with a slight ammonia odor used in cutting oils.

isosceles triangle: A triangle having two sides of equal length.

isothermal annealing: Complete transformation at constant temperature of an austenitized ferrous alloy to relatively soft ferrite carbide aggregate and pearlite.

isothermal quenching: A process resembling austempering in that the steel is first rapidly quenched from above the transformation point down to a temperature that is above that at which martensite begins to form and is held at this temperature until the austenite is completely transformed into bainite. The constant temperature to which the piece is quenched and then maintained is usually 450°F/232°C or above. The process differs from austempering in that after transformation to a bainite structure has been completed, the steel is immersed in another bath and is brought up to some higher temperature, depending on the characteristics desired, and is maintained at this temperature for a definite period of time, followed by cooling in air. Thus, tempering to obtain the desired toughness or ductility takes place immediately after the structure of the steel has changed to bainite and before it is cooled to atmospheric temperature.

isothermal transformation: A change in phase or structural transformation which occurs at constant temperature.

isotropic: A term used to describe the properties at any point in a body are the same, regardless of the direction in which they are measured. For example, cast metals and extruded plastics tend to be isotropic so that samples cut in any direction within a cast body tend to have the same physical properties. For a comparison of terms See anisotropic.

ISS: Abbreviation for Iron and Steel Society.

Izod Test: A single-blow impact test of a material's ability to resist the propagation of a crack. The specimen (usually notched, mounted vertically, and fixed at one end) is subjected to a sudden blow of 120 ft-lb at a velocity of 11.5 ft/sec by a hammer mounted on the end of a pendulum. The specimen is broken by the weight of the pendulum; the energy absorbed, as measured by the subsequent rise of the pendulum, is a measure of impact strength or notch toughness of the material.

J

J: Symbol for joule. Symbol for moment of inertia.

japan: A hard, glossy, dark-colored varnish used on metallic objects. Japans are normally applied in successive applications and dried by baking in an oven at relatively high temperatures.

Japanese Standards Association: An international standards organization located at 1-24 Akasaka, 4-chome, Minato-ku, Tokyo 107, Japan. Sets the Japanese Industrial Standards (JIS) and publishes standards including metals and welding filler metals.

japanning: The process of applying a japan varnish to an object. See japan.

jewelry bronze: Alloys nominally containing 87.5% copper and 12.5% zinc.

jig: A work-holding device that places the workpiece in proper position and holds it securely. The jig can contain guides and hardened steel bushings for aligning, guiding, and supporting various cutting tools for drilling, reaming, tapping, etc. For comparison, see fixture.

jig borer: A machine (or a task of NC machining centers) for accurately locating, aligning, and boring hole in jigs, fixtures, dies, for centering, drilling, step boring, counterboring reaming, contouring, etc.

JIS: Abbreviation for Japanese Industrial Standards.

JIT: Abbreviation for *just-in-time*, as in just-in-time manufacturing.

Jo blocks: Short for Johansson Gage blocks invented in 1895 by Carl E. Johansson in Sweden. Precision blocks made from hardened steel or other hard material.

Considered the universal standard of precision measurement for science and industry throughout the world, they provide practical standards for measuring accurately in millionths of an inch, an impossible feat prior to their availability.

joggle: An offset surface consisting of two adjacent, continuous or nearly continuous short-radius bends of opposite curvature.

josephinite: A naturally occurring iron-nickel alloy found in the state of Oregon.

joule (J): The international (SI) unit of work, energy, and quantity of heat.

journal bearing: Also known as a sleeve bearing. Cylindrical or ring-shaped bearings designed to carry radial loads. The most widely used and simplest journal bearings are cast bronze that are lubricated by oil or grease, and porous bronze that are impregnated with oil or have an oil reservoir in the housing. Carbon graphite and high-load carrying plastic bearings are often used to replace metal bearings.

JSA: Abbreviation for the Japanese Standards Association.

K

k: Abbreviation for kilo- (10^3), as in kilometer. Symbol for thermal conductivity.

K: (1) Symbol for the element potassium; it stands for the Latin *kalium*; (2) abbreviation for the Kelvin temperature scale; (3) symbol for computer memory, in approximate number of thousands (or 1,024) of bytes or words.

K.A.: An alloy of aluminum. Similar to Duralumin®.

karat: Another name for carat as used in the United States to denote the fineness of gold. See also carat.

Kelvin: The temperature scale used in certain engineering calculations such as the change in volume of a gas with temperature. Absolute temperatures are expressed either in degrees Kelvin, corresponding respectively to the centigrade and Fahrenheit scales. Temperatures in Kelvins are obtained by adding or subtracting 273 to the centigrade temperature depending if above or below 0°C.

kerf: The width of a cut produced during a cutting process, as the slot or passageway cut by a saw.

kerosene: Also kerosine. A clear, oily, liquid; a mixture of hydrocarbons. Used as a fuel, solvent, cleaner, and sometimes as a cutting fluid for cutting metals, threading and turning aluminum alloys. This flammable liquid must be stored in approved and properly labeled containers. Never use it near open flames, sparks, or temperatures above its flash point (about 100-150°F/37.7-65.5°C).

key: A small square, rectangular, or other shaped piece of metal which locks locked into both a keyway in a shaft and a slot in a gear or pulley. Once locked, the positive means is provided for transmitting torque and preventing the gear or pulley from spinning on the shaft.

key leash: A rubber or plastic device that attaches to drill cord to help prevent chuck key loss.

keyseat: An axially located rectangular groove in a shaft or hub.

keyway: A groove, usually in a shaft and the piece that is to be fastened to it, in which a key is driven. The keyway is usually square or rectangular, but can be round or other shaped.

kg: Abbreviation for kilogram.

kgf: Abbreviation for kilogram force.

kibbled: Broken up into small pieces, usually of approximately 1 cm.

killed steel: Steel that has been thoroughly deoxidized with a strong deoxidizing agent such as silicon, ferrosilicon, manganese, or aluminum, in order to insure that it solidifies quietly in the mold while the mixture is maintained at melting temperature until bubbling ceases. The purpose of the process is to reduce oxygen content and allow solidification without any evolution of gas due to an absence of a reaction between carbon and oxygen.

kilo-, kilo: (1) Prefix in the SI metric system for 1,000. Prefix for 10^3; (2) symbol (k or K) e.g., 1km=1 kilometer=1,000 meters; (3) kilogram.

kilogram (kg): The standard unit of mass in the SI metric system. 1 kg = 1,000 g = 2.2046 lb = 15,432 grains.

kinetic energy: Energy possessed by a body due its motion; it is equal to the product of one half of the mass of the body by the square of its velocity (= $mv^2/2$ where *m* is the mass, and *v* is the velocity).

kip: A load of 1,000 lb.

Kirksite®: A proprietary trade name for a zinc-based alloy used principally in low-production dies.

kish: A name given to crystalline graphite from molten iron, found on the walls of iron furnaces.

Kleanrol®: A registered trademark of DuPont for a soldering flux crystal based on ammonium chloride and zinc chloride.

knife bandsaw blade: See bandsaw.

knife file: Shaped like a knife, this file is double cut and tapered in width and thickness. Used for filing grooves, notches, and narrow slots.

knockout: A mechanism, sometimes called an ejector, used to release blanks from a die.

Knoop hardness test: Microhardness determined from the resistance of metal to indentation by a pyramidal diamond indenter. The indenter has edge angles of 172° 30' and 130°, making a rhombohedral impression with one long and one sort diagonal.

knurling: A lathe process using a knurling tool to impress a rough pattern design on the peripheral surface of a work blank, generally used to provide a non-slip surface for gripping, handling, or turning cylindrical pieces by hand or to improve appearance of the workpiece.

knurling tool: A lathe tool containing hard patterned steel rollers that is mounted in the tool-post and forced against the cylindrical metallic surface of a work blank with considerable pressure. As the work blank and the knurling tool are mechanically rotated together, each roller, under pressure, cuts itself into the work blank. As one wheel contains right-hand and the other left-hand grooves, a diamond-shaped pattern is produced.

Kroll process: A widely used process for obtaining titanium metal. Titanium tetrachloride is reduced with magnesium metal at red heat and atmospheric pressure, in the presence of an inert gas blanket of helium or argon. Magnesium chloride and titanium metal are produced. Essentially the same process is also used for obtaining zirconium.

Kronaplate®: A proprietary trade name of the E/M Corporation (Great Lakes Chemical Corporation) for a line of lubricating greases.

Krupp's disease: Brittleness exhibited by steels following tempering and reheating.

K.S. magnet steel: Alloys nominally containing 35% cobalt, used for short magnets.

L

ladle: A refractory-lined receptacle equipped with a spout used for transporting molten metal from the furnace to ingot molds, from which it is poured. Large ladles may contain lugs for handling by a crane.

lamination: Metal defects involving the separation of two or more layers, sometimes found in wrought, rolled, or forged metals. These separations are generally aligned parallel to the worked surface and may be the result of non-metallic inclusions, segregation elongated and made directional by working.

land: (1) A drill term used to describe the peripheral portion of the drill body between adjacent flutes; (2) a gear term: the top land is the top surface of a tooth, and the bottom land is the surface of the gear between the fillets of adjacent teeth.

land width: A drill term used to describe the distance between the leading edge and the heel of the land measured at a right angle to the leading edge.

lanthanum: CAS number 7439-91-0. Soft, malleable, lustrous, silvery-white or gray metal. An alloying element. One of the most-abundant rare-earth metals. Decomposes in cold water, slowly producing lanthanum hydroxide and flammable hydrogen gas. *Danger:* Pyrophoric, the finely divided material will catch fire at room temperature.

Symbol: La
Physical state: Gray solid
Periodic Table Group: IIIB
Atomic number: 57
Atomic weight: 138.9055
Valence: 3

Density (g/cc): 6.15
Specific heat: 6.65
Melting point: 1,688°F/920°C
Boiling point: 6,267°F/3,464°C
Source/ores: Monazite sand, bastnasite, seawater
Oxides: La_2O_3
Crystal structure: Hexagonal (α); f.c.c. (β); b.c.c. (γ)

lap: (1) A material or another piece of work containing an imbedded abrasive that can be used for lapping or honing other and harder materials. A workpiece to be lapped will have a very small amount of material (0.0001-0.002 in) left to be removed; (2) a surface defect caused by folding a fin or sharp corner of a metal and then rolling or forging them into the surface, without welding them; (3) a term applied to seams that lap each other.

lap joint, riveted: one of two kinds of riveted joint. In the lap-joint, the plates overlap each other and are held together by one or more rows of rivets. The other rivet joint is the *butt joint*.

lapping: A precision grinding process for polishing and/or accurately finishing the surface of an object by abrasion with a fine material such as abrasive pastes or abrasive flour suspended in a liquid medium. Also, an object can be used such as cloth, lead, copper, cast iron, plastic, or close-grained hard wood, having very fine abrasive particles rubbed, rolled, or otherwise embedded in its surface.

lard oil: Colorless to yellowish liquid with a peculiar odor. It is an animal oil, derived by cold-pressing lard (hog fat) and used in metal cutting compounds and lubricants. It may be mixed with mineral oils to lower its relatively high cost. Lard oil is combustible and subject to spontaneous heating. It can also breed bacteria that can cause skin irritation and dermatitis.

laser: A term derived from the initial letters *l*ight *a*mplification by *s*timulated *e*mission of *r*adiation. A controlled beam of intense, high-energy, and concentrated monochromatic (one color) light of constant and uniform wavelength, frequency, and phase traveling in a series of parallel rays (coherent light) that can be accurately focused. Laser beams are used in industry for cutting, welding, drilling, machining, surface treatment, and marking. *Safety note:* Coherent laser light can be 100,000 times higher in energy density than equivalent-power incoherent light and can cause irreparable damage to the eyes. Proper shielding and the use of laser-proof safety glasses is essential at all times.

laser beam machining (LBM): A limited-production machining process using a laser that can heat up and vaporize the surface layer of any material. The removal rate of material is relatively minimal and the process is limited to hole drilling and the cutting of thin material. See also laser.

laser beam welding (LBW): A welding method using a concentrated, high-energy beam of light. Used primarily for joining relatively thin materials. See also laser.

LaserMike®: A registered trademark of the Techmet Company for an optical micrometer that is highly accurate and uses an inert gas helium-neon beam.

latent heat: See heat of fusion.

lathe: The most basic of all metalworking machine tools. A turning machine capable of producing cylindrical and conical parts by spinning the circumference of the workpiece against the cutting edge of a single-point cutting tool that is mounted on a carriage that may be fed manually or automatically along the work while machining. Also used for boring holes more accurately and/or larger than those produced by drilling or reaming.

lathe center: A lathe accessory with a 60° cone point that is incorporated into a lathe headstock or tailstock and used to support the workpiece.

lathe tool: Any cutting instrument held in the tool post of a lathe.

lattice: Also called the atomic lattice arrangement. The structural arrangement of atoms in a crystal. Accurate information is obtained by X rays, which are diffracted by the lattice at various angles. As the atoms are from 1.5 to 3 Å apart in most crystals, the lattice acts as a diffraction grating.

Lautal®: A proprietary trade name for a hard aluminum alloy nominally containing 4-5% copper, 1.5-2% silicon, 0.4-0.7% iron, and small amounts of manganese, magnesium or other metals.

lay: A term describing the direction of the predominant surface pattern, ordinarily determined or caused by the production method used.

layout: The transfer of information containing lines and geometrical shapes from a working drawing to a metal surface of a workpiece.

LC: Abbreviation for locational clearance fit. See also fits, description.

lead: CAS number 7439-92-1. Soft, dull gray metal. An alloying element. Lead has many commercial uses including solder, low-melting alloys, batteries.

Symbol: Pb	**Density (g/cc):** 11.35
Physical state: Gray metal.	**Melting point:** 622°F/328°C
Periodic Table Group: IVA	**Boiling point:** 3164°F/1740°C
Atomic number: 82	**Source/ores:** Galena, anglesite, cerussite,
Atomic weight: 207.21	pyromorphite, mimetesite
Valence: 2,4	**Oxides:** PbO, Pb_2O, Pb_2O_3, Pb_3O_4
	Crystal structure: f.c.c.

lead: (1) In thread design, used to describe the axial distance between two consecutive points of intersection of a helix by a line parallel to the axis of the cylinder on which it lies, i.e., the distance that a threaded part advances axially as it is rotated one revolution (or one complete turn) around the circumference. On a single lead screw or tap, the lead and pitch are identical. On a double lead screw or tap, the lead is twice the pitch, etc; (2) in gear design, the distance a helical gear or worm would thread along its axis in one revolution if it were free to move axially.

lead angle: On a straight thread, the lead angle is the angle made by the helix of the thread at the pitch line with a plane perpendicular to the axis. On a taper thread, the lead angle at a given axial position is the angle made by the conical spiral of the thread with the perpendicular to the axis at the pitch line.

lead, arsenical: Also known as F3 alloy. Contains 99.65 lead, 0.15 arsenic, 0.10 tin, 0.10 bismuth and sometimes small amounts of copper and/or tellurium.

lead-base babbitt: Babbitt metal nominally containing 75-85% lead, 10-15% antimony, and 5-10% tin. Some examples include SAE 13, containing 85% lead, 10% antimony, and 5% tin. SAE 14 containing 75% lead, 15% antimony, and 10% tin. Alloy 8 containing 80% lead, 15% antimony, and 5% tin. Other lead-base babbitt materials include arsenical lead babbitt. See also Babbitt and arsenical lead babbitt.

lead bath tempering: A process for heating steel in connection with tempering, as well as for hardening. The bath is first heated to the temperature at which the steel should be tempered; the preheated work is then placed in the bath long enough to acquire this temperature, after which it is removed and cooled. As the melting temperature of pure lead is about 620°F, tin is commonly added to it to lower the temperature sufficiently for tempering.

lead, calcium: Alloy containing 99.972% lead, 0.028 calcium. Used for cable sheathing and creep-resistant sheet or pipe.

lead coating: Coatings of lead or lead alloys can be deposited on base metals in various ways, including dipping into the molten lead following application of a tin layer to the base metal. The intermediate tin layer is applied by electroplating or spraying and increases the bond of the lead coating.

lead dross: Another name for lead scrap or waste.

leaded brass: Alloys nominally containing 61-72% copper, 24-37% zinc, 1% lead, and 0-3% tin.

leaded commercial bronze: See commercial bronze, leaded.

leaded Muntz metal: See Muntz metal, leaded.

leaded naval brass: See naval brass, leaded.

leaded nickel brass: See nickel brass, leaded.

leaded red brass: See red brass, leaded.

leaded semi-red brass: See red brass, leaded.

leaded tin bronze: See tin bronze, leaded.

leaded yellow brass: See yellow brass, leaded.

lead float: A single-cut file with large straight rows of teeth or cutting edges running across its surface, used for filing soft materials such as lead and babbitt. A similar file with curved teeth is the *vixen file.*

lead fluoborate: [$B_2F_8\cdot Pb$] Used in metal finishing and as an electroplating solution for coating metallic objects with lead.

lead glance: Another name for galena, a primary ore of lead and secondary ore of silver.

lead hammer: A hammer with a soft lead alloy head, used for striking finished surfaces that would be dented or otherwise damaged by a hammer with a hardened head. As a pure lead head would deform too quickly due to its softness, the head contains a small amount of antimony.

Lead Industries Association, Inc. (LIA): 295 Madison Avenue, New York, NY 10017. Telephone: 800/922-LEAD; 212/578-4750. FAX: 212/684-7714. WEB: http///:www.leadinfo.com

lead-soap lubricants: An "extreme-pressure lubricant" (EP) made from lead salts that are treated with an alkali and with fats. They are hard at low temperatures, viscous at medium temperatures, and somewhat fluid on heating by friction, and are not suited for high speeds.

lead solder alloys: There are various lead-base solder alloys including the following: Alloy 5B, ASTM B32-60T containing 95% lead and 5% tin; Alloy 20B,

ASTM B32-60T containing 80% lead and 20% tin; Alloy 50A or 50B, ASTM B32-60T containing approximately 50% lead and 50% tin and small amounts of impurities. Alloy 50A contains 0.12% (max.) antimony and Alloy 50B contains 0.20-0.50% (max.) antimony. Both 50A and 50B contain maximum amounts of the following: 0.25% bismuth, 0.08% copper, 0.02% iron, 0.005% aluminum, 0.005% zinc, 0.02% arsenic, and 0.08% other impurities.

lead thread: A thread design term used to describe that portion of the incomplete thread that is fully formed at the root but not fully formed at the crest that occurs at the entering end of either an external or internal thread.

lead variation, involute spline: The lead tolerance for the total spline length applies also to any portion thereof unless otherwise specified. The variation of the direction of the spline tooth from its intended direction parallel to the reference axis, also including parallelism and alignment variations. *Note*: Straight (nonhelical) splines have an infinite lead. These variations will cause clearance variations and therefore reduce the effective clearance.

lead wool: Fine strands of metallic lead used for pipe joint packing.

lean ore: An ore containing low metal content.

least material condition (LMC): The condition in which a feature of size contains the least amount of material within the stated limits of size, for example, maximum hole diameter and minimum shaft diameter.

LED: Abbreviation for Light Emitting Diode.

LEED: Abbreviation for low-energy electron diffraction.

left-hand thread: A thread is a left-hand thread if, when viewed axially, it winds in a counterclockwise and receding direction. All left-hand taps are stamped "L" or "LH." Right-hand taps are unmarked.

LEL: Abbreviation for lower explosive limit.

length, file: The distance from the heel to the point.

length of action, gear: The distance on an involute line of action through which the point of contact moves during the action of the tooth profile.

length of complete thread: The axial length of a thread section having full form at both crest and root, but also including a maximum of two pitches at the start of the thread which may have a chamfer or incomplete crests.

length of engagement, involute spline (L_q): The axial length of contact between mating splines.

length of thread engagement: The length of thread engagement of two mating threads is the axial distance over which the two threads, each having full form at both crest and root, are designed to contact. See also length of complete thread.

letter drills: Twist drills with letter designations from A to Z [with diameters form 0.234 in (5.94 mm) to 0.413 in (10.49 mm)]. On charts containing inch-sized drill sizes, the letter drills begin where number drills end.

leveling: (1) A plating term describing the capacity of a nickel coating to cover surface irregularities of the base metal. This is achieved by the addition of brighteners to the plating bath. See also brightener; (2) a process used to flatten or "even out' deformities in sheet metals by slight tensile straining or by feeding them through slightly offset rollers.

Levelume®: A registered trademark of M & T Harshaw (division of Atochem) for a bright, high-leveling nickel process. Prepared from nickel sulfate, nickel chloride, boric acid, and organic addition agents. Used for automotive trim, plumbing fixtures.

Levol's alloy: A silver alloy nominally containing 71.9% silver and 28.1% copper.

Lewis metal: An alloy of 50-50% tin and bismuth, with a melting point of 280°F/138°C, used for holding and sealing die parts.

LH: Abbreviation for left hand thread.

Li: Symbol for the element lithium.

life, bearing: For an individual bearing, the number of revolutions (or hours at some given speed) that a bearing runs before the first evidence of fatigue develops in the material of either the rings (or washer) of any of the rolling elements.

light-curable adhesives: Light-curing systems use a unique curing mechanism. The adhesives contain photoinitiators that absorb light energy and dissociate to form radicals. These radicals then initiate the polymerization of the polymers, oligomers, and monomers in the adhesive. The photoinitiator acts as a chemical solar cell, converting the light energy into chemical energy for the curing process. Typically, these systems are formulated for use with ultraviolet light sources but some products have been formulated for use with visible light sources. One of the biggest benefits that light-curing adhesives offer to the manufacturer is the elimination of the work time to work-in-progress trade-off, which is embodied in most adhesive systems. With light-curing systems, the user can take as much time as needed to position the part without fear of the adhesive curing. Upon exposure to the appropriate light source, the adhesive then can be fully cured in less than a minute, minimizing the costs associated with work in progress. Adhesives that utilize light as the curing mechanism are often one-part systems with good shelf life.

light metal: engineering terminology for one of the low-density metals (with a specific gravity less than 3 that is strong enough for use in construction such as aluminum, magnesium, titanium, beryllium, or their alloys.

limit: (1) a border, perimeter, or boundary; (2) In mathematics, the value upon which an infinite series converges. Thus, the series $1+1/2+1/4 + 1/8 + 1/16. . .$has a limit of 2. See also allowance.

limit gages: Tools used in the manufacture of interchangeable parts that are required to be machined to size and degree of precision so that in assembly little or no fitting will be necessary. A commonly used limit gage may be double ended and have a "go" end. When the work is reduced to the correct size, one end of the gage will pass over it, but the other "no go" end will not. By using a gage for testing the minimum and maximum size, every part must come within the limits of the gage. This allowance or limit is made to conform to whatever amount experience has shown to be correct for the particular class of fit required. Limit gages are also known as go and no-go gages.

limits of size: The limits of size are the applicable maximum and minimum sizes.

linear: Pertaining to the dimension of length.

linear expansion: The lengthwise expansion of a material due to heat.

line of action, gear: The path of contact in involute gears. It is the straight line passing through the pitch point and tangent to the base circles.

liner bushings: Liner bushings are provided with and without heads and are permanently installed in a jig to receive the renewable wearing bushings. They are sometimes called master bushings.

linnaeite: A principal ore of cobalt, found in Canada, Zaire, and Zambia.

lip angle: See lip relief angle.

Lipowitz's metal: A fusible (low-melting) alloy nominally containing 50% bismuth, 27% lead, 13% tin, and 10% cadmium. It has a melting point of 150°F/66°C, used for making automatic sprinklers.

lips, two flute drill: The cutting edges extending from the chisel edge to the periphery.

lips, three or four flute drill (core drill): The cutting edges extending from the bottom of the chamfer to the periphery.

lip relief: A drill term used to describe the axial relief on the drill point.

lip relief angle: A drill term used to describe the axial relief angle at the outer corner of the lip. It is measured by projection into a plane tangent to the periphery at the outer corner of the lip. Lip relief angle is usually measured across the margin of the twist drill.

liquid rosin: See tall oil.

liquidus: The lowest temperature at which a metal or alloy is completely liquid.

lithium: An alkali metal element. The lightest solid element; lighter than lead, harder than other alkali metals. An alloying element with aluminum, copper, lead, magnesium, and zinc; also used in silver solders batteries; deoxidizer in copper and copper alloy production; a scavenging agent in inert gases. A source of tritium. Must be kept in airtight containers to prevent reaction with air.

Symbol: Li	**Density (g/cc):** 0.534
Physical state: Silvery metal	**Melting point:** 357°F/181°C
Periodic Table Group: IA	**Boiling point:** 2,426°F/1,330°C
Atomic number: 3	**Source/ores:** Spodumene, lepidolite, petallite,
Atomic weight: 6.941	amblygonite, triphylite, lithia, mica, other min.
Valence: 1	**Oxides:** LiO_2, Li_2O_2
	Mohs hardness: 0.6

lithium grease: A grease based on lithium stearate or lithium soap. It is water-resistant and stable when heated above its melting point (about 428°F/220°C) and re-cooled. See also lubricating grease.

lithium silicon alloy: An alloy of lithium and silicon. A fire or explosive risk in powder form. Reacts with water forming flammable gases.

live center: A lathe term use to describe the headstock center that turns with the spindle or the tailstock center that turns on ball bearings. See also center.

LMC: Abbreviation for least material condition. See minimum material condition.

LN: Abbreviation for locational interference fit. See also fits, description.

load: The total pressure acting on a surface; thus, if an engine piston having an area of 300 square inches is subjected to a steam pressure of 100 pounds *per square inch*, then the load, or total pressure on the piston is 300 x 100=30,000 pounds.

loading: A condition that prevents an abrasive material from functioning properly because although the abrasive grains are still sharp, the spaces between them are partly or entirely filled with compacted dust from the workpiece.

Lo-Ex®: An aluminum alloy having a low coefficient of expansion and containing 12-14% silicon, 2-2.5% nickel, 1% magnesium, 1% copper, 0.7% iron, 0.5% manganese. The alloy is solution treated, quenched in hot water, and precipitated treated for 4-16 hours, used in internal combustion engines, etc.

London Standard Silver: Silver of 92.5% purity.

longitudinal feed: In lathework, the principal direction or movement of the cutting tool along the workpiece, parallel to the bed of the lathe.

locational fit: See fits, description.

locators: See adjustable locators, conical locators, equalizing locators, expanding locators, retractable locators, spring-loaded locators.

lost wax casting: A centuries-old method of casting sculptures and jewelry. Adapted to industry as precision investment casting for the production of precision metal parts and intricate castings to close tolerances and accuracy. The process involves the making of a wax prototype in a metal mold. The prototype is dipped and dried multiple times in a ceramic glaze, thus building up a coating or "investment " which is fired in an oven, removing (melting out) the wax, leaving a cavity that becomes a mold for liquid metal.

lotus metal: An alloy containing 75% lead, 15% antimony and 10% tin, used as a bearing metal.

low angle point: Generally refers to drill points having an included angle of 60° or 90°. The low angle reduces the effective rake at the outer periphery of the cutting lip, which tends to prevent cracking when drilling soft materials and grabbing on breakthrough in low tensile strength, non-ferrous materials.

low brass: Alloys nominally containing 80% copper and 20% zinc.

low carbon steel: Steel without enough carbon (content usually below 0.03%) to allow it to harden to any great extent when heated to a specific temperature and quenched in water, oil, or brine. Where resistance to wear is required, low carbon steel may be heat treated to increase its strength or case hardened by cyaniding or by carburizing and hardening.

lower deviation (ISO term): In thread design, the algebraic difference between the minimum limit of size and the basic size. It is designated EI for internal and ei for external thread diameters.

Lower Explosive Limit (LEL): The lowest concentration (lowest percentage of the substance in air) of a vapor or gas that will produce a flash of fire when an ignition source (heat, arc, or flame) is present. At lower concentrations, the mixture is too "lean" to burn. Compare with Upper Explosive Limit (UEL).

lowest point of single tooth contact: The smallest diameter on a spur gear at which a single tooth of one gear is in contact with its mating gear. Often referred to as *gear set*, contact stress is determined when a load is placed at this point on the pinion.

low silicon bronze: Alloys nominally containing copper and silicon. Low silicon bronze, Type B contains 1.5% silicon.

low-zinc brass: The generic term for brass alloys nominally containing less than 20% zinc. These alloys are easily formed and have resistance to stress-corrosion cracking.

lozenge: See diamond chisel.

LT: Abbreviation for transition clearance or interference fit. See also fits, description.

lube-oil additive: See lubrication oil additives.

lubricant: Also known as lubricating agent. A substance that reduces friction between moving solid surfaces.

lubricant, liquid: A wet lubricant such as an oil, or semiliquid grease.

lubricant, solid: A compound or material that is crystalline in nature, lamellar-structured, usually about 40 micron in size, and produces an lubricating effect by shearing into thin, flat plates, which readily slide over one another. Solid lubricants include barium stearate, boron nitride, graphite, hafnium disulfide, mica, molybdenum diselenide, molybdenum disulfide, talc, tantalum disulfide, thorium disulfide, titanium disulfide, tungsten diselenide, and zirconium disulfide.

lubrication: The introduction of a substance of low viscosity that reduces friction between two adjacent solid surfaces, one of which is in motion.

lubrication oil additive: A material added to lubricants to impart special qualities, such as low pour point, viscosity index improver, detergent properties, oxidation stability, foam reducers, and resistance to high operating temperatures.

lubricity: (1) Describes the load-bearing characteristics of a material under relative motion. Materials with good lubricity, such as some plastics, have low coefficients of friction with other materials (or sometimes with themselves) and no tendency to gall; (2) the ability of a substance to act as a lubricant.

Lucalox®: A proprietary trade name for a strong, transparent refractory produced from the high-temperature firing of compressed alumina.

Lucerno: A nickel alloy nominally containing 62.5-67.9% nickel, 27.5-30% copper, 2.2-5% manganese, and 2.4-2.5% iron, used for electrical resistance.

Lurgi metal: A lead base alloy nominally containing 96.3-97.3% lead, 2-3% barium, 0.4% calcium, and 0.3% sodium, used as a bearing metal.

luster: The reflection of light from the surface of an object or material, and the appearance of the surface by that reflected light, as in metallic luster.

M

M: Symbol for mega or million (10^6).

M (*italicized*)*:* Symbol for (1) moment of force; (2) mutual inductance.

m: Symbol for meter.

m^2: Symbol for square meter.

m^3: (1) Symbol for cubic meter; (2) milli- (10^{-3}).

machinability: A term used to describe the ease with which an engineering material can be machined, worked, or cut. It is based on various factors including rate of metal removal, tool life, surface roughness, and power consumption. Most common metals are rated in terms of machinability using B1112 steel at 100%, as the comparison baseline. By comparison, carbon and chromium steels are rated from about 45 to 60, while aluminum is rated from 300 to 2,000. Generally, as the machinability rating is increased, the cutting speed is also increased.

machinability factor: Also called machininability rating. A percent indicator of the machinability, or ease in machining or cutting various workpiece materials. See also machinability.

machine bolt: A bolt containing a square head, and unlike the carriage bolt, having no square portion on the shank, and fitted with either a square or hexagonal nut. Machine bolts are machined straight from the point to the head and are usually threaded for about three-fourths of their length. The unthreaded portion is the same diameter as the threaded part. In order to tighten a machine bolt a wrench is put on the head while the nut is held stationary.

machining center: Also called N/C machining center. A multipurpose machine tool that performs many of the different operations traditionally performed on several basic machine tools; is usually controlled by numerically coded instructions stored in a computer.

machine reamers: Also called *chucking reamers*. Usually made of high-speed steel or having cemented carbide tips, these reamers are designed for use on drill presses, lathes, vertical milling machines, etc. See also reamers.

machine reaming: A drill press operation used to make light finishing cuts to a predrilled and/or bored, undersize holes with a tool called a *fluted machine reamer*. Machine reaming procedures are called *one-step* or *two-step*. For one-step, machine reaming holes are generally drilled 1/64 in (0.0156 in) undersize. The two-step method is used to produce finer reamed holes having closer tolerance and better finish than that produced with the one-step and holes are generally drilled 1/32 in smaller, rough reamed to about 0.002 in to 0.005 in undersize, followed by finish reaming. The feed for machine reaming should be faster than used for drilling and a cutting fluid should always be used. The speed for machine reaming should be slower, perhaps by one-half to two-thirds that of drilling.

machine reference point: A term used in numerical control (N/C) operations using absolute dimensioning. It is the point on a machine from which all tool movements are measured and programmed.

machine steel: Easily machined steel containing 0.2-0.3% carbon.

machining tolerance, involute spline (m): The permissible variation in actual space width or actual tooth thickness.

machinist's hammer: See ball-peen hammer.

Macht metal: A copper alloy containing 57-60% copper, 38-43% zinc, and sometimes 2% iron, used for casting and forging.

Mackenite metals: A range of heat resistant alloys, consisting of nickel-chromium or nickel-chromium-iron.

macro-: A prefix indicating "large" and sometimes "long."

macro-etching: The metallurgical process of etching the surface of a metal surface to accentuate gross structural details or the distribution of impurities and defects for observation by the naked eye or at low power magnification, up to about 10-15 diameters.

macrostructure: The structure of metals and alloys as revealed by examination of the surface of a machine polished specimen, that may or may not be etched, with the naked eye, with a magnifying glass or, other low magnification, up to about 10-15 diameters. See also macro-etching.

mafic: A material found in the earth's crust, mainly magnesium and iron silicates.

magazine: A chute-like bin for storing parts in a uniform position, used to feed work to a press or other production machine.

mag-lith: A alloy of magnesium and lithium, used as a structural metal in space vehicles.

magnaflux: A nondestructive, magnetic-particle inspection technique used to reveal surface cracks and other defects in iron, steel, and ferrous alloys. One technique involves magnetizing a specimen followed by dusting with finely divided magnetic particles. A similar method involves immersing the specimen in low-viscosity oil containing magnetic dust. These magnetized particles are attracted to, and outline, the pattern of any magnetic-leakage fields created by flaws on, or just below, the surface.

magnalite: Alloys of aluminum and magnesium containing 1-10% magnesium, and possibly up to 4% copper and/or up to 2% nickel.

magnalium: Alloys of magnesium and aluminum nominally containing 88-98% aluminum, 1-10% magnesium, and sometimes 1.5-2% copper.

magnesite: [$MgCO_3$] *A* natural magnesium carbonate, used as a refractory lining.

magnesium: CAS number: 7439-95-4. When alloyed with small amounts of aluminum, rare earths, manganese, thorium, zinc, or zirconium, this element produces alloys with high strength-to-weight ratios and unexcelled machinability. Used in sand and permanent mold casting, especially for structural materials.

Note: When machining or grinding magnesium, it is important to periodically stop the machine and carefully sweep or brush out the chips or powder that can be ignited by the slightest spark, causing a fire with extremely high heat. Do not use water-based cutting fluids on magnesium. Water reacts with hot magnesium, releasing explosive hydrogen. It also reacts violently with oxidizers and forms heat, friction-, or shock-sensitive explosive mixtures with acetylene compounds, methyl alcohol, and other compounds that may be found in a machine shop environment. On fires use only Class D extinguishers, or smother with special powders (G-1), powdered talc, dry sand, dry clay, crushed limestone, dry graphite.

Symbol: Mg
Physical state: Silvery metal
Periodic Table Group: IIA
Atomic number: 12
Atomic weight: 24.3050
Valence: 2
Density (g/cc):1.738
Melting point: 1,200°F/649°C
Boiling point: 1,994°F/1,090°C
Source/ores: Dolomite, magnesite, carnalite, kieserite, epsomite, carnallite, asbestos, talc, meerschaum, and other minerals; seawater
Oxides: MgO, MgO_2
Crystal structure: h.c.p.

magnesium dust: Finely powdered metal used in pyrotechnics.

Magnesium-Monel®: A registered trademark for an alloy containing 50% magnesium and 50% monel metal.

magnesium-thorium alloys: Magnesium alloy containing 2-3% thorium. to improve ductility and structural stability.

magnesium welding rod: Alloys of magnesium nominally containing 5.5-9% aluminum, 1-3% zinc, and 0.1-0.2% manganese. Also available are non-aluminum alloys: Alloy M1A containing 98.71% magnesium, 1.2% manganese, 0.09% calcium; Alloy HK31A containing 96.05% magnesium, 3.25% thorium and 0.7% zirconium; Alloy EZ33A containing 9.3% magnesium, 3.0% rare earths, 2.7% zinc, and 0.5% zirconium.

magnetic: Having magnetic properties; or, pertaining to magnetic substances or magnets; or, possessing magnetism.

magnetic chuck: A permanent magnet or electromagnetic table-like device for holding flat, ferrous-containing objects (with as large an area in actual contact as possible), used principally for surface grinding, and sometimes used on lathes, planers, and milling machines.

magnetic-particle inspection: A nondestructive method of inspection for determining the existence and extent of possible defects in ferromagnetic materials. The piece to be tested is magnetized and coated with dry magnetic powder or iron dust, or with a low-viscosity oil containing finely divided particles of iron. These dust particles are attracted to, and outline, the pattern of any magnetic-leakage fields created by flaws on, or just below, the surface. See also magnaflux.

magnetism: A characteristic property of substances (as iron) which exhibit a spatial polarization (depending upon their relative positions) and either attract or repel each other, or another magnetic substance.

magnetite: [Fe_3O_4] lodestone. Black, dense, magnetic mineral, dull to metallic luster. Contains 72.4% iron, often with titanium or magnesium. Hardness 5.5-6.5.

mainframe: A centrally located, often large, computer capable of simultaneous multiple tasks.

Maintain®: A registered trademark of the Aqualon Co. for a line of products for cleaning, inhibiting corrosion, and coating architectural copper, brass, and bronze.

major circle, involute spline: The circle formed by the outermost surface of the spline. It is the outside circle (tooth tip circle) of the external spline or the root circle of the internal spline.

major clearance: A thread design term used to describe the radial distance between the root of the internal thread and the crest of the external thread of the coaxially assembled designed forms of mating threads.

major cone: A thread design term used to describe the imaginary cone that would bound the crests of an external taper thread or the roots of an internal taper thread.

major cylinder: A thread design term used to describe the imaginary cylinder that would bound the crests of an external straight thread or the roots of an internal straight thread.

major diameter: (1) *thread:* The largest inside (internal) or outside (external) diameter of a workpiece measured between the crests of an external thread, or roots of and internal thread. On a straight thread, the major diameter is that of the major cylinder. On a taper thread the major diameter at a given position on the thread axis is that of the major cone at that position. See also major cylinder and major cone; (2) *involute spline* [$D_{o \ (external \ spline)}$, $D_{ri \ (internal \ spline, \ root)}$]: The diameter of the major circle.

major diameter fit, involute splines: Mating parts for this fit contact at the major diameter for centralizing. The sides of the teeth act as drivers. The minor diameters are clearance dimensions.

malleability: A measure of a metal's ability to endure hammering or rolling without breaking, fracturing, or returning to its original shape.

malleablizing: A process for annealing white cast iron in which some or all of the combined carbon is transformed wholly or partly into graphite or free carbon. In some instances, part of the carbon is removed completely. See also temper carbon.

mandrel: An accurately ground cylindrical bar made in various diameters and lengths and slightly below standard size at one end, but with a slight taper (approximately 0.0005 per inch) along the length. Thus, a mandrel of suitable diameter can be driven with sufficient force into the bore of the work to withstand the cutting action without slipping when the mandrel is mounted between the lathe centers. At each end, the mandrel is reduced in diameter and provided with a flattened portion, upon which the driving dog grips. For special diameters, expanding mandrels are sometimes used, these being adjustable for diameters within a small range. As the accuracy of the work machined depends on the mandrel running true between the lathe centers, any damage to its center ends should be avoided. After a wheel or similar work has been bored in a chuck, it is often finished on a mandrel supported between the lathe centers.

manganese: CAS number: 7439-96-5. A hard, brittle, silvery metal. In steel, small amounts (to about 0.60%), manganese is added to reduce brittleness and to improve forgeability. Larger amounts of manganese increases hardness, permitting oil quenching for nonalloyed carbon steels, thus reducing deformation, although with regard to several other properties, manganese is not an equivalent replacement for the regular alloying elements. In steel, manganese has a neutralizing effect on sulfur. In cast iron it combines with the sulfur and goes off in the slag and produces harder and stronger castings. An oxidizer or reducing agent, depending on valence.

Symbol: Mn
Physical state: Silvery solid
Periodic Table Group: VIIB
Atomic number: 25
Atomic weight: 64.938
Valence: 2,3,4,6,7

Density (g/cc): 7.43 (α); 7.29 (β); 7.18 (γ)
Melting point: 2,273°F/1,245°C
Boiling point: 3,564°F/1,962°C
Source/ores: Pyrolusite, psilomelane, ryptomelane, manganite, rhodochrosite
Oxides: MnO, Mn_3O_5, Mn_2O_3, MnO_2, MnO_3
Rockwell harness: 35
Crystal structure: b.c.c. (α, β, δ); f.c.c. (γ)

manganese boron alloy: Alloys of manganese and boron and usually containing 60-65% manganese, used to harden steel (in trace quantities), in making brass, bronze, and other alloys, as degasifying and deoxidizing agents, and to increase the conductivity of copper.

manganese brass alloy: Alloys nominally containing 50-65% copper, 30-40% zinc, 1-4% manganese, 0-3% iron, 0-2% tin, 0-2% nickel. Some alloys may also contain aluminum in small amounts.

manganese bronze alloys: Alloys nominally containing of 55-60% copper, 1.5% tin, 0.9-2.0% iron, 0.7-1.0% aluminum, 0.3-0.9% manganese, and 0.4% lead,and the remainder zinc. One alloy contains 58.5% copper, 39% zinc, 1.4% iron, 1% tin, and 0.1% manganese. High strength manganese bronzes with moderate machinability contains 64% copper, 5.25% aluminum, 3% iron, 3.75% manganese, 0.2 tin, 0.2 lead, and 23.6% zinc and another contains 58% copper, 1.5% manganese, 1.25% iron, 1% aluminum, 1% tin, 0.4 lead, and 36.85% zinc. More difficult to machine is an alloy containing 64% copper, 7.5% aluminum, 3% iron, 3.75% manganese, 0.2 tin, 0.2 lead, and 21.35% zinc. Following is a description of some casting alloys: Alloy 7A (60,000 psi) also know as high strength yellow brass, 59% copper, 37% zinc, 1.25% iron, 0.75% tin, 0.75% aluminum, 0.5% manganese. Alloy 8A (65,000 psi) 57.5% copper, 39.25% zinc, 1.25% iron, 1.25% aluminum, 0.25% manganese. Alloy 8B (90,000 psi) also known as high strength yellow brass, 64% copper, 24% zinc, 5% iron, 5% aluminum, 4% manganese. Alloy 8C (110,000 psi) also known as high strength yellow brass, 64% copper, 26% zinc, 3% iron, 5% aluminum, 4% manganese.

manganese phosphate: Also known as manganous phosphate [$Mn_3(PO_4)_2 \cdot 7H_2O$]. A reddish-white powder. Used for conversion coating of steels, aluminum, and other metals.

manganese steel: See austenitic manganese steel.

manganese-titanium alloys: Alloys nominally containing manganese, titanium, aluminum, iron, silicon. Used as a deoxidizer in steels and nonferrous alloys.

manganic: A general term for a nickel alloy containing a small amount of magnesium.

manganism: Chronic intoxication caused by manganese and its derivatives.

manganous sulfate: [MnS] A green crystalline solid used in steel making.

Man-Made®: A registered Trademark of the General Electric Company for its manufactured diamond superabrasive products.

Mapp gas: Common name for a colorless, liquefied, stabilized industrial fuel gas used for cutting, welding, brazing, heat treating, and metallizing. Although highly flammable, mapp gas is as versatile as acetylene, and considered safer.

Mar-aging® or Maraging®: A trademark of the International Nickel Co for a precipitation hardening treatment for iron-nickel alloys.

Maraging® steel: A steel alloy nominally containing 18-25% nickel, less than 0.03% carbon, and sometime with the addition of aluminum, cobalt, niobium, or titanium. and hardened by aging at about 842-932°F/450-500°C.

margin: (1) A drill term used to describe the cylindrical portion of the land, which is not cut away, to provide clearance; (2) in riveting, a term used to describe the distance from the edge of the plate to the center line of the nearest row of rivets.

martempering: A hardening process in which the steel is first rapidly quenched from some temperature above the transformation point down to some temperature (usually about 400°F/204°C) just above that at which martensite begins to form. It is then held at this temperature for a length of time sufficient to equalize the temperature throughout the part, after which it is removed and cooled in air. As the temperature of the steel drops below the transformation point, martensite begins to form in a matrix of austenite at a fairly uniform rate throughout the piece. The soft austenite acts as a cushion to absorb some of the stresses which develop as the martensite is formed. The difficulties presented by quench cracks, internal stresses, and dimensional changes are largely avoided, thus a structure of high hardness can be obtained. If greater toughness and ductility are required, conventional tempering may follow. In general, heavier sections can be hardened more easily by the martempering process than by the austempering process. The martempering process is especially suited to the higher-alloyed steels.

martensite: A chief microconstituent of hardened steel alloys and carbon tool steels having carbon contents up to 0.9%. It is an interstitial supersaturated solid solution of carbon or cementite [Fe_3C] in β-iron, or a very fine-grained α-iron with carbon or cementite in atomic or molecular dispersion. Its microstructure has a body-centered tetragonal lattice and is characterized by an acicular, or needle-like, pattern. It is the product of rapid cooling from above the Ac_3, such as the cooling in water.

master bushings: See liner bushings.

master gages: (1) Any type of gage that is used to check the accuracy of other gages; (2) a gage whose gaging dimensions represent as exactly as possible the physical dimensions of the workpiece or component.

matte: (1) A roughened surface which reflects light diffusely, as the surface of unpolished metal; (2) a general term used to describe the intermediate, impure product containing metal sulfides obtained during smelting, the process of recovering metals from their ores. Common examples are copper matte and nickel matte.

maximum material condition (MMC): A thread design term used to describe the condition where a feature of size contains the maximum amount of material within the stated limits of size. For example, minimum internal thread size or maximum external thread size.

maximum material limit (MML): (1) The condition in which a feature of size contains the maximum amount of material with the stated limits of size, for example, minimum hole diameter and maximum shaft diameter; (2) a maximum material limit is that limit of size that provides the maximum amount of material for the part. Normally it is the maximum limit of size of an external dimension or the minimum limit of size of an internal dimension. *An example of exceptions*: An exterior corner radius where the maximum radius is the minimum material limit and the minimum radius is the maximum material limit. See also minimum material limit.

MCAA: Abbreviation for the Mechanical Contractors Association of America.

mean: An average. An intermediate value between two or more values.

measure: To determine the dimension or dimensions of an object.

measured profile: A term related to the measurement of surface texture, a representation of the profile obtained by instrumental or other means. When the measured profile is a graphical representation, it will usually be distorted through the use of different vertical and horizontal magnifications, but shall otherwise be as faithful to the profile as technically possible.

measured surface: A representation of the surface obtained by instrumental or other means.

Mechanical Contractors Association of America (MCAA): A trade association of contractors in the piping, heating and air-conditioning business. It sponsors the National Certified Pipe Welding Bureau. Formerly the Heating, Piping and Air-Conditioning Contractors National Association. Located at 5530 Wisconsin Avenue N.W., Washington, D.C. 20015. See also National Certified Pipe Welding Bureau (NCPWB).

mechanical properties: The characteristics of materials when subjected to outside forces such as elasticity, elongation, hardness, ductility, fatigue limit, and tensile strength, thereby indicating the suitability of materials for various mechanical applications.

medium pitch cutter: A cutter with approximately twice as many teeth as its diameter size in inches.

medium carbon steels: Steel with a carbon content of about .20% to .60%. Where resistance to wear is required, medium carbon steels may be heat treated increase its strength or case hardened by cyaniding or by pack hardening, or quenched directly from the furnace.

Meehanite®: A registered trade name for cast irons of various analysis, usually containing a large percentage of scrap metal to which is added calcium silicide [CaSi] while the iron is melted. This produces a close-grained, fine graphite, pearlitic structure that is superior and tougher than cast iron. These irons can be given heat-, abrasion-, acid-, salt-, alkali-, and corrosion-resisting properties, and can also be chill-cast and heat-treated. Chilling can be carried to any depth and hardness up to 500 Brinell. One formula contains high nickel content, about 14% with copper and chromium, that is practically free from growth and will not scale excessively at temperatures up to 1,500°F/816°C.

Melonite®: A proprietary product of the Kolene Corp for an anhydrous molten salt bath used to produces nitride surfaces on carbon and alloy steels.

melting point of solids: The temperature at which solids become liquid or gaseous.

Mercers liquor: A solution containing potassium ferricyanide. Used for etching metal.

mercury: CAS number: 7439-97-6. Liquid metallic element. Alloys of mercury are termed almalgums. Salts or mercury are extremely toxic.

Symbol: Hg	**Density (g/cc):** 13.456
Physical state: Silvery liq.	**Specific heat:** 0.03325
Periodic Table Group: IIB	**Melting point:** −38°F/−39°C
Atomic number: 80	**Boiling point:** 675°F/357°C
Atomic weight: 200.59	**Source/ores:** cinnabar
Valence: 1,2	

Meta Bond®: A registered trademark of International Rustproof Co. for microcrystalline zinc phosphates. Used for the preparation of steel, galvanized steel, and aluminum for decorative and protective finishing. Application by spray or immersion.

Metal Construction Association (MCA): 1101 14th Street NW, Suite 1100, Washington, DC 20005. Telephone: 202/371-1243. FAX: 202/371-1090.

metal dye: An organic dye suitable for use on metals such as aluminum or steel.

Metal Fabricating Institute (FMI): 650 Race Street, Rockford IL 61101. Telephone: 815/965-4031.

metal fatigue: The local deformation of metals caused by repeated or fluctuating stresses.

Metal Finishing Suppliers Association (MFSA): 801 North Cass, Suite 300, Westmont, IL 60559. Telephone: 630/887-0797. FAX: 630/887-0799.

metal, heavy: A metal of high specific gravity.

metallic: Pertaining to or containing a metal. Possessing the properties of a metal.

metallic glass: See glass, metallic.

metallic soaps: See heavy metal soaps.

metal, light: A metal of low specific gravity.

metallizing: A process for coating objects with thin layers of metal by spraying them with molten metals or by means of vacuum deposition. The thickness of such films may vary from 0.01-3 mils.

metallography: The science dealing with the properties and characteristics of metals and metallic compounds and mixtures, especially in regard to the study of their surface by scientific methods such as microscopical or chemico-microscopical methods.

metallurgy: The science that deals with metals theory and the processing of metals, from their recovery as ores to their purification, alloying, and fabrication into usable industrial products.

Metal Treating Institute (MTI): 1550 Roberts Drive, Jacksonville Beach FL 32250. Telephone: 904/249-0448. FAX: 904/249-0459.

meter: A unit of linear measure (length) in the S.I. metric system, intended to be one ten-millionth part of the earth's quadrant from equator to pole. 1 meter=1,000 mm=3.28083 feet=1.093611 yards =39.37 in.

Metglas®: A trademark of Allied-Signal Inc. for a metallic glass, an amorphous metal alloy used in transformer coils. See glass, metallic.

2-methyl-3-butyn-2-ol: $[C_5H_8O]$ A colorless liquid used as an electroplating brightener.

methyl ricinoleate: A colorless liquid used as a lubricant and as a cutting oil additive. Combustible.

metric drills: Twist drills ranging in diameter size from 0.15 mm (0.006 in) to 100 mm (3.94 in). In the smaller sizes, the diameter differs by only 0.01 mm (0.0004 in). As the size increases, diameter difference increases progressively to 0.02 mm, 0.05 mm, 0.1 mm, and then to 0.5 mm. See also drills.

metric ton: A standard unit of weight measure in the S.I. metric system. 1,000 kg=2204.6 pounds=1.1 tons(US).

metrology: The science concerned with systems of measurement.

Mg: Symbol for the element magnesium.

microcrystalline: Also known as crytocrystalline. Crystallizing in microscopic crystals.

microfractography: The microscopic study of microfractures on the surfaces of metals.

microfracture: Also known as microcrack. A fracture or crack of microscopic size.

micrometer: (1) A unit of length in the SI system equal to one millionth of a meter, or one micron, sometimes denoted by μm, or by $\imath$. See also micron; (2) a measuring instrument for calibrating small lengths under a microscope, or used with a telescope or laser for fine focusing.

micrometer, caliper: A direct-contact instrument used to take precision measurements of small lengths by means of rotation of a screw of fine pitch. The measurements are taken by means of a curved "C" or "U" shaped frame having at its open extremities a micrometer at one end and a hardened anvil at the other. The micrometer has a vernier device with a threaded spindle having a barrel divided into graduations so as to measure slight degrees of rotation. The barrel contains an extremely fine screw (40 pitch, a screw with 40 threads per inch on inch micrometer or 50 threads per 25 millimeters on a metric micrometer). Micrometers are widely used and designed for inside, outside, and depth measurements and are available in a variety of shapes and sizes and generally measure in thousands of an inch (1/1,000) or ten-thousands of an inch (1/10,000), or hundreds of a millimeter (1/100 mm) or two thousands of a millimeter (2/1000). The standard outside micrometer is used the most

micron: A unit of length equal to one millionth (0.000001) of a meter or approximately 0.00004 inch.

microprobe: See electron beam microprobe analyzer.

microstructure: The structure of surfaces of metals as revealed by microscopy.

MIG (metal inert gas) welding: Also known as gas metal arc welding (GMAW). A welding process using a consumable wire electrode that is automatically fed through a welding gun with inert gas shielding.

mike: (1) A term officially adopted by ANSI for a micro-inch (0.000001 or 10^{-6} inch); (2) the nickname for a micrometer.

mil: A measure of thickness, especially of wire products. 1/1000 in. (0.001 or 10^{-3}).

mildewcide: A mildew preventer. A chemical compound used to prevent the growth of parasitic fungi in cutting fluids and on organic materials such as woods, textiles, leather, paper, etc. commonly used compounds include metal salts of copper, mercury, zinc, and cresols, benzoic acid, formaldehyde, phenols and organic substances such as *p*-dichlorobenzene, *p*-chloro-3,5-dimethylphenol, and benzylhexadecyldimethylammonium chloride.

mild steel: Carbon steel containing a maximum of about 0.25% carbon.

mill file: A hand file that is always single-cut and usually tapered from the heel to the point. Used for general deburring and draw- and lathe-filing to produce a fine, smooth, surface. Also preferred for finishing brass and bronze.

milli-: Prefix meaning 10^{-3} unit, or 1/1000th part.

milliliter (ml): One thousandth of a liter. The volume occupied by i gram of water at 4°C/40°F and one atm. The actual difference between a milliliter and a cubic centimeter (cc) is so small (1 ml = 1.0000027 cm^3) that the two are used interchangeably, except for the most precise sceintific calculations.

millimeter (mm): 10^{-3} of a meter equal to 0.03937 of an inch.

millimeter drills: See metric drills.

milling: A machining operation in which metal is removed by bringing the workpiece into contact with a horizontal or vertically mounted rotating, multiple-tip cutting tool called a mill or milling cutter. Milling on the larger and more powerful horizontal machine can be accomplished in the conventional mode where the cutter rotates in a direction opposite of the table feed, or by climb milling where the cutter rotates in the same direction as the table feed. See also climb milling, conventional milling, helical milling.

milling arbor: A machine shaft that is inserted in the milling machine spindle that both drives and holds a rotating milling cutter.

milling cutter: Also known as a *mill*. A rotating tool containing one or multiple teeth which engages the workpiece and removes material as the workpiece moves past the cutter. Milling cutters can be termed *horizontal* with teeth (cut or inserted) on the periphery or *vertical* (usually held in a vertical spindle) with teeth on both the end and periphery.

milling cutter geometries: Terms used to describe the four basic milling cutter geometries: double positive, double negative, shear angle and high shear.

mineral oil: Also referred to as straight oil. A mixture of liquid hydrocarbons from petroleum. Mineral oils are used as cutting fluids for light-duty cutting operations, especially with free-cutting (easily cut) steels and nonferrous metals. Recommended for power sawing carbon steels, malleable iron, wrought iron, stainless steels, tool steels, high-speed steels, aluminum alloys, brass, bronze, and magnesium alloys; for drilling, reaming, threading, turning, and milling aluminum alloys, copper-base alloys, and magnesium alloys; and grinding gray cast iron, aluminum alloys, and magnesium alloys. Other kinds of mineral oils have better lubricating properties such as those containing fatty acids and fatty-oils, including fish oil, lard oil, and sperm oil (e.g., *mineral-lard* oil); chemical additives such as *sulfurized and chlorinated* mineral oils used for their anti-weld properties; and for severe cutting operations including tapping and broaching. Mineral oil is also used as a defoaming agent, protective coating, and release agent. See also sulfurized and chlorinated mineral oil, mineral-lard oil.

mineral-lard oil: A mixture of lard derived from animal fats and mineral oil. One of the basic straight mineral cutting oils used for medium-duty cutting operations. See also mineral oil.

minimum bore diameters: See thread milling.

minimum material condition (MMC): Also known as least material condition

(LMC): The condition where a feature of size contains the least amount of material within the stated limits of size. For example, maximum internal thread size or minimum external thread size.

minimum material limit: A minimum material limit is that limit of size that provides the minimum amount of material for the part. Normally it is the minimum limit of size of an external dimension or the maximum limit of size of an internal dimension. *An example of exceptions*: An exterior corner radius where the maximum radius is the minimum material limit and the minimum radius is the maximum material limit. See also maximum material limit.

minor circle, involute spline: The circle formed by the innermost surface of the spline. It is the root circle of the external spline or the inside circle (tooth tip circle) of the internal spline.

minor clearance: A thread design term describing the radial distance between the crest of the internal thread and the root of the external thread of the coaxially assembled design forms of mating threads.

minor cone: A thread design term describing the imaginary cone that would bound the roots of an external taper thread or the crests of an internal taper thread.

minor cylinder: A thread design term describing the imaginary cylinder that would bound the roots of an external straight thread or the crests of an internal straight thread.

minor diameter: Also root diameter. (1) *thread:* The smallest inside (internal) or outside (external) diameter of a workpiece measured between roots of an external thread, or crests of an internal thread. On a straight thread the minor diameter is that of the minor cylinder. On a taper thread the minor diameter at a given position on the thread axis is that of the minor cone at that position. See also minor cylinder and minor cone; (2) *involute spline* (D_{re}, D_i): is the diameter of the minor circle.

mischmetal: The primary commercial form of mixed rare-earth metals containing

40-75% cerium, with lanthanum, neodymium, praseodumium, etc., and sometimes 1-5% iron. Used for pyrophoric alloys and in ferrous, nonferrous, and magnetic alloys.

miscibility: The ability of two or more substances to mix completely in all proportions and stay mixed under normal conditions.

mist: Liquid droplets suspended in air.

MMC: Abbreviation for maximum material condition.

Mn: Symbol for the element manganese.

Mo: Symbol for the element molybdenum.

Mobilmet®: A registered trademark of the Mobil Corporation for its mineral oil-based cutting oils that are suitable for most machining applications. Mobilmet® Omega is extremely heavy-duty oil for hard-to-machine metals; not suitable for use with copper and copper alloys. Mobilmet® Omicron is a dual- or tri-purpose cutting/lubricating oil for automotive screw machine shops that will not stain copper, brass or other nonferrous metals.

modified profile: A term relating to the measurement of surface texture, a measured profile where filter mechanisms (including the instrument datum) are used to minimize certain surface texture characteristics and emphasize others.

module, gear: Ratio of the pitch diameter to the number of teeth. Ordinarily, module is understood to mean ratio of pitch diameter in millimeters to the number of teeth. The English module is a ratio of the pitch diameter in inches to the number of teeth.

modulus (or coefficient) of elasticity: Also known as *elastic modulus, coefficient of elasticity* and *Young's modulus (E)*. It is a measure of the rigidity (softness or stiffness) of metal. A coefficient of elasticity representing the ratio of stress to strain as a metal is deformed under tension or dynamic load. Also stated as the load per unit of section divided by the elongation or contraction per unit of length. Wrought iron and steel have a well-defined elastic limit, and the modulus within that limit is nearly constant. See also Young's modulus.

$$E = \frac{Force}{SectionalArea} \times \frac{Originallength}{Extension}$$

Mohs harness scale: An empirical scale is used in mineralogy to classify the harness of materials. Although the original system consisted of a standard series of minerals having values ranging from 1-10, the list has been expanded to 15 with the inclusion of several synthetic materials between the original 9 and 10 positions. In this system talc has a harness number of 1, and diamond has a hardness number of 15. The standard series of minerals are used to determine and rate materials based on the ability of each material to just scratch the one directly below it in the series.

moisture-cured polyurethane adhesives: These adhesives start to cure when moisture from the atmosphere diffuses into the adhesive and initiates the polymerization process. In general, these systems will cure when the relative humidity is above 25%, and the rate of cure will increase as the relative humidity increases. The dependence of these systems on the permeation of moisture through the polymer is the source of their most significant process limitations. As a result of this dependence, depth of cure is limited to between 0.25 (6.35 mm) and 0.5 in. (12.7 mm). Typical cure times are in the range of 12 to 72 hours. Applications for moisture-cured polyurethane adhesives include windshield bonding in automobile bodies, bonding of metals, glass, rubber, thermosetting, and thermoplastic plastics.

molecular weight: weight (mass) of a molecule based on the sum of the Atomic weights of the atoms that make up the molecule.

mold: A hollow, negative cavity, form that is filled with molten metal, molding powders, or rubber under heat and pressure to form an article into a desired shape. Dies for rubber and plastics are called molds.

molding methods: See cement molding, dry-sand molding, green-sand molding, shell molding, vacuum molding.

molding sand: See foundry sand.

molybdenite: [MoS_2] Also known as molybdenum glance. Native molybdenum sulfide. Principal ore of molybdenum.

molybdenum: CAS number: 7439-98-7. Pure metal is soft, lustrous, and silvery. An alloying element in steels and cast iron. In small amounts it can improve certain metallurgical properties of alloy steels such as deep hardening, toughness, and heat-resistance. Used in combination with other alloys such as chromium or nickel, or both to promote structural uniformity in heavy sections of castings. It is often used in larger amounts in certain high-speed tool steels to replace tungsten, primarily for economic reasons, often with nearly equivalent results.

Symbol: Mo
Physical state: Gray powder
Periodic Table Group: VIB
Atomic number: 42
Atomic weight: 95.94
Valence: 2,3,4,5,6

Density (g/cc): 10.2
Specific heat: 0.062
Melting point: 4,730°F/2,610°C
Boiling point: 10,040°F/5,560°C
Source/ores: Molybdenite; wolfenite; by-product of copper production
Oxides: Mo_2O_3, MoO_2, MoO_3, Mo_2O_5
Crystal structure: b.c.c.

molybdenum aluminide: A cermet that is capable of being flame-sprayed.

molybdenum boride: Used as brazes to join molybdenum, tungsten, tantalum, and niobium parts; in cutting tools.

molybdenum disilicide: [$MoSi_2$] a cermet. Used as a high-temperature coating.

molybdenum ditelluride: [$MoTe_2$] A solid lubricant, available in powder form

molybdenum glance: See molybdenite.

molybdenum nickel: An alloy nominally containing 73-75% molybdenum and 25-27% nickel.

molybdenum silicide: An alloy containing 60% molybdenum, 30% silicon, and 10% iron, used for introducing molybdenum into steel.

molybdenum steel: Alloys for high-speed tool steels nominally containing 99.75-83,25% iron, 0.20-15% molybdenum, 0.05-1.75% carbon. One common type (M10) contains 8% molybdenum, 4% chromium, and 2% vanadium.

4-79 Moly Permalloy®: A registered trademark for an alloy nominally containing 79% nickel, 17% iron, and 4% molybdenum, having high permeability at low field strength, and high electrical resistance.

moment of force: See torque.

Mond process: A refining process for the production of nickel involving the heating of mixed ores to 122-176°F/50-80°C in a stream of carbon monoxide [CO] gas. Nickel forms nickel carbonyl [$Ni(CO)_4$], which passes off as a vapor while other oxides (other than nickel) are reduced to the metallic state.

monel metal: An alloy of copper and nickel and a small percentage of iron. Melting point 2,480°F/1,360°C. Tensile strength approximately 70,000 lb/in². Weight = 0.32 lb/in³.

monel 400: An alloy nominally containing 63-70% nickel, 25-37% copper, 0-2.5% iron, 0-2% manganese, 0-0.5% silicon, and 0-0.3% carbon.

Monimax®: An alloy nominally containing 49% iron, 48% nickel, and 3% molybdenum, having high permeability at low field strength and high electrical resistance.

mosaic gold: See ormolu.

mossy zinc: A powdered zinc material formed by pouring molten zinc into water.

mottled iron: That grade of cast iron in which half the carbon is combined and half separates out as carbon and graphite.

mrr: Abbreviation for metal removal rate.

MTI: Abbreviation for Metal Treating Institute.

*mu (*M or μ*):* The twelfth letter of the Greek alphabet. Also used as a symbol to denote the prefix micro-.

muffle furnace: A furnace or kiln in which the steel or other materials being heated are kept out of direct contact with the flame or heat source, the combustion being effected by heat reflected from the walls of the furnace.

multiple spindle drill press: A production drilling machine containing a number of spindles, each holding a drill, and all run at once, together. Used to drill multiple holes in a workpiece at the same time.

multiple-start thread: Also known as a multi-start thread. A thread in which the lead is an integral multiple, other than one, of the pitch. Stated more simply, the multi-start thread has more than one entry point on a given workpiece, and always has a more severe helix angle. See also start.

Mumetal®: A patented alloy nominally containing 77% nickel, 16.5% iron, 5% copper, 1.5% chromium, having high permeability at low field strength and high electrical resistance.

Muntz metal: An alloy containing 60% copper and 40% tin and having the highest tensile strength of all the brasses.

Muntz metal, leaded: Alloys nominally containing 60% copper, 39.4% tin, 0.6% lead.

n: (1) The symbol for spindle speed; (2) symbol for *nano-*. SI system prefix for 10^{-9}; (3) symbol for unspecified number; (4) symbol for rotational frequency (*n*-italic).

N: (1) Symbol for Avogadro's number, (6.023×10^{23}); (2) symbol for the element nitrogen; (3) symbol for newton.

Na: Symbol for the element sodium.

NADCA: Abbreviation for North American Die Casting Association.

naphthalene: $[C_{10}H_8]$ A white crystalline solid used in metal cutting fluids.

National Association of Metal Finishers (NAMF): 112 J. Elden Street, Herndon VA 20170. Telephone: 703/709-8299. Web: http://www.namf.org

National Certified Pipe Welding Bureau (NCPWB): 1385 Picard Drive, Rockville, MD 20850-4329. Telephone: 301/869-5800; 800/556-3653. FAX 301/990-9690. WEB: http://www.mcaa.org A division of the Mechanical Contractors Association of America, Inc (MCAA). Its purpose is to develop and test procedures and, through its local chapters, to establish pools of people qualified to weld under these procedures.

National Electrical Manufacturers Association (NEMA): 1300 N. 17th Street, Suite 1847, Rosslyn VA 22209. Telephone: 703/841-3200. FAX: 703/841-5900. WEB: http://www.nema.org An industry association of manufacturers of electrical machinery. Publishes standards and industry statistics including welding.

National Fire Protection Association (NFPA): 1 Batterymarch Park, Quincy MA 02269-910.Telephone: 1-800-344-3555. Web: www.nfpacatalog.org

An international membership organization established to promote the science and improve the methods of fire protection and prevention, and establish safeguards against loss of life and property by fire. Best known in industry for the National Fire Codes and the National Electrical Code®. Among these codes is the NFPA 704M, the code for classifying substances according to their fire and explosion hazard (as they might be encountered under fire or related emergency conditions) using the familiar diamond-shaped label (see also fire diamond) or placard with appropriate color, numbers, or symbols.

National Lubricating Grease Institute (NLGI): 4635 Wyandotte Street, Kansas City, MO 64112. Telephone: 816/931-9480. FAX: 816/753-5026. WEB: http://www.nlgi.com

National Metal Spinners Association (NMSA): Box 358, Farmingdale NY 11735. Telephone: 516/249-2468. FAX: 516/249-2599.

National Ornamental and Miscellaneous Metals Association (NOMMA): 532-A Forest Parkway, Forest Park GA 30297. Telephone: 404/363-4009. FAX: 404/366-1852. WEB: http://www.nomma.org

National Tooling and Machining Association (NTMA): 9300 Livingston Road, Fort Washington MD 20744. Telephone: 800/248-NTMA. FAX: 301/248-7104. WEB: http://www.ntma.org

National Welding Supply Association (NWSA): 1900 Arch Street, Philadelphia, PA 19103. Telephone: 215/564-3484. FAX: 215/564-2175. WEB: http://www.nwsa.com An industry association of welding supply distributors.

naval brass: Alloys nominally containing 60% copper, 39.25% zinc, and 0.75% tin. See also admiralty brass.

naval brass, leaded: Alloys nominally containing 60% copper, 39.5% zinc, 1.75% lead, and 0.75% tin. See also admiralty brass.

Navy M bronze, alloy 2A: Also know as Navy M, steam or valve bronze. Alloys nominally containing 88% copper, 4.5% zinc, 1.5% lead, and 6% tin

Nb: Symbol for the element niobium.

N/C programming codes: Also called N/C coding systems. Instructions that are provided to N/C control systems by codes representing characters in numeric form such as ASCII (American Society for Computer Information Interchange) code or EIA (Electronics Industries Association) code.

NCPWB: Abbreviation for National Certified Pipe Welding Bureau.

Nd: Symbol for the element neodymium.

neck: A drill term used to describe the section of reduced diameter between the body and the shank of a drill.

NEMA: Abbreviation for National Electrical Manufacturers Association.

Neochel®: A registered trademark of M & T Harshaw (division of Atochem) for a wetting agent used in copper plating. A liquid replacement for Rochelle salt with proprietary additives.

neodymium: CAS number: 7440-00-8. A rare-earth element of the lanthanide (cerium) group. It has little structural strength but has value as an alloying element and used as scavenger in refining ferrous alloys.

Symbol: Nd	**Density (g/cc):** 7.007
State: Silvery white	**Melting point:** 1,875°F/1,024°C
Periodic Table Group: IIIB	**Boiling point:** 5,486°F/3,030°C
Atomic Number: 60	**Ores:** Monazite sands, allanite, bastnasite
Atomic Weight: 144.24	**Oxides:** $Nd_2(SO_4)_3$
Valence: 3	**Crystal structure:** Hexagonal, body-centered cubic

nevyanskite: A native alloy of iridium and osmium with allied metals. Also a name used to describe the alloy iridosmine when it contains a high content of the element iridium. See also iridosmine.

NFPA: Abbreviation for National Fire Protection Association.

Ni: Symbol for the element nickel.

nib: See diamond dresser.

Nichrome®: A registered trademark for alloys nominally containing 55-80% nickel, 7-24% iron, 16% chromium, 0.1% carbon and sometimes smaller amounts of copper, manganese, molybdenum, and silicon. It is used principally for electric resistance metals and offers good resistance to seawater and wet, sulfurous environments.

nickel: CAS number 7440-02-0. An alloying element for steel, copper, brass, etc. Generally in combination with other alloying elements, particularly chromium, nickel is used to improve the strength, toughness and, to some extent, the wear resistance of tool steels. Nickel eliminates hard spots in castings imparting added density to the metal in thicker and softer sections, and makes the metal more fluid and easier to cast. Very high nickel alloys have special thermal and magnetic properties. Used in electrical devices, protective coatings.

Symbol: Ni	**Density (g/cc):** 8.88
Physical state: Silvery metal	**Melting point:** 2,650°F/1,455°C
Periodic Table Group: VIII	**Boiling point:** 4,950°F/2,732°C
Atomic number: 28	**Source/ores:** chalcopyrite, garnierite, penlandite
Atomic weight: 58.69	**Oxides:** NiO, Ni_2O_3, Ni_3O_4
Valence: 2,3	**Crystal structure:** f.c.c.

nickel alloy, base (ASTM B127): Also known as Alloy 400. An alloy nominally containing 63-70% nickel, 25-37% copper, 0-2.5% iron, 0-2% manganese, 0-0.5% silicon, and 0-0.3% carbon.

nickel alloy base (ANSI 687): A nickel-based cobalt alloy containing 47-59% nickel, 17-20% cobalt, 13-17% chromium, 4.5-5.7% molybdenum, 3.7-4.7% aluminum, 3-4% titanium, 0-1 iron, and 0-0.1% carbon.

nickel alloy, base (ASTM B344): An alloy nominally containing 57-62%, nickel, 22-28% iron, 14-18% chromium, 0.8-1.6% silicon, 0-1% manganese, and 0-0.2% carbon.

nickel-aluminum bronze: Alloys nominally containing 78-81% copper, 4.5-5.5 nickel, 3.5-5.5% iron, 9-10.3% aluminum., 0.5-1.0% manganese, 0.01% lead, and 0.5% other.

nickel aluminide: A cermet material that can be flame sprayed.

nickel ammonium chloride: [$NiC_{12} \cdot NH_4Cl$] A green crystalline solid or yellow powder used in electroplating.

nickel ammonium sulfate: [$O_8S_2 \cdot Ni \cdot 2H_4N$] A dark blue-green crystalline solid used in electroplating.

Nickel Babbit®: A proprietary trade name for high-speed bearing alloys containing copper-tin-nickel.

nickel brass: Nickel silver alloys nominally containing 50-55% copper, 34-43% zinc, 2-15% nickel, and 0-18% lead.

nickel brass, leaded: Nickel silver alloys nominally containing 57% copper, 20% zinc, 12% nickel, 9% lead, 2% tin, and having good machinability. Following is a description of some casting alloys: Alloy 10A, also known as 12% Nickel-silver and Benedict metal, containing 57% copper, 20% zinc, 12% nickel, 9% lead, 2% tin. Alloy 10B, also known as 16% Nickel-silver, containing 60% copper, 3% tin, 5% lead, 16% zinc, and 16% nickel.

nickel bronze: Nickel silver alloys nominally containing 50-86% copper, 20-30%

nickel, 8-25% tin. Others contain 11-18% zinc, and 0-18% lead. Following is a description of some casting alloys: Alloy 11A, also known as leaded nickel bronze and 20% nickel silver, contains 64% copper, 4% tin, 4% lead, 8% zinc, and 20% nickel. Alloy 11B, also known as leaded nickel bronze and 25% nickel silver, contains 66.5% copper, 5% tin, 1.5% lead, 2% zinc, and 25% nickel.

nickel cyanide: [C_2N_2Ni] Yellow-green powder used in electroplating.

nickel, electroless: A rust inhibiting, metal coating which is deposited in a dipping bath by chemical reduction, without application of an electric current. The advantage of this process is that uniform coatings can be formed in tubes, deep holes, etc. Electroless nickel coating appears to be a better inhibitor than black oxide or polymer chrome but with an added cost of approximately 20%.

nickel-gallium alloy: An alloy nominally containing 60% nickel and 40% gallium.

Nickel-Lume®: A registered trademark of M & T Harshaw (division of Atochem), Inc. for a bright nickel electroplating process using nickel sulfate, nickel chloride, boric acid, and organic addition agents.

nickel nitrate: [$N_2O_6 \cdot Ni$] A green crystalline solid used in nickel electroplating.

nickel phosphate: [$Ni_3(PO_4)_2 \cdot 7H_2O$] A light green powder used in electroplating.

nickel rhodium: Alloys nominally containing 23-80% rhodium and 20-75% nickel, and sometimes small amounts of cobalt, copper, iridium, iron, molybdenum, palladium, platinum, or tungsten.

nickel silver: Alloys having a silvery appearance and containing nominally 55-65% copper, 17-18% nickel, 17-27% zinc. Nickel silver 65-18 (also known as Nickel silver Alloy A) contains 65% copper, 18% nickel, and 17% zinc. Nickel silver 55-18 (also know as Nickel silver Alloy B) contains 55% copper, 27% nickel, and 18% zinc. Nickel silver 65-12 (also known as Nickel silver Alloy D and German Silver) contains 65% copper, 12% nickel, and 23% zinc.

nickel-zirconium steel: A tough iron alloy with a carbon content of 0.4%, and containing nickel (2%), zirconium (0.35%), manganese (1%), silicon (15%), used for armor plate.

Nicomo®12: A registered trademark of W.R. Grace, Inc. for a hydrodesulfurization catalyst containing nickel, cobalt, and molybdenum or alumina.

Nimonic® alloys: A proprietary trade name for line of nickel-base alloys available from Wiggin Alloys Ltd. nominally containing 10-21% chromium, 5-37% iron, 1-1.5% silicon, 0.2-4% titanium, 0.08-0.15% carbon, and sometimes cobalt, niobium, molybdenum, titanium, and aluminum, copper, or both.

Nimonic® AP1: A proprietry name of Wiggin Alloy Ltd for a nickel-based cobalt alloy containing 47-59% nickel, 17-20% cobalt, 13-17% chromium, 4.5-5.7% molybdenum, 3.7-4.7% aluminum, 3-4% titanium, 0-1 iron, and 0-0.1% carbon.

niobium: CAS number: 7440-03-1. Also known as columbium. An alloying element in nickel, stainless steels, and cobalt-base superalloys. A stabilizing agent in stainless steels. Also used in cermets, magnetic, and conducting alloys. The name niobium has been officially in use for more than 50 years. The previous name, columbium (symbol Cb) is still used as an alternate.

Symbol: Nd	**Density (g/cc):** 8.4 to 8.57
Physical state: Soft metal	**Melting point:** 4,474°F/2,468°C
or steel-gray crystals.	**Boiling point:** 8,901°F/4,927°C
Periodic Table Group: VB	**Source/ores:** columbite; tin by-product
Atomic number: 41	**Oxides:** Nb_2O_5
Atomic weight: 92.91	**Crystal Structure:** Body centered cubic
Valence: 2,3,4,5	

niobium carbide: [NbC] Melting point approximately 6,332°F/3,500°C. Used for cemented carbide tipped tools, special steel alloys.

niobium tin: [Nb_3Sn] Alloy used for special wire and superconducting magnets.

niobium titanium: Alloy used for special magnetic devices.

Ni-O-Nel®: A registered Trademark for an acid-resistant alloy nominally containing approximately 40% nickel, 31% iron, 21% chromium, 3% molybdenum, less than 2% copper, and traces of carbon, silicon, and manganese.

Niranium®: A chrome-cobalt alloy nominally containing 64% chromium, 29% cobalt, 4.3% nickel, 2% tungsten, and small amounts of carbon, aluminum, and silicon.

Ni-Resist®: An alloy nominally containing approximately 74-80% iron, 12-15% nickel, 5-7% copper, 1.5-4% chromium. Heat resistant, nonmagnetic, and weldable, used in automobile engine valves.

nitride: A binary compound of a metal and nitrogen.

nitride treatment: See nitriding.

nitriding: A treatment that produces an extremely hard, abrasion-resistant surface on steel parts. It also retards the tendency of softer materials to cling or load on machine tools. An case hardening process in which an alloy of special composition (nitroalloy) is heated in an atmosphere of ammonia (or in contact with nitrogenous material) at a specific temperature for an extended period of time depending on the case depth required. The ammonia breaks down into nitrogen and hydrogen and the nitrogen reacts with the steel. Surface hardening is produced by the absorption of nitrogen and quenching is not required. The major advantage to this process is that parts can be quenched and tempered, then machined, prior to nitriding, because only a little distortion occurs during nitriding. This treatment is recommended for tools that are used for ferrous, non-ferrous, and non-metallic materials which are abrasive and have loading characteristics.

nitrites: Rust inhibiting chemical agents added to cutting fluids.

nitrogen: A gaseous element. CAS number: 7727-37-9. Used for chilling in aluminum foundries and the bright annealing of steel

Symbol: N	**Density:** 1.2506 g/L at STP
Physical state: Gas	**Melting point:** $-346°F/-210°C$
Periodic Table Group: VA	**Boiling point:** $-321°F/-196°C$
Atomic number: 7	**Sources:** Liquid air
Atomic weight: 14.007	**Oxides:** NO, N_2O, NO_2, N_2O_3, N_2O_5
Valence: 1,2,3,4,5	**Crystal structure:** cubic (α); h.c.p. (β)

Nivco®: A registered trademark of Westinghouse Corp. for an alloy nominally containing 78% cobalt, 20% nickel and small amounts of iron, titanium, and zirconium.

NLGI: Abbreviation for National Lubricating Grease Institute.

NMSA: Abbreviation for National Metal Spinners Association.

noble gas: Refers to the six noble gases constitute group VIIIA in the periodic table, also known as noble gas group, the inert gas group, and rare gases. The gases of this group are helium, neon, argon, krypton, xenon, and radon. The last three gases are not inert and form compounds with chlorine, fluorine, nitrogen and oxygen.

noble metal: A term applied generally to the gold, platinum, and palladium families of the periodic table, and are generally considered to be gold, silver, platinum, palladium, iridium, rhenium, mercury, ruthenium, and osmium. The term has no reference to their commercial value. These metals are relatively inert (e.g., either completely unreactive or reacts only to a limited extent with other elements), not readily oxidized, and highly resistant to corrosion.

nominal clearance, involute spline: The actual space width of an internal spline minus the actual tooth thickness of the mating external spline. It does not define the fit between mating members, because of the effect of variations.

nominal profile: A term related to the measurement of surface texture, a profile of the nominal surface; it is the intended profile (exclusive of any intended roughness profile).

nominal size: The designation used for the purpose of general identification.

nominal surface: The intended surface contour (exclusive of any intended surface roughness), the shape and extent of which is usually shown and dimensioned on a drawing or descriptive specification.

NOMMA: Abbreviation for National Ornamental and Miscellaneous Metals Association.

nondestructive testing (NDT): Testing techniques used by quality control inspectors to determine the presence of internal defects likely to cause operational failure without actually causing damage to the sample or materials tested. Some tests are performed during manufacturing and others are performed after parts have been in service for a period of time, such as during maintenance or overhaul. Such tests are usually carried out by using harness instruments such as the Rockwell, Brinell and Shore testers, and by fluorescent penetrant inspection, magnetic particle inspection, radiographic inspection, ultrasonic inspection, pulse-echo inspection, eddy current inspection, infrared scanning inspection, and x-radiation inspection.

nonferrous metal: Metals containing no iron.

nonmetallic: Not of a metallic nature.

Norbide®: A registered trademark of the Norton Company, Inc. for a line of products made of boron carbide [B_4C], a wear-resistant, synthetic abrasive.

normalizing: An annealing process for ferrous alloys that relieves internal stress and removes the induced strains caused by the effects of other heat-treatment processes, making the structure more uniform, and refining grain size. The iron-based alloy is heated to a temperature above Ac_1, the transformation range, and subsequently cooled in still air to a temperature substantially below the transformation range. This process returns the alloy to a "normal" condition following forging.

normal pitch: In gear design, the distance between adjacent teeth measured on the pitch cylinder at right angles to the tooth spirals.

normal plane, gear: A plane normal to the tooth surfaces at a point of contact, and perpendicular to the pitch plane.

North American Die Casting Association (NADCA): 9701 W. Higgins Road, Suite 880, Rosemont IL 60018. Telephone: 847/292-3600. FAX: 847/292-3620. WEB: http://www.diecasting.org

notched point drills: Heavy web drills containing two additional positive rake cutting edges extending toward the center of the drill. Made for drilling tough alloy materials, this design can assist in chip control, reduce the machine torque required, and allows the point to withstand the higher thrust loads. These notched points or secondary cutting lips extend no further than half the original cutting lip and can be incorporated on both 118° and 135° included point angles, making them suitable for drilling a wide variety of materials.

notching (depth of cut): See depth of cut notch (DOCN).

notch sensitivity: A measure of the ease with which a crack progresses through a material from an existing notch, crack, or sharp corner.

NTMA: Abbreviation for National Tooling and Machining Association.

number drills: Also called wire gage drills. Twist drills designated from No. 97 to No. 1. [with diameters from 0.0059 in (.14986 mm) to 0.228 in (5.79 mm)]. *Note:* the larger the number, the smaller the drill. See also drills, letter drills, drill types.

NWSA: Abbreviation for National Welding Supply Association.

O

O: Symbol for the element oxygen.

Ω, ω: See omega.

oblique triangle: A common name for both acute triangles and obtuse triangles.

obtuse angle: If the inclination of the arms of an angle are more than a right angle (90° but less than a straight line 180°, also known as a straight angle since it corresponds to a rotation through half a full circle), the angle is called obtuse.

obtuse triangle: A triangle having one angle larger than 90°. Both obtuse- and acute-angled triangles are known under the common name of oblique-angled triangles. The sum of the three angles in every triangle is 180°.

OD: Abbreviation for outside diameter.

OEM: Abbreviation for Original Equipment Manufacturer.

oil bath tempering: A heat-treatment process whereby the work is immersed in oil heated to the correct temperature depending on type of steel, carbon content of the steel, hardness required, and toughness required. It is important that the temperature of the oil be uniform throughout and that the work be immersed long enough to reach this temperature. The best method for tempering is to submerge the work before heating the oil, so that it is heated along with the oil. After the pieces tempered are removed from the oil bath, they should be immediately dipped in a tank of caustic soda, followed by a bath of hot water. This final step will remove all oil that might adhere to the work. Used extensively for tempering steel tools, especially in quantity. Cold steel plunged into a heated tempering bath can cause cracking.

oil hole drills: Twist drills containing one or more holes that allow oil or cutting fluid fed from the machine spindle to the drill's cutting edges, to lubricate, cool, and force the chips out of the drilled hole, used on high production screw machines.

Olsen ductility test: A qualitative test for assessing the ductility of strip and sheet metals performed by forcing a hemispherical-shaped plunger or hardened ball into the metal and measuring the depth at which fracture occurs in the cupping.

omega (Ω, ω): The twenty-fourth letter of the Greek alphabet. Symbol for angular speed, angular frequency, solid angle (ω). Symbol for ohm (Ω).

opacity (or transparency): A measure of the amount of light transmitted through a given material under specific conditions. Measures are expressed in terms of haze and luminous transmittance. Haze measurements indicate the percentage of light transmitted through a test specimen that is scattered more than 2.5° from the incident beam. Luminous transmittance is the ratio of transmitted light to incident light.

open-hearth furnace: A reverberatory furnace used for steel production.

open-hearth process: In steel production, a large open crucible containing iron ore, scrap iron, and pig iron is subjected to hot gases that oxidize out the excess carbon and other impurities. The open-hearth process, which uses air, has been supplanted by the basic oxygen process (BOP) in which pure oxygen is injected into molten iron.

open washer: Also known as "C"-washers, made with one open side, so they can be slipped under a nut or taken off, thereby avoiding the necessity of entirely removing the nut.

orange peel: A pebble- or alligator-grain surface effect encountered in the forming of metals from stock having coarse grains.

ordinary steel: See carbon steel.

ore: An aggregate of natural minerals mixed with earthy matter, such as matrix or gangue, from which useful and valuable substances, usually metals, can be profitably extracted. The principal ores are usually metal-oxides, -carbonates, -silicates, -sulfides, -arsenides, -antimonides, and -halides.

ormolu: Also known as mosaic gold. French meaning ground gold. An alloy containing equal parts of copper and zinc, used in inexpensive jewelry and as an imitation of gold in gilding brass or bronze ornaments often associated with French decorative arts of the 18th century.

Os: Symbol for the element osmium.

oscillating grinding: Another name for reciprocating grinding.

osmiridium: An alloy of osmium and iridium nominally containing 40-80% iridium and 20-60% osmium. If the osmium content is high this alloy may be termed siserskite.

osmium: CAS number: 7440-03-1. Lustrous, silvery metal. An alloying element and catalyst. Used hardener for iridium and platinum, instrument pivots.

Symbol: Os	**Density (g/cc):** 22.48
Physical state: Silvery metal	**Specific heat:** 0.0311
Periodic Table Group: VIII	**Melting point:** 5,529°F/3,054°C
Atomic number: 76	**Boiling point:** 9,081°F/ 5,027°C
Atomic weight: 190.2	**Sources:** By-product of nickel production;
Valence: 2,3,4,5,6,8	occurs in platinum ores
	Oxides: OsO, Os_2O_3, OsO_2, OsO_4
	Crystal structure: h.c.p.

osram: An alloy of osmium and tungsten.

out-of-roundness, involute spline: The variation of the spline from a true circular configuration. Heat treatment and deflection of thin sections may cause out-of-roundness, which increases index and profile variations. Tolerances for such conditions depend on many variables and are therefore not tabulated. Additional tooth and/or space width tolerance must allow for such conditions.

out-of-truth: See out-of-round.

outside calipers: A two-legged steel instrument with the hardened steel ends of each bent inward. Their size is measured by the greatest distance they can be opened between the two points. Used to measure widths and thickness of workpieces and the outside diameters of circular objects.

overaging: Aging under conditions of time and temperature greater than those required to obtain maximum change in a desired property, resulting in the property being altered in the direction of the initial value. See aging.

overall length: A drill term used to describe the length from the extreme end of the shank to the outer corners of the cutting lips. It does not include the conical shank end often used on straight shank drills, nor does it include the conical cutting point used on both straight and taper shank drills.

overcut, file: The first series of teeth put on a double-cut file.

overheated: A metal is said to have been overheated if, after exposure to an unduly high temperature, it develops an undesirably coarse grain structure but is not permanently damaged. The structure damaged by overheating can be corrected by suitable heat treatment or by mechanical work or by a combination of the two. In this respect it differs from a burnt structure, which is permanently damaged.

overheating: The exposure of a metal or alloy to such a high temperature that it acquires a coarse-grained structure, causing its physical properties to be impaired. Unlike *burning*, which results in permanent damage, an overheated metal may be restored by further heat treating, by mechanical working, or by combination of working and heat treating.

Oxanal®: A registered trademark for dyes for coloring anodized aluminum.

oxidation: The original meaning meant simply combination with oxygen. The term in its broadest sense means a chemical reaction whereby electrons are transferred The reverse of reduction. Oxidation and reduction occur simultaneously, and the substance that gains electrons is termed the oxidizing agent.

oxidizer: See oxidizing agent.

oxidizing agent: Also known as oxidizer. Substances that yields oxygen readily, thereby causing ignition or explosion of other, usually organic, materials or compounds. Perchlorates, peroxides, permanganates, chlorates, and nitrates are examples of oxidizers. These materials are dangerous storage hazards and must be kept away from powdered metals, combustible materials, organic substances, fuels, cutting fluids, solvents, etc.

oxyacetylene welding: A welding process using a mixture of oxygen and acetylene. The oxyacetylene flame is applied with a torch to the butted ends or edges of the metal pieces to be joined or bonded together.

oxygen free copper: See copper, deoxidized.

P

P: Symbol for the element phosphorus.

P: Symbol for power.

p: Symbol for pressure.

Π, π: See pi.

Φ, φ: See phi.

Ψ, ψ: See psi.

palladium: CAS number: 7440-05-3. An alloying metal that does not tarnish in air. The most reactive of the platinum group. Used for electrical relay contacts, Alloys with ruthenium makes "white" gold for use in jewelry, making special resistance alloys, aircraft spark plugs, high-temperature solders, protective coatings.

Symbol: Pd	**Density (g/cc):** 12.02
Physical state: Silvery solid	**Melting point:** 2,826°F/1,552°C
Periodic Table Group: VIII	**Boiling point:** about 7,200°F/3,980°C
Atomic number: 46	**Source/ores:** By-product of copper and zinc
Atomic weight: 106.42	production
Valence: 2,3,4	**Crystal structure:** Face-centered cubic.
	Mohs hardness: 4.8

palm oil: Yellow-brown, buttery solid at room temperature. Used as a cutting-tool lubricant; terne plating, in hot-dipping of tin coatings to prevent oxidation while it is cooling. Combustible.

parallel clamp: Also known as a toolmakers' clamp. A clamp consisting of two pointed jaws with a screw passing through the center of each jaw and another screw at the flat or opposite end of the jaws. Used to hold parts together or to hold small parts in the pointed end by tightening the screw at the flat end of the jaws which should be kept parallel to the workpiece surfaces.

parallel thread: See screw thread.

parallelism variation, involute spline: The variation of parallelism of a single spline tooth with respect to any other single spline tooth.

Parkes process: A process used to separate silver from lead by adding molten zinc to the lead-silver mixture, which is heated to above the melting point of zinc. It is a process that must be repeated several time to separate out all of the silver.

partial thread: See vanish thread.

parting: (1) An operation that uses a parting tool performed on a lathe or screw machine to remove or separate a finished part from a length of chuck- or collet-held stock; (2) a sheet-metal operation that produces two or more parts from a common stamping.

parting tool: See cutoff tool.

passivation: The changing of the chemically active surface of certain metal such as iron, chromium, and related metals to a much less reactive state following treatment with strong oxidizing agents, like nitric acid [HNO_3], or when oxygen forms an oxide coating during electrolysis.

patenting: A heat-treatment process applied to medium- or high-carbon steel in wire making prior to the wire drawing or between drafts. It consists of heating to a temperature above the transformation range, followed by cooling to a temperature below that range in air or in a bath of molten lead or salt maintained at a temperature appropriate to the carbon content of the steel and the properties required of the finished product.

patina: The ornamental and/or corrosion-resistant green coating that slowly forms on the surface of copper and copper alloys such as bronzes, and other metals that are exposed to the atmosphere or from suitable chemical treatment. See also verdigris.

Pattinson process: A process used to separate silver from lead by melting the silver-lead alloy that is permitted to cool slowly. The lead, that has separated from silver or that has low silver content, separates out as crystals. This crystalline material is removed, leaving a residual lead which is rich in silver. The operation is repeated using a series of melting pots, and the silver recovered.

Pb: Symbol for the element lead, derived from *plumbum*, Latin for "lead."

PCBN: Abbreviation for polycrystalline cubic boron nitride.

PCD: Abbreviation for polycrystalline diamond.

Pd: Symbol for the element palladium, adapted from Pallas, and asteroid named for a Greek goddess.

peak: A term related to the measurement of surface texture, the point of maximum height on that portion of a profile that lies above the centerline and between two intersections of the profile with the centerline.

peak-to-valley height: A term related to the measurement of surface texture, the maximum excursion above the centerline plus the maximum excursion below the centerline within the sampling length. This value is typically three or more times the roughness average.

pearlite: A lamellar aggregate of ferrite (a name applied to α- and β-iron, and sometimes to Δ-iron) and cementite (iron carbide [Fe_3C]), often occurring in annealed steel, carbon steels, and gray cast iron, and pig iron. In steelmaking, the eutectic between ferrite and cementite.

peen: (1) The ball, wedge, or cone shaped end of a peening hammer; (2) to hammer with the peen of a hammer. See also peening.

peening: Also known as swaging, the process of expanding, stretching, or spreading metal by hammer blows, shot impingement, or rolling the surface. Used in metal-stretching operations such as flattening the end of a rivet, straightening bars by stretching the short or concave side, compressing or spreading babbitt to fit tightly in a bearing, etc.

perforating: The piercing or punching of identical multiple holes arranged in a regular pattern.

periodic system: A scientific arrangement of the chemical elements in a systematic grouping wherein elements having like properties occur in relative positions, as in horizontal or vertical sequence, so that properties of known elements, can be deduced from their positions in this arrangement. The arrangement of the elements is based on the Atomic number of each element.

periodic table of elements: An arrangement of the periodic system in the form of a table.

Permalloy®: A registered trademark for a family of magnetic alloys nominally containing 78.23-78.5% nickel, 18-21.5% iron, 0.04% carbon, 0.37% cobalt, 0.035% sulfur, 0.03% silicon, 0.22% manganese, 0.1% copper, sometimes up to 3% molybdenum, and small amounts of phosphorus. These alloys have high permeability at low field strength and high electrical resistance.

45 Permalloy®: An alloy nominally containing 55% iron and 45% nickel and annealed at 1,920°F/1,049°C, having high permeability at low field strength and high electrical resistance.

Permalume G®: A registered trademark of M&T Chemicals for a nickel electroplating process.

permanent set: The alteration in form that occurs when a metallic piece is stressed beyond its elastic limit and deformation occurs, the piece being either stretched, crushed, bent, or twisted, according to the nature of the strain.

personal protection equipment (PPE): Safety equipment designed to protect parts or all of the body from workplace hazards. Such protective equipment includes chemical resistant clothing, gloves, respirators, and eye protection.

pewter: Tin alloys nominally containing 89-95% tin, 6-7% antimony, and 1-3% copper; in the past, it sometimes contained 0-15% lead. Lead is an undesirable element in pewter items used for food and may make the surface very dull. White metal and Britannia metal are also of this general composition. Pewter has been used for useful and decorative items for the home such as candlesticks, porringers, drinking mugs, vases, trays, and tea and coffee services.

PFI: Abbreviation for Pipe Fabrication Institute.

pH: A symbol representing the concentration of hydrogen ions (H+) in aqueous solution. This logrithmic scale is expressed as a numerical value usually between 0 and 14. A pH of 7 indicates a neutral or noncorrosive substance. A pH between 0 and 7 indicates greater acidity. A pH between 7 and 14 indicates greater alkalinity. A pH of 0 (very acid), or 14 (very basic) are highly corrosive. The symbol is useful in the identification of the appropriate type of protective equipment necessary for handling a chemical material. The pH of an electroplating bath will be in the range of 6.5 - 5.0.

phase: Any of the forms in which matter can exist depending on the number of atoms or molecules per unit volume: gaseous, liquid, or solid. Gasses and liquids are amorphous (without structure) while solids such as metals are crystalline in nature are tightly packed and strongly bound together. The best known substance to demonstrate the three phases is water, which can be easily converted to solid (ice), liquid, or vapor.

phenolic resin: A resin made from phenol.

phi (Φ, φ, or φ): The twenty-first letter of the Greek alphabet. Symbol for angle (φ). Symbol for heat flow rate. Symbol for fluidity (φ). Symbol for critical angle ($φ_c$). Symbol for function (of) (φ). Symbol for magnetic flux (Φ).

phosphates: Corrosion inhibiting chemical agents added to cutting fluids.

phosphatizing: See costellizing.

phosphinic acid: [H_3O_2P] Colorless crystalline solid or oily liquid used in electroplating.

phosphomolybdic acid: [$H_3PMo_{12}O_{40}·xH_2O$] A yellow crystalline solid used in electroplating.

phosphor bronze: A tin bronze that has been deoxidized by the addition of up to 0.5% phosphorus. Relatively hard, strong, and corrosion resistant. Has good cold-work properties and high strength. Available in the following grades: Grade A containing 5% tin; Grade C containing 8% tin; Grade D containing 10% tin; Grade E containing 1.25% tin, and free-cutting phosphor bronze containing 88% copper, 4% lead, 4% tin, and 4% zinc.

phosphorus: CAS number: 7723-14-0. A nonmetallic element. Allotropes: white (or yellow), red, and black. Exhibits phosphorescence at room temperature. It contributes to hardenability, strengthens low carbon steels, and adds some corrosion resistance. In cast, iron phosphorus lowers the melting point and increases fluidity of the molten iron. Also it reduces tensile strength and makes harder and brittle castings. In steel, phosphorus enhances its strength, adds to the hardness of the plate and thus makes it abrasion resistant, but brittle. Steel containing a high content of phosphorus is particularly weak in shock or vibratory conditions and may be considered a harmful impurity. Red phosphorus is a reddish-brown to violet-red amorphous powder that is obtained from white phosphorus by heating with a catalyst at 464°F/240°C. It is much less reactive than white phosphorus. White phosphorus is a dangerous fire risk and ignites spontaneously in air @ 86°F/30°C. Must be stored under water and away from heat. Phosphorus improves the machinability in high sulfur steels, but must be limited to less than 0.05% to obtain plasticity.

Symbol: P
Physical state: Solid
Periodic Table Group: VA
Atomic number: 15
Atomic weight: 30.97376
Valence: 1,3,4,5

Density (g/cc): 1.82 (white @20°C); 2.34 (red)
Melting point: 111°F/ 44°C (white)
Boiling point: 536°F/280°C
Source/ores: Fluorapatite, phosphate rock, phosphorite(found as nodules on the ocean floor)
Oxides: O_5P_2
Crystal structure: cubic (α- or γ-white, red); orthorhombic (black); rhombohedral (β-white)

phosphotungstic acid: [$H_3PW_{12}O_{40}\cdot xH_2O$] White to slightly yellow crystalline solid or powder used in electroplating.

physical properties: The physical phenomena of a material which remain unchanged so long as there is no change in molecular composition or chemical nature; i.e., color, density, tensile strength, flexural strength, impact strength, water absorption, weight, mass, electrical conductivity, heat conductivity, toughness, thermal expansion, freezing point, and boiling point. Mixing metals to form alloys; or, the presence in a metal of very low percentage of other elements (not necessarily metals such as carbon in iron) can profoundly affect the original metals' physical properties.

pi (Π, π): The sixteenth letter of the Greek alphabet. Symbol for the mathematical constant 3.141592; the ratio of the circumference of a circle to the diameter.

pickling: The process of descaling and removing surface oxides from metallic surfaces by soaking them in strong acids such as sulfuric or hydrochloric, or by electrochemical reaction.

Pidgeon process: A process producing high-purity magnesium metal from dolomite or magnesium oxide by treating it with ferrosilicon [FeSi] under vacuum at high temperature (about 2,100°F/1,149°C).

pig iron: The basic raw material for cast iron and steel. A compound of about 93% pure iron ore with about 3.5% carbon, and small amounts of silicon, sulfur, phosphorus, and manganese that are obtained from smelting iron ores in a blast furnace and run into gridiron-shaped molds in the open air and slowly cooled either by spraying or running through a tank of water. The cooled metal takes the form of ingots or semi-cylindrical bars, attached at right angles to a longitudinal bar, thus resembling a sow and pigs. Commercial pig iron is produced when iron ore is charged in a blast furnace, mixed with limestone as a flux, and melted down with either charcoal, coke, or anthracite coal as fuel.

pillar file: A narrow, rectangular file that may be parallel or tapered, usually double cut with one or two uncut ("safe") edges. Used for narrow work such as grooves or keyways.

pill press: A press for making small pieces or compacts made by powder metallurgy techniques. See also compacts.

pin expansion test: A test for tubes to gauge the expansion capacity or for uncovering the presence of longitudinal weaknesses such as cracks, performed by forcing a tapered pin into the tube's open end.

pinion: of two gears which mesh together, the one with the smaller number of teeth is called the pinion, and the other is the gear. An exception occurs in worm gearing, where the gear with the smaller number of teeth is called the worm.

pin punch: A hand punch made of hardened tool steel, with a long, straight nose and a flat, straight "point" that is used to drive straight pins, cotter pins, taper pins, and keys, etc. It is normal practice to loosen the pin with a drift punch followed by use of the pin punch to drive them out. This punch is available in various sizes.

pipe: (1) The central cavity deformation in the top of an ingot, or the feeder head, caused by contraction during solidification; (2) a cast or wrought tubular metal conduit having walls thick enough to accept a standard pipe thread; otherwise, it may be called a tube.

Pipe Fabrication Institute (PFI): 1326 Freeport Road, Pittsburgh, PA 15238. An association of the pipe-fabricating industry.

pipe-forming dies: See tube-forming dies.

Pipe Line Contractors Association (PLCA): 2800 Republic National Bank Building, Dallas, TX 75201. An industry association of contractors and builders of long underground pipelines.

pipe wrench: Also known as a Stillson wrench. A wrench containing serrated adjustable jaws used to grip cylindrical parts and pipes. The L-shaped adjustable jaw slides loosely in a sleeve that allows the user to increase the grip on the workpiece by adding pressure to the handle. Adding pressure to the handle causes the serrated edges of the jaws to cut into the part being turned and may damage plated or other finished surfaces.

pitch: (1) The distance from the center of one screw thread, gear tooth, rivet heads, to the center of the next, or serration of any kind such as teeth per inch in a saw blade; (2) in thread design, term used to describe the distance measured from one point on a thread to the corresponding point on the next thread measured parallel to the axis. The pitch equals one divided by the number of threads per inch.(3) in gear design, the distance between similar, equally-spaced tooth surfaces, in a given direction and along a given curve or line. The single word "pitch" without qualification has been used to designate circular pitch, axial pitch, and diametral pitch, but such confusing usage should be avoided.(4) involute spline (P/P_s): A combination number of a 1:2 ratio indicating the spline proportions; the upper or first number is the diametral pitch, the lower or second number is the stub pitch and denotes, as that fractional part of an inch, the basic radial length of engagement, both above and below the pitch circle.

pitch circle: (1) *gear:* A circle the radius of which is equal to the distance from the gear axis to the pitch point; (2) *involute spline:* the reference circle from which all transverse spline tooth dimensions are constructed.

pitch cone: A thread design term used to describe an imaginary cone of such apex angle and location of its vertex and axis that its surface would pass through a taper thread in such a manner as to make the widths of the thread ridge and the thread groove equal. It is, therefore, located equidistantly between the sharp major and minor cones of a given thread form. On a theoretically perfect taper thread, these widths are equal to one-half the basic pitch. See also axis of thread and pitch diameter.

pitch cylinder: A thread term used to describe an imaginary cylinder of such diameter and location of its axis that its surface would pass through a straight thread in such a manner as to make the widths of the thread ridge and groove equal. It is, therefore, located equidistantly between the sharp major and minor cylinders of a given thread form. On a theoretically perfect thread these widths are equal to one-half the basic pitch. See also axis of thread and pitch diameter.

pitch diameter: (1) *thread:* on a straight thread the pitch diameter is the diameter of the pitch cylinder. On a taper thread the pitch diameter at a given position on the thread axis is the diameter of the pitch cone at that position*; (2) *gear:* The diameter of the pitch circle. In parallel shaft gears the pitch diameters can be determined directly from the center distance and the numbers of teeth by proportionality. Operating pitch diameter is the pitch diameter at which the gears operate. Generating pitch diameter is the pitch diameter at which the gear is generated. In a bevel gear the pitch diameter is understood to be at the outer ends of the teeth unless otherwise specified. (See also reference to standard pitch diameter under pressure angle.); (3) involute spline (*D*): is the diameter of the pitch circle.
Note: When the crest of a thread is truncated beyond the pitch line, the pitch diameter and pitch cylinder or pitch cone would be based on a theoretical extension of the thread flanks.

pitch line: The generator of the cylinder or cone specified in pitch cylinder and pitch cone.

pitch plane, gear: In a pair of gears, the plane perpendicular to the axial plane and tangent to the pitch surfaces. In a single gear it may be any plane tangent to its pitch surface.

pitch point: (1) *gear:* This is the point of tangency of two pitch circles (or of a pitch circle and a pitch line) and is on the line of centers. The pitch point of a tooth profile is at its intersection with the pitch circle; (2) *involute spline:* is the intersection of the spline tooth profile with the pitch circle.

pits: Cavities in a metal surface that are the result of localized corrosion, usually confined to a small area.

pitting: The formation of small sharp cavities in a metal surface by nonuniform electrodeposition or by corrosion. Pitting can be non-progressive or destructive. Non-progressive pitting occurs during the initial operation of a machine because of surface fatigue at roughness peaks. As the peaks wear, the surface becomes smoother and hence the load becomes more uniformly distributed. Pitting stops when the stress becomes less than the endurance limit of the material. Destructive pitting, however, is progressive and continues until the surface disintegrates. Pitting is a common mode of failure for gear teeth.

plain carbon steel: See carbon steel.

plain milling: A machining operation that produces flat surfaces with a plain milling cutter mounted on a horizontal milling machine arbor.

plain milling cutter: A broad category of cylindrical cutters with straight or helical teeth having a helix angle generally up to 60° on the circumference or periphery only, and available in a wide variety of widths and diameters. Plan milling cutters can be sharpened on a cutter and tool grinder by grinding the land on the periphery.

plane bearing: A term use to describe journal or sleeve bearings to distinguish them from journal, sleeve and other antifriction or rolling element bearings.

plain mandrel: See mandrel.

plane of rotation, gear: Any plane perpendicular to a gear axis.

planing: A machining operation for the removal of material from large, flat, plane surfaces using cutting tools fixed in a planing machine, above a reciprocating table on which work to be planed is carried. The table moves horizontally, back-and-forth, under the tool head containing the cutting tool, which can be fed across the table. Planing is used principally for workpieces that are too large for shapers. See also shaper.

planishing: Means to make smooth. An operation for producing a smooth surface finish by flattening the work using rapid series of overlapping light blows delivered by highly polished dies or by a hammer designed to be used against planishing stakes. Also accomplished by rolling in a planishing mill, usually done to level an uneven surface and to remove burrs and wrinkles, but sometime with the object of reducing the thickness.

planishing hammer: A hand tool having a flat, smooth, finely polished face.

plastic deformation: The permanent deformation of metals resulting from applied stresses and strains beyond the material's elastic limits. The deformation results in dislocations and distortions of the crystal structure.

plasticity: The property of a substance that permits it to accept a permanent change in shape without fracture or rupture. The inverse of elasticity. A material that tends to stay in the shape or size to which it has been deformed has high plasticity.

plastic metal: A general term used to describe bearing metals containing 78-85% tin, 9-15% antimony, and 5-11% copper.

plated ware: Objects such as eating utensils made of a heavy base metal of copper or other metals and covered with a thin layer of silver. From the 1740s the English perfected a technique in Sheffield to manually fuse the plate to copper and from the 1840s on, the silver was chemically electroplated on the base metal.

platen: (1) The flat surface of a press on which work is placed; (2) the sliding component or ram of a power press.

platinum: CAS number: 7440-06-4. A silver-white metallic alloying element, soft. lustrous. The least rare of the platinum group metals. Used in jewelry, automobile exhaust gas catalytic converters, jewelry, electrical contacts, brushes, thermocouples, anodes, surgical wire, bushings, electroplating, permanent magnets.

Symbol: Pt	**Density (g/cc):** 21.5
Physical state: Silvery metal	**Melting point:** 3,217°F/1,769°C
Periodic Table Group: VIII	**Boiling point:** 8,185°F/4,430°C (approx.)
Atomic number: 78	**Source/ores:** A by-product of copper and nickel
Atomic weight: 195.08	production; in platinum ores
Valence: 2,4	**Brinell hardness:** 97, annealed; (Vickers: 42)
	Crystal structure: f.c.c.

platinum bronze: An alloy nominally containing 90% nickel, 9% tin, and 1% platinum.

platinum cobalt alloys: Alloys nominally containing 40-77% platinum and 23-60% cobalt. Used to produce the most powerful permanent magnets.

platinum gold: Alloys containing widely varying amounts of copper, platinum, gold, silver, and zinc. In some alloys the precious metal content can range up to 70%.

platinum iridium alloys. Alloys containing 5-35% iridium. As the iridium is increased, the thermal conductivity is decreased and hardness of the alloy increases, along with electrical resistivity, and resistance to chemical attack.

platinum metals: Any of the group of six metals in group VIII of the periodic table: ruthenium, rhodium, palladium, osmium, iridium, and platinum. Also known as transition metals.

platinum-nickel alloy: Alloys nominally containing 80-95% platinum and 5-20% nickel have good strength at high temperatures.

platinum-rhodium alloys: Alloys nominally containing 3.5-40% rhodium. Used in high-temperature operations under oxidizing conditions; vessels, windings in high temperature furnaces, and components of gas-turbine aircraft engines.

platinum-tungsten alloy: Alloys containing 2-8% tungsten. Having high resistance to lead contamination this alloy was originally developed for spark plug electrodes in aircraft engines.

Plat-Iron®: A proprietary trade name of the SCM Corp. for high-purity electrolytic iron powder and reduced iron oxide powder, used in welding rod coatings, sintered structural parts, and oil-less bearings.

PLCA: Abbreviation for Pipe Line Contractors Association.

plug gage: Also known as an internal gage. One whose outside measuring surfaces may be straight plug gages, tapered plug gages, thread plug gages, and special plug gages, used any cross-sectional shape such as square holes. Plug gages contain a length of metal very accurately ground to the correct size and fitted with a handle. Used for internal measuring, to verify the specified uniformity of holes. The taper plug gage contains gaging limits indicated by marks or flats ground on the end of the gage in such positions that the distance the gage enters the work can be observed. They may be made in *go* and *no go* dimensions. (a) Plain plug gage is used to check the size limits of straight cylindrical holes; (b) tapered plug gages are used to check the fit, size, and amount of taper of tapered holes; (c) threaded plug gages are used to check the size limits and fit of internal screw threads. The "opposite" type of gage is a ring gage or *external* gage.

plug tap: A tool used as a starting tap on easily cut metals. It is designed with 3 or

4 threads at the front end of the land that are tapered to distribute the cutting action over several teeth as the internal threads are gradually cut.

plumb- or plumbo-: Containing lead.

plumber's solder: See solder.

PMA: Abbreviation for Precision Metalforming Association (PMA).

point: (1) Drill term used to describe the cutting end of a drill made up of the ends of the lands, the web, and the lips. In form, it resembles a cone, but departs from a true cone to furnish clearance behind the cutting lips; (2) a file term used to describe the front end of a file; the end opposite the tang.

point angle: A drill term used to describe the angle included between the lips projected upon a plane parallel to the drill axis and parallel to the cutting lips. For general-purpose drills, included point angle of 118° is the most commonly used. The point angle can range from 135° for hard materials to 150° for very hard materials such as manganese steel. Drills used for softer material such as molded plastics, bakelite, and hard rubber, should have an included angle of 90°, and for even softer materials, the angle can range down to 60°.

point-to-point system (N/C): A numerical control positioning system for machining operations programmed along the X- and Y-axis. In most point-to-point N/C systems the cutting tool or workpiece is rapidly moved in straight lines from one programmed point to another.

polar additives: See additives, cutting oils.

polar film: See additives, cutting oils.

polishing: A process of flexible grinding, used to smooth the surface of a workpiece by rubbing with abrasive particles imbedded in, or attached to, a soft cushioned matrix or flexible backing in the form of a wheel or belt that is moving or rotating. When the abrasive material is applied loosely, the process is called buffing. See also buffing.

pollucite: [$Cs_4Al_4Si_9O_{26} \cdot H_2O$] A natural cesium aluminum silicate. Source of cesium, used in welding materials.

polycrystalline: Materials made of aggregates of fine crystals bonded together.

polycrystalline diamond (PCD): An ultra-hard cutting tool material (substrate) consisting of fine, synthetic crystals of diamond bonded together to form a solid material that may be brazed to steel shanks or clamped in a throw-away carbide insert carrier. Primarily used to machine non-ferrous materials or for light cuts on abrasive materials at very high speeds, they last from 10-450 times as long as cemented and coated carbide tools.

porcelain enamel: An inorganic vitreous coating composed of clays and silicates that are sprayed and bonded to metal by fusion above 799°F/426°C. These finishes are hard, abrasion- and chemical-resistant, smooth and glasslike, and easy to clean. These coatings have been used on metal eating and drinking utensils since the 19th century.

porcelain glaze: A hard, glassy, heat-resistant coating for metals. Consisting of clays, feldspar, lime, and quartz they are brushed or sprayed on the metal surface and subsequently fired in a kiln at temperatures high enough to fuse the coating into the metal. See porcelain enamel.

porosimeter: A scientific instrument used to determine the porosity of solid to liquids, both liquids and gases.

porosity: The property of containing pores or minute channels or open spaces in a metal. Also used to describe the proportion of the total volume occupied by such pores or spaces.

postheating: A heat treatment for weld-metal including the heat-affected zone around the weld, applied immediately following welding for the purpose of tempering, for relieving stress, or for providing a controlled rate of cooling to prevent formation of a hard or brittle structure.

pot annealing: See box annealing.

potassium: CAS number: 7440-09-7. A soft white metal, silvery when cut . It has many uses including the manufacture of heat-exchange alloys. *Warning:* A storage hazard. Difficult to extinguish in fire. Store in inert atmospheres, such as argon or nitrogen, under liquids that are oxygen free, such as toluene or kerosene.

Symbol: K	**Density (g/cc):** 0.862
Physical state: White metal	**Melting point:** 147°F/64°C
Periodic Table Group: IA	**Boiling point:** 1,400°F/760°C
Atomic number: 19	**Source/ores:** Sylvite, sylvinite, carnallite
Atomic weight: 39.0983	**Crystal structure:** b.c.c.
Valence: 1	

potassium copper cyanide: Also known as potassium cuprocyanide. [$C_2CuN_2 \cdot K$] White crystalline solid used in electroplating. Highly poisonous.

potassium cyanide: [KCN] A white granular powder or crystalline mass used in gold and solver ore extraction, electroplating, steel nitriding and hardening.

potassium dichromate: [$K_2Cr_2O_7$] Bright, yellowish-red, transparent crystals used as a corrosion inhibitor, metal treatment, in electroplating, and for making brass pickling compositions.

potassium ferricyanide: [$K_3Fe(CN)_6$] A red or yellow crystalline solid used for tempering steel and in electroplating.

potassium ferrocyanide: [$K_4Fe(CN)_6 \cdot 3H_2O$] A lemon-yellow crystalline solid used for tempering steel. It is a lemon-yellow crystalline solid or powder.

potassium hydroxide: [KOH] A white crystalline solid used in electroplating.

potassium stannate: [$K_2OSn \cdot 3H_2O$] A white to slightly yellow crystalline solid used in tin electroplating.

pour point: The temperature at which an alloy is cast.

powder coating: A process using powdered thermosetting resins such as acrylics, epoxies, and polyesters that are applied metal objects by electrostatic spraying techniques. Reportedly, this process minimizes the pollution problems encountered with solvent-based sprayed coatings.

powdered metals (danger!): In general, finely dispersed powdered metals are a dangerous fire hazard. Many powdered metals suspended in air can ignite spontaneously and/or explode when exposed to flame or sparks. They may react violently with oxidizers such as chlorine and fluorine. Do not use water to fight fire; use earth, dry sand, dolomite, powdered graphite, sodium chloride, etc. See supplier's MSDS for specific instructions for storage, handling, disposal, and response to spills and fire.

powder metallurgy: The production of objects to desired shape from finely divided metals or clays by various methods including compressing them, possibly with heat, into precision dies or molds, or drawing into wire. See also sintering.

power: In mechanics, is the product of force times distance divided by time; it measures the performance of a given amount of work in a given time. It is the rate of doing work and as such is expressed in foot-pounds per minute, foot-pounds per second, kilogram-meters per second, etc. The metric SI unit is the watt, which is one joule per second.

PPE: Abbreviation for personal protection equipment.

Pr: Symbol for the element praseodymium.

praseodymium: A metallic element, one of the rare earth elements of the lanthanide group. A soft, crystalline solid, similar in color to iron. It has some specialized applications in metallurgy. An ingredient in misch metal.

Symbol: Pr **Melting point:** 1,686°F/919°C
Physical state: Silvery metal **Boiling point:** 5,468°F/3,020°C
Periodic Table Group: IIIB **Source/ores:** Monazite sands, bastnasite
Atomic number: 59 **Oxides:** Pr_2O_3
Atomic weight: 140.90765 **Crystal structure:** h.c.c. (α); b.c.c. (β)
Valence: 3,4

precipitation hardening: Hardening caused by the precipitation of a component in a supersaturated solid solution. See also age hardening.

precision gage blocks: See gage blocks, Jo blocks, angle gage blocks, rectangular.

Precision Metalforming Association (PMA): 6363 Oak Tree Blvd. Cleveland OH 44131-2556. Telephone:216/901-8800. FAX: 216/901-9190. E-mail: metalforming..com WEB: http://www.pma.org

preheating: A general term applied to any heating operation, usually at moderate temperature, that is employed immediately prior to subsequent thermal or mechanical treatment such as austenitizing when hardening many of the tool steels, and structural steels of high-hardenability. High-hardenabilty steels are often preheated prior to welding to avoid cracking and hard spots. However some nonferrous alloys are preheated to a high temperature for extended periods of time, in order to homogenize the structure prior to working.

press: A general term for various machines used for varied operations employing pressure to accomplish the work required. Presses range in size and capacity from small bench presses to massive machines capable of exerting more than 1,000 tons of pressure. Presswork includes cold pressing, hot pressing or stamping, blanking, punching, drawing, bending, forming, coining, embossing, swaging, and shearing operations, many of these processes being sometimes combined into a single operation. The finished job is called a *stamping* if flat, and a *pressing* if shaped and formed. See also blanking.

press, double-action: A press having a stationary bed and two slides or rams, one working inside the other, and carries two punches working independently.

press fit: An interference fit that requires the aid of a press to assemble the components.

press fit bushings: Press fit wearing bushings to guide the tool are for installation directly in the jig without the use of a liner and are employed principally where the bushings are used for short production runs and will not require replacement. They are intended also for short center distances.

press forging: An operation for working hot or cold metal between dies in a forging press.

press, hydraulic: A press actuated by a hydraulic cylinder and piston.

press, single-action: A press with a stationary bed, a single slide or ram, and carrying one punch.
Press-brake (bending brake): An open-frame press for bending, cutting, and forming; usually handling relatively long work in strips.

pressure angle: (1) *gear:* The angle between a tooth profile and a radial line at its pitch point. In involute teeth, pressure angle is often described as the angle between the line of action and the line tangent to the pitch circle. Standard pressure angles are established in connection with standard gear-tooth proportions. A given pair of involute profiles will transmit smooth motion at the same velocity ratio even when the center distance is changed. Changes in center distance, however, in gear design and gear manufacturing operations, are accompanied by changes in pitch diameter, pitch, and pressure angle. Different values of pitch diameter and pressure angle therefore may occur in the same gear under different conditions. Usually in gear design, and unless otherwise specified, the pressure angle is the standard pressure angle at the standard pitch diameter, and is standard for the hob or cutter used to generate the teeth.

The operating pressure angle is determined by the center distance at which a pair of gears operates. The generating pressure angle is the angle at the pitch diameter in effect when the gear is generated. Other pressure angles may be considered in gear calculations.

In gear cutting tools and cutters, the pressure angle indicates the direction of the cutting edge as referred to some principal direction. In oblique teeth, that is helical, spiral, etc., the pressure angle may be specified in the transverse, normal, or axial plane. For a spur gear or a straight bevel gear, in which only one direction of cross-section needs to be considered, the general term pressure angle may be used without qualification to indicate transverse pressure angle. In spiral bevel gears, unless otherwise specified, pressure angle means normal pressure angle at the mean cone distance; (2) *involute spline (ö):* is the angle between a line tangent to an involute and a radial line through the point of tangency. Unless otherwise specified, it is the standard pressure angle.

prick punch: A hardened tool steel punch having a slender point with a 30° to 60° included angle, used in layout work to mark scribed or layout lines with small indentations, to provide center marks for divider points when laying out arcs and circles, or for hole locations before drilling.

primes: A term used in the metal industry to describe metal products, principally sheet and plate, that are free from visible surface defects, and of the highest quality.

Princes metal: A pewter-colored alloy containing 84.8% tin and 15.2% antimony. Other types contain 61-83% copper and 17-39% zinc. These latter alloys are brasses.

principal reference planes, gear: These are a pitch plane, axial plane, and transverse plane, all intersecting at a point and mutually perpendicular.

process annealing: An imprecise term for a process commonly applied in the sheet and wire industries, in which an iron-base alloy is heated to a temperature close to, but below, the lower limit of the transformation range and allowed to cool. The process softens the metal for further cold working and to improve workability. For the term to be meaningful, the condition of the material and the time-temperature cycle used must be stated.

profile: A term related to the measurement of surface texture, the contour of the surface in a plane perpendicular to the surface, unless some other angle is specified. See also nominal profile, measured profile, modified profile.

profile gage: Made for checking the shape of specially formed jobs such as those produced by form-tools in the lathe.

profile grinding: Another name for form grinding.

profile variation, involute spline: Any variation from the specified tooth profile normal to the flank. The reference profile, from which variations occur, passes through the point used to determine the actual space width or tooth thickness. This is either the pitch point or the contact point of the standard measuring pins. Profile variation is positive in the direction of the space and negative in the direction of the tooth. Profile variations may occur at any point on the profile for establishing effective fits. Positive profile variations affect the fit by reducing effective clearance. Negative profile variations do not affect the fit but reduce the contact area.

profilometer: An electronic roughness-measuring instrument, used for the measurement of surface irregularities.

progression notch: See French notch.

proof circle: A technique used in drilling operations to accurately position a drill using a scribed circle that is slightly smaller than the hole to be drilled.

1,2-propanediamine:[$C_3H_{10}N_2$] A colorless liquid used in electroplating.

propeller bronze: See aluminum bronze.

protective coating: A film or thin layer of metal applied to a substrate primarily to inhibit corrosion, and secondarily for decorative purposes. Metals such as nickel, chromium, copper, and tin are electrodeposited on the base metal; paints may be sprayed or brushed on. vitreous enamel coatings that require baking are also used. Zinc coatings are applied by a continuous bath process in which a strip of ferrous metal is passed through molten zinc.

Protina®: A registered trademark of Aqualon Co. for line of products for cleaning, inhibiting corrosion, and coating copper, brass, or bronze artwork and statuary.

protractor: An instrument for setting off, reading, or measuring angles. In its simplest form it is a graduated semi-circle, marked out to 180° along the circumference. At the center of the protractor's straight base is a starting point such as a hole or notch which acts as a reference, the angle being read in degrees from the circumference.

protrusion: Another name for projection.

psi (Ψ, ψ) or *psi* : The twenty third letter of the Greek alphabet. Symbol for electric flux (ψ). Abbreviation for pounds per square inch (lb/in^2), a measurement of pressure. In the metric system pressure is measured by kilometers per square centimeter (km/cm^2).

Pt: Symbol for the element platinum.

punch: (1) A small hand tool with a straight or tapered end that is struck with a hammer, used for various purposes including punching holes in sheet metal, layout marking, to drove out tapered or cotter pins and rivets, and to transfer the location of a hole; (2) The male die part, usually the upper member; (3) to die-cut a hole in metal sheet or plate.

punching: A press operation involving the punching of holes of any shape in the blank, and the piece removed is waste. This differs from blanking in that the flat piece removed is either the finished workpiece or the first step of a finished workpiece. See also blanking.

punch press: A machine made for punching holes or forming metal. It contains a ram with a punch that is moved against or through a die held in the frame on a bed or bolster. Also, a general name applied to any mechanical press.

push fit: Fits that require slight manual effort to assemble the parts. Suitable for detachable or locating parts, but not for moving parts. As the clearance is small or negligible, this is termed a transition fit. See fits, description.

putty powder: A soft abrasive made from tin oxide.

PVD: Abbreviation for physical vapor deposition (coated carbide).

pyrite: Also known as iron pyrite and fools gold. See also fools gold.

Pyrobrite®: A registered trademark of M & T Harshaw (division of Atochem) for a bright high-leveling copper electroplating process using copper pyrophosphate trihydrate, potassium pyrophosphate, ammonium hydroxide, and organic addition agents.

pyrochlore: An ore of niobium.

pyrolusite:[MnO_2] Black manganese dioxide, a native manganese ore.

Pyromet® Alloy 625: A proprietary trade name of Carpenter Technology for a non-magnetic, corrosion- and oxidation-resistant nickel alloy with high strength and toughness.

pyrometer: A meter for measuring the heat generated in a furnace used for heat treatment of metals.

Q

Q: Symbol for quantity of electrical charge; (2) symbol for quantity of heat; (3) symbol for quantity of light.

q: Symbol for heat flow rate although the symbol Φ is preferred.

QC: Abbreviation for quality control.

quadrant: Quarter of a circle.

quad: A unit of energy equal to 10^{15} btu.

quadrilateral: Referring to a four-sided figure.

quality control (QC): The maintenance of a specific level of quality during the manufacturing process, to ensure a consistently good product. Techniques include physical sampling and statistical quality control, the gathering and analyzing of quality data, and action are taken if the quality deviates by a specified amount.

quart: A unit of liquid or dry measure equal to 0.25 gallon (U.S.) = 0.9464 liter liquid; 1.1012 liters dry.

quench: Rapid cooling of metals or alloys by immersion in cold water or oil.

quench annealing: Annealing an austenitic alloy by solution heat treatment.

quench hardening: The process of hardening a ferrous alloy by austenitizing and then cooling rapidly enough so that a substantial amount of the austenite transforms to martensite.

quenching: The operation of rapid cooling hot metal in a suitable medium such as liquids (water, brine, oil), gasses (air), or solids. An essential part of the tempering process, especially for steels and alloys. When the quench time of molten metal is extremely short (less than a second), the final product will have an amorphous or glasslike structure, because crystallization will not occur.

quick-change toolholder: A lathe toolholder made to hold four or more different cutting tools, containing a patented locking system that enable the operator to change tools fast and accurately.

quickening liquid: An electroplating solution of mercury (II) nitrate or cyanide.

quicklime: Calcined limestone; calcium oxide [CaO]. Used as a refractory, flux in steel manufacture, carbon dioxide absorbent.

quicksilver: A synonym for mercury. See also mercury.

quill-type punch: A small, fragile punch mounted in a protective shouldered sleeve or quill.

quintal: A unit of weight equal to 100 kg in the metric system and 100 lb in the avoirdupois system.

R

Ra: Abbreviation for roughness average.

R&D: Acronym for research and development.

rack: (1) A gear whose teeth lie in a straight line instead of a circle. A worm has a pitch plane lying parallel to the center lines of worm and gear touching the pitch cylinder of the worm wheel; (2) a metal bar cut with gear teeth so that a circular gear can mesh with it to convert rotary into reciprocating motion or vice versa; (3) a gear with teeth spaced along a straight line, and suitable for straightline motion. A basic rack is one that is adopted as the basis of a system of interchangeable gears.

radial drill: A drilling machine for large, heavy work. It has a movable arm containing the spindle that can be set at different distances from the post or column and rotated around it, offering great flexibility in dealing with massive or cumbersome workpieces.

radian: Also called a rad. An arc of a circle, equal in length to the radius; an angle at the center of a circle, formed by 2 radii cutting off such an arc. One rad = $180°/\pi$ = $57.29578°$. The SI unit for plane angles.

radius and fillet gage: It consists of a holder containing a series of thin stainless steel or stainless steel blades or "leaves" of varying size, somewhat like a pocket knife. Each thin steel blade has the corresponding radius accurately formed in it. Used for checking the size and accuracy of internal and external radii of rounded corners. The majority of gages are made with an external and internal radius of the same size on one blade. External and internal radii are also known as male and female (a female radius being called a fillet), and convex and concave radii.

rake: (1) *gear:* Any deviation of a straight cutting face of the tooth from a radial line. *Positive rake* means that the crest of the cutting face is angularly advanced ahead of the balance of the face of the tooth. *Negative rake* means that the same point is angularly behind the balance of the cutting face of the tooth. *Zero rake* means that the cutting face is directly on the center line; (2) *drill:* The angle between the flute and the workpiece. An angle less than 90° is called a *positive rake*, and a rake angle more than 90° is a *negative rake*. A slightly negative rake is preferred for drilling softer materials such as brass or plastics.

raker set: A tooth set having one tooth bent to the right, one straight tooth in the middle, followed by one tooth bent to the left. Perhaps the most widely used blade set in metalworking, it is recommended for cutting most contours and the majority of cutting where the shape and thickness of the stock is consistent. See also set and wave set.

rancid: Having the characteristic, tainted odor of spoiled, or about-to-spoil, oily substances, due to the formation of free fatty acids; as, rancid cutting fluid.

Raney nickel: Also known as Raney alloy. A finely-divided nickel made from an alloy of 50-50% aluminum and nickel and a caustic soda solution. Must be stored under water or alcohol; ignites spontaneously in air.

rare earth: An oxide of one of the fifteen related elements in group IIIB of the periodic table having atomic numbers 57 (lanthanum) to 71 (lutecium), inclusive. The elements themselves constitute the rare earth metals, possessing very similar chemical properties: Lanthanum, cerium, praseodymium, neodymium, promethium, samarium, europium, gadolinium, terbium, dysprosium, holmium, erbium, thulium, ytterbium, lutetium.

rare gas: Any of the six gases composing the extreme right-hand group of the periodic table, namely helium, neon, argon, krypton, xenon, and radon. They are preferably called noble gases or (less accurately) inert gases. The first three have a valence of zero and are truly inert, but the others can form compounds to a limited extent.

Rareox®: A registered trademark of W.R. Grace, Inc. for cerium oxide used for high speed polishing.

rasp cut: A file tooth arrangement that is formed individually by raising a series of individual rounded teeth, usually not connected, from the surface of the file blank with a sharp narrow, punch-like cutting tool. The rasp file is used with a relatively heavy pressure on soft substances for fast removal of material.

ratchet: A mechanism containing a gear with triangular-shaped teeth adapted to be engaged by a pawl which either imparts intermittent motion to the ratchet or else locks it against backward movement, permitting operation in one direction only. This tool allows the handle to be moved backwards, without force, in the opposite direction in preparation for subsequent strokes used to tighten or loosen a fastener such as a bolt or screw, without having to remove it from the fastening device.

ratio of gearing: Ratio of the numbers of teeth on mating gears. Ordinarily the ratio is found by dividing the number of teeth on the larger gear by the number of teeth on the smaller gear or pinion. For example, if the ratio is 2 or "2 to 1," this usually means that the smaller gear or pinion makes two revolutions to one revolution of the larger mating gear.

rat-tailed file: A tapered round file.

rawhide mallet: Used to deliver non-marring blows from a large striking face, and ideal for working soft metals such as sheet metal, copper, and brass. The mallet head is made from animal hides that are compressed under hydraulic pressure and seasoned for durability.

Rayox®: A registered trademark for titanium dioxide.

RC: Abbreviation for running or sliding clearance fit. See also fits, description.

Re: Symbol for rhenium.

reamers: Multi-edged, fluted, rotary cutting tools used to bring a hole to accurate size and to produce a good surface finish on existing cylindrical or tapering holes, whether a cast or cored hole, or one made by a drill or boring bar. There are two main types of reamer, one of which is parallel for making straight holes and the other tapered for making tapered holes. It is possible still further to divide reamers by flute design, for while some have straight flutes, others have spiral flutes (also called helical flutes) and resemble a drill, but often of much greater pitch. In general, the spiral-flute type is preferred for accurate work, since there is less tendency to *chatter* when in use and, therefore, a finer, smoother finish is produced. It should be noted that the spiral flutes are left-handed, although the reamer is turned in a right-hand or clockwise direction. Still further sub-dividing, some reamers are made to a taper for their full length, and others which are tapered for only about half the length of the flutes, the remainder being parallel. Hand reamers are generally made of carbon steel or high-speed steel, and their cutting edges are backed off in the same manner as those of twist drills to give suitable clearance. When buying reamers, care should be taken to order the correct pattern since many of them have a tapered or Morse shank for fitting a drill press, milling machine, or lathe. Machine reamers are also called *chucking reamers*, and the most common types are called *rose* and *fluted* . Reamers for hand use have a cylindrical shank with a square end to engage with the hand wrench. See also reaming.

reaming: A machining process used for a number of applications, the chief of which are: To enlarge existing holes; to make a parallel hole into a tapered hole; and to bring existing holes accurately to size; and to produce a good surface finish. In some respects, hand reamers are similar to screw-cutting taps, for they cut away metal from the inside of a hole. Additionally, they are held in a wrench of similar type to that used for taps. Reaming is often performed on drill presses, milling machines, lathes and other equipment. This often called *machine reaming*.

recess: A groove set below the normal surface of a workpiece.

recrystallization: A process for refining the microstructure existing in cold worked metals and solid solutions by replacing the distorted structure with one that is new and strain-free, usually accomplished by prolonged annealing.

recrystallization annealing: Annealing cold-worked metal to produce a new, undistorted, strain-free grain microstructure without phase change.

recrystallization temperature: The lowest temperature at which complete recrystallization of a cold worked metal occurs during prolonged annealing. The temperature is a function of the metal and its purity, the extent of the deformation and time.

re-cut: A worn-out file which has been re-cut and re-hardened after annealing and grinding off the old teeth.

Redalloy®: A proprietary trade name for a brass (copper alloy) containing 85% copper, 14% zinc, and about 1% tin.

red brass: Highly corrosion resistant brass alloys nominally containing 85-87.5% copper and 12.5-15% zinc. The term is general and used for several different types of brass including those containing more than 80% copper, usually with 1-5% lead, often with 3-5% tin. Other alloys termed red brass include those with 75-85% copper, up to 20% zinc, and sometimes very small amounts of lead and tin. Due to their resistance to atmospheric corrosion and dezincification, red brasses are widely used for decorative purposes and in plumbing and piping.

red brass, leaded: Leaded semi-red brass alloys nominally containing 71-85% copper, 5-15% zinc, 5-7% lead, 3-5% tin, and having good machinability.

red metal: A general term used to describe brasses containing a minimum of 80% copper. It is also commonly used for any copper-base metal.

red mud: A by-product sludge from the Bayer Process of alumina from bauxite for aluminum production, containing 30-60% iron oxide [Fe_2O_3].

reducing agent: (1) A substance that is readily oxidized; (2) a compound, element, or force that is capable of effecting a reduction reaction under the proper conditions. Chemical reduction may occur in the following ways: (a) acceptance of one or more electrons from another substance; (b) removal of oxygen from a compound; (c) addition of hydrogen to a molecule. Also known as a reducer.

reduction: Any operation that decreases the size of a unit or material by any of several methods, e.g., grinding, crushing, etc. In metallurgy, the physical separation of an ore into desirable economically viable materials and undesirable impurities (gangue) by any of several methods, i.e., flotation, roasting, elutriation, screening, milling, and magnetic means.

reference cut: A machining term for the initial cut that is used to set the base for subsequent cutting measurements.

reference dimension: A dimension, usually without tolerance, used for information purposes only. It is considered auxiliary information and does not govern production or inspection operations. A reference dimension is a repeat of a dimension or is derived from other values shown on the drawing or on related drawings.

reference gage: A fixed gage made of hardened steel, used for testing the accuracy of inspection instruments such as gages and micrometers.

reflex angle: Used to describe any angle greater than 180°.

Refractaloy®: A trademarked product of Westinghouse Electric Corp. for an alloy containing nickel, cobalt, chromium, molybdenum, and iron.

Refractaloy® 26: A trademarked product of Westinghouse Electric Corp. for an alloy containing nickel, cobalt, chromium, molybdenum, iron and titanium.

refractory metal: A metal with an extremely high melting point; one that resists heat and is slow to soften. In the broad sense, it refers to metals having melting points above the range of cobalt, iron, and nickel.

regardless of feature size (RFS): The term used to indicate that a geometric tolerance or datum reference applies at any increment of size of the feature within its size tolerance.

resilience: The quality of elasticity; the property of springing back or recoiling upon removal of pressure, as with a spring which can be strained to the extreme limit again and again without rupture or developing a permanent set.

resinoid: Any organic, polymeric liquid that, when converted to its final state for use, becomes solid. (ASTM) Any material that contains thermosetting synthetic resin. Drying oils, raw or partially polymerized linseed oil, and partially condensed phenol-formaldehyde are considered resinoids.

resinoid bond: See resin bond.

resinoid grinding wheel: A kind of elastic bond grinding wheel that often uses potassium fluoborate [KBF_4] as a grinding aid. See also elastic bond grinding wheel.

Resistance Welder Manufacturer Association (RWMA): 1900 Arch Street, Philadelphia, PA 19103. An professional association for manufacturers of resistance welding equipment. Provides standards and dissemination of technology for equipment and procedures.

Resistox®: A registered trademark of the SCM Corp. for stabilized grades of copper powder assaying at greater than 99% copper with a density 8.9 and apparent density range of 2.0-3.5 g/cm3. Marketed in several grades of various particle sizes. Fabrication of porous bearings, sintered ferrous machine parts, metal friction surfaces, electric brushes, electrical contacts.

resolution: See discrimination.

resulfurized steel: Steel containing sulfur that was added following refining, to improve machinability.

rethreading die: A hardened steel tool with a square or hexagonal nut-like shape for use with a common wrench, used to repair stripped or damaged threads.

reverberatory furnace: An ore-roasting kiln having a long, shallow hearth and curved or sloping roof shaped to deflect the flame and radiate the heat downward onto the charge, ore, or material being treated. After passing over the charge, the heat is vented out of the furnace.

Rexalloy®: A proprietary name for a nonferrous cast alloy used as brazed tips on tool shanks, as removable tool bits, as inserts in toolholders and milling cutters.

RFS: An abbreviation for "regardless of feature size."

Rh: Symbol for the element rhodium.

rhenium: CAS number: 7440-15-5. Resists oxidation and corrosion, but slowly tarnishes in moist air. Retains its crystalline structure all the way to its melting point. Not attacked by seawater. Alloyed with tungsten and molybdenum for electronic products, plating of metals by electrolysis and vapor-phase deposition.

Symbol: Re	**Density (g/cc):** 21.02
State: Silvery-gray metal	**Melting point:** 5,755°F/3,180°C
Periodic Table Grp.: VIIB	**Boiling point:** 10,650°F/5,627°C
Atomic Number: 75	**Source/ores:** Molybdenite, from flue dusts of
Atomic Weight: 186.207	molybdenum smelting
Valence: -1 through 7; 4, 6,	**Oxides:** Re_2O_7
and 7 are most common,	**Crystal structure:** Hexagonal, close-packed
the last being the most stable	

Rhine metal: An alloy nominally containing 97% tin and 3% copper.

rho (P, ρ): The Greek letter. the italic ρ is a symbol for mass density, rdflection coefficient.

rhodanizing: Plating with rhodium, usually on silver.

rhodium: CAS number: 7440-16-6. Metallic element of the platinum group. Used as an alloying element with platinum for high temperature thermocouples, furnace windings, electrical contacts, jewelry, catalyst, optical instrument mirrors, and electrodeposited coatings for metals.

Symbol: Rh	**Density (g/cc):** 12.41
Physical state: Gray metal	**Melting point:** 3,571°F/1,966°C
Periodic Table Group: VIII	**Boiling point:** ca 8,130°F/4,500°C
Atomic number: 45	**Ores:** Rhodite; occurs with platinum.
Atomic weight: 102.90550	**Crystal structure:** f.c. cubic
Valence: 3	**Brinell hardness:** 101 (Vickers: 122)

rhodite: A naturally occurring alloy consisting of 57-65% gold and 35-43% rhodium.

rhodonite: An ore of manganese.

rifflers (files): Name given to small files that are curved upwards at the ends into an arc and used to reach the bottom of a sinking, for filing the inside of curved castings, and similar work.

right triangle: A triangle having one angle that is a right or 90° angle.

right-hand thread: A thread is a right-hand thread if, when viewed axially, it winds in a clockwise and receding direction. A thread is considered to be right-hand unless specifically indicated. In other words, right-hand taps are unmarked while left-hand taps are stamped "L" or "LH."

rimmed steel: Also called rimming steel. A partially deoxidized, low-carbon steel containing sufficient iron oxide that reacts with the carbon in the steel to give continuous evolution of carbon monoxide [CO] while the ingot is solidifying, resulting in the formation of an outer circumferential rim of substantially pure metal virtually free from carbon, other impurities, or voids. The carbon monoxide bubbles and impurities are trapped in the center portion or core of the ingot . The sheet and strip material produced from the surface material of rimmed steel is highly suitable for deep drawing and pressing.

ring gage: Also known as an external gage or collar gage. A cylindrical shaped gage with a hole of the exact size specified for the part being measured. The ring gage is used to externally measure surfaces that may be cylindrical or conical in form, and should slip easily, without force, over the part being measured and there should be no noticeable side-by-side movement. The inside measuring surfaces of a ring gage may be (a) plain ring gages, used to test the external dimension limits of straight round parts; (b) tapered ring gages, used to test the size and fit of a taper; (c) thread ring gages, made to test the fit and pitch diameter limits of external screw threads. They are made in *go* and *no go* dimensions. The no-go ring is identified by a groove around the peripheral face. The "opposite" type of gage is a plug gage or *internal* gage.

ring test: A grinding wheel safety inspection procedure for detecting cracks or other damage that may pose a health hazard. This test should be performed directly prior to mounting a new or used grinding wheel because cracks in abrasive wheels are frequently not visible to the naked eye. If the wheel is not too heavy it can usually be suspended from its hole on a mandrel and tapping it with a non-metallic implement. Heavy wheels should be mounted on a wheel balancing stand or may be allowed to stand upright (vertically) on their peripheral face on a clean board (if the floor is hard) while performing this test. Tap the wheel on the right and on the left of the vertical centerline, about 45° (from the centerline) and 1 or 2 in. from the peripheral edge. The test should be repeated after rotating the wheel 45°). A vitrified wheel will emit a clear metallic, bell-like tone, a ring. A dead or dull sound being indicative of a possible crack in the wheel. The crack may not be visible with the naked eye. Resin bonded wheels do not emit the same metallic tone as vitrified wheels. The ring test is not applicable to mounted wheels, small wheels (4 in. diameter and smaller), plugs and cones, threaded wheels, etc.

rivet: A pin used in a pre-formed hole for joining two or more pieces of metal. A rivet is made with a head formed on one end, and the shank or tail at the other end. After the rivet is put into place, the shank is treated with direct blows from a hammer, rivet-set, or other rivet-setting device, clinching the rivet and drawing the riveted members close together. Rivets are generally of the same metal as the parts to be joined. They are made with heads of various shapes for different purposes; are specified according to their length, diameter, and shape of head. The length is normally measured from the underside of the head along the shank or tail. When the faces of a joint must be smooth and flush, it is necessary to use rivets of the countersunk-head type. The countersunk rivet is not as strong as the button- or conical-headed type, and a greater number of rivets is necessary to insure strength.

riveting: A method for joining together two or more metal plates or pieces. Riveting is especially suited to light construction, for attaching handles to objects, for forming pivots which act as pivots on which parts revolve or swivel, making strong angle joints using angle iron or plate, attaching fittings such as hinges to metal plates, and for joining the ends of metal formed into cylinder or band. In heavier work, riveting is used for joining girders to stanchions and in other constructional engineering where it would be impossible to use a welded joint.

Basically there are two kinds of riveted joints, the lap-joint and the butt-joint. In the ordinary lap-joint, the plates overlap each other and are held together by one or more rows of rivets. In the butt-joint, the plates being joined are in the same plane and are joined by means of a cover plate or butt strap, which is riveted to both plates by one or more rows of rivets. The term single riveting means one row of rivets in a lap-joint or one row on each side of a butt-joint; double riveting means two rows of rivets in a lap-joint or two rows on each side of the joint in butt riveting. Joints are also triple and quadruple riveted. Lap-joints may also be made with inside or outside cover plates. The holes to receive the rivets in the various plates are generally drilled, but in thin plates they may be punched. When fastening thin plate, it is particularly important to maintain accurate spacing to avoid buckling. In either case the rivet should be a tight push fit or light driving fit in the hole.

In all types of construction where large rivets are necessary they are first heated to red. Not only does this facilitate the panning-over of the rivet, but results in a firmer joint. This is because the rivet contracts on cooling, and so pulls the two members more closely together. In hot riveting, the diameter of the rivet is increased by heating and therefore the rivets must be smaller initially than the holes that receive them. In jobs that must be made water and steam tight by riveting, the edges of the rivet heads and the edges of the plates are generally burred down with a caulking tool, which is similar in shape to a chisel, but with a flat end. Caulking is a very skillful operation, and if done badly may tend to open the plates instead of closing them. Fullering is a similar process, but is perhaps less difficult. The fullering tool has its end made equal in thickness to the plates, the ends of which are often beveled to about 80° to facilitate the process.

roasting: An oxidation process involving heating metals in the presence of air or oxygen to effect a chemical change that results in the conversion of natural metal sulfide ores to oxides. Used as a facilitating initial step to smelting or recovery of non-ferrous metals such as copper, lead, zinc, etc.

Rockwell hardness test: A measure of relative hardness of the surface of a material based on the indentation made by a 1/16, 1/8, or 1/4 in standard steel ball penetrator, or a conical diamond cone, called a brale indenter, with an apex angle of 120°. The steel balls are used on soft, non-ferrous metals and read on the "B" scale; the diamond point used on harder materials and read on the "C" scale (used particularly for steel and titanium). The hardness is indicated on a dial gage graduated in the Rockwell-B (RB) and Rockwell-C (RC) hardness scales. The results are reported by using numbers to denote the pressure in kilograms, and letters scale to denote the ball or diamond producing a given indentation. The harder the material, the higher Rockwell number will be.

roll angle, gear: The angle subtended at the center of a base circle from the origin of an involute to the point of tangency of the generatrix from any point on the same involute. The *radian* measure of this angle is the tangent of the pressure angle of the point on the involute.

roller bearing: An antifriction bearing that contains loose hardened rollers that convert sliding friction into rolling friction. The rollers fit between a race and journal.

roll pins: Fasteners made from steel sheet that is rolled into a cylindrical shape, chamfered on the ends, and hardened by spring tempering, used to hold mechanical parts together. Roll pins are made slightly oversized for the hole into which they are driven, thus allowing the springing action of the cylinder to hold tightly.

Roman bronze: A copper alloy nominally containing 10% tin.

Rompel's alloys: A copper alloy nominally containing 18% lead, 10% zinc, and 10% tin.

Ronfusil Steel®: A proprietary trade name for a steel containing 12% manganese.

Ronia metal: A brass alloy nominally containing small amounts of cobalt, manganese, and phosphorus.

root circle, gear: A circle coinciding with or tangent to the bottoms of the tooth spaces.

root diameter: See minor diameter.

root diameter, gear: Diameter of the root circle.

root, thread: The bottom surface of the thread which joins the tow sides or flanks of adjacent thread forms and is immediately adjacent to the cylinder or cone from which the thread projects. The root of and external thread is at its minor diameter, while the root of an internal thread is at its major diameter.

root truncation: A thread design term used to describe the radial distance between the sharp root (root apex) and the cylinder or cone that would bound the root.

roscoelite: A species of mica and source of vanadium.

rose reamer: Rough machine end-cutting tools, designed to cut only on the end and are particularly adapted to rough-reaming of cored or drilled holes 0.003 to 0.010

in undersize. They have a 45° end cutting chamfer to do all of the cutting. The chip grooves or flutes are ground cylindrical and contain no cutting edges, providing room for cutting fluids and chip evacuation.

rosin oil: A thick, water-white to brown liquid. Odorless and insoluble in water. Made by distilling and fractionation of rosin above 680°F/360°C. Combustible; a fire risk when heated. Used as a lubricant for iron bearings.

rosin spirit: A hydrocarbon mixture that is often used as a turpentine substitute.

Ross alloys: A bronze alloys nominally containing 68% copper and 32% tin.

rottenstone: Another term for tripoli, an abrasive made from decayed and disintegrated limestone. See tripoli.

rouge: A very fine iron oxide [Fe_2O_3], used as an abrasive. The finest rouges are prepared from safflower (carthamus, African or American saffron).

rough-cut file: A coarse file having approximately 20 teeth per inch.

rough machining: Machining without concern to finish, usually to be followed by a subsequent operation to refine the workpiece.

roughness (R): Describes the finely-spaced micro-geometric irregularities of the surface texture, usually including those irregularities which result from the inherent action of the production process. These are considered to include traverse feed marks and other irregularities within the limits of the roughness sampling length.

roughness average value (Ra): Also known as arithmetic average (AA) and centerline average (CLA). It is the arithmetic average of the absolute values of the measured profile height deviations taken within the sampling length and measured from the graphical centerline. Roughness average is expressed in micrometers or one millionth of a meter (0.000001 meter).

roughness average value (Ra) from continuously averaging meter readings (measurement of surface texture): So that uniform interpretation may be made of readings from stylus-type instruments of the continuously averaging type, it should be understood that the reading that is considered significant is the mean reading around which the needle tends to dwell or fluctuate with a small amplitude. Roughness is also indicated by the root-mean-square (rms) average, which is the square root of the average value squared, of a series of measurements of deviations from the roughness centerline, expressed in microinches. A roughness-measuring instrument calibrated for rms average usually reads about 11% higher than an instrument calibrated for arithmetical average. Such instruments usually can be recalibrated to read arithmetical average. Some manufacturers consider the difference between rms and arithmetic average to be small enough that rms on a drawing may be read as arithmetic average for many purposes.

roughness sampling length: A term related to the measurement of surface texture, the sampling length within which the roughness average is determined. This length is chosen, or specified, to separate the profile irregularities which are designated as roughness from those irregularities designated as waviness.

roughness spacing: A term related to the measurement of surface texture, the average spacing between adjacent peaks of the measured profile within the roughness sampling length.

round file: A file that is generally single cut. When parallel they are called *parallel round*; tapered round files are described as *rat-tailed*. Used for enlarging holes, filing fillets, and concave radii.

round-nose chisel: Also known as a round-nose chisel. A cold chisel with a rounded cutting edge ground back at an angle, used for roughing out small concave surfaces of filleted (rounded) corners, oil grooves, concave surfaces, and similar work.

rpm: Abbreviation for revolution per minute.

RR 53 alloys: Aluminum alloys used in die casting nominally containing 91.85% aluminum, 2.25% copper, 1.3% nickel, 1.5% magnesium, 1.5% iron, 1.5% silicon, 0.1% titanium.

Ru: The symbol for the element ruthenium.

rubber-based solvent cements: Adhesives made by combining one or more rubbers or elastomers in a solvent. These solutions are further modified with additives to improve the tack or stickiness, the degree of peel strength, flexibility, and the viscosity or body. Rubber-based adhesives are used in a wide variety of applications such as contact adhesive for plastics laminates like counter tops, cabinets, desks, and tables.

rubber-pad forming process: See Guerin process.

rules: Following is a description of various kinds of rules are used in the machine trades, including flat rules, hook rules, short (holder) rules, slide caliper rules, shrink rules, and flexible rules.

Flat rules are made of tempered steel and contain graduations cut with great accuracy. Although the most common plain flat rule is 6 in. long, they are made in a variety of lengths and the thicknesses, which can varies from 1/64 in. to 1/20 in. Some flat rules contain end graduations used for measuring the depth and width of grooves, recesses and countersinks.

The hook rule is one having a rigidly attached or adjustable sliding hook projecting from the zero division at one end. The sliding hook makes it possible to make accurate measurements when it is not convenient for the user to see the starting edge of the rule, such as through the hubs of pulleys, against shallow shoulders or round corners of a workpiece. The hook rule is made in many sizes. The narrow hook rule is made for measuring very small holes such as 3/8 in.

The short or holder rule can be used for measuring small spaces, where other rules can not fit. It can be held at any angle and is convenient for measuring keyways, recesses, and for general tool and die work. It consists of a set of small tempered steel rule of various lengths that are graduated on both sides together with a holder which is a split chuck adjusted by a knurled nut at the top of the holder.

The rules come in various lengths, i.e., ¼, ⅜, ½, ¾, and 1 in. and the inches are graduated into 32nd and 64^{th}s.

The slide caliper rule is and used to make internal and external measurements, for making quick measurements on tubing, small rods, sheet stock, etc. The jaws of the slide caliper rule contain "nibs" that can be inserted into small holes (as small as 1/8 in. diameter) and a clamp nut locks the slide for a particular measurement.

The shrink rule is a tempered steel rule used by pattern makers to make allowances for the contraction or shrinkage of castings. Since the rate of shrinkage changes with different metals each shrink rule has the shrink allowance stamped on it. The shrink rule is graduated like any ordinary rule, but is actually slightly longer. The additional length is proportionally throughout its length. *For example*, steel shrinks about 1/4 in. to the foot, so the shrink rule for iron would be 12-1/4 in. long. Shrink rules are marked "shrink 1-4" to foot," "shrink 1-8" to foot," etc. Although there are published standard tables for the shrinkage of different metals, castings will shrink (more or less) depending on thickness, the quality of the casting materials, and the process of molding and cooling.

The flexible rule can be a self-supporting steel tape rule or the entirely flexible steel tape that can be mounted in a spring-loaded case, making it retractable for ease of handling.

runner: In sand casting, a horizontal channel for carrying molten metal from sprue to the gate.

running and sliding fit: See fits, description.

running balancing: See dynamic balancing.

runout, thread: Unless otherwise specified, runout refers to circular runout of major and minor cylinders with respect to the pitch cylinder. Circular runout, in accordance with ANSI Y14.5M, controls cumulative variations of circularity and coaxiality. Runout includes variations due to eccentricity and out-of-roundness. The amount of runout is usually expressed in terms of *full indicator movement* (FIM).

Ruolz alloy: A kind of nickel silver.

Ruselite®: A proprietary trade name for a corrosion-resistant aluminum alloys nominally containing 94% aluminum, 4% copper, 2% chromium, and 2% molybdenum.

rust: A reddish-brown corrosion product (primarily ferric oxide) of iron and ferrous alloys. The reaction occurs most rapidly in a moist atmosphere.

Rust-Tap®: A proprietary trade name for an oil and rust preventative used for tapping all metals except aluminum.

rust, white: A form of steel corrosion caused by excess chloride formed during the galvanizing operation.

ruthenium: CAS number: 7440-18-8. A metallic element of the platinum group. Used as an alloying element for hardening platinum and palladium for jewelry, making corrosion-resistant alloys, electrodeposited coatings, the oxide is used to coat titanium anodes in electrolytic production of chloride; the dioxide serves as an oxidizer in photolysis of hydrogen sulfide.

Symbol: Ru
Physical state: Silvery metal
Periodic Table Group: VIII
Atomic number: 44
Atomic weight: 101.07
Valence: 3,4,5,6,8

Density (g/cc): 12.41
Melting point: 4,190°F/3,900C°
Boiling point: 4,190°F/2,310C°C
Source/ores: By-product of Ni refining; found in Free state
Oxides: RuO_2, RuO_4
Brinell hardness: 220 (cast); Vickers: 220
Crystal structure: h.c.p.

rutile: Natural titanium dioxide (TiO_2). A source of titanium. Used for steel deoxidizing, welding rod coatings.

RWMA: Abbreviation for the Resistance Welder Manufacturer Association.

S: Symbol for the element sulfur.

sacrificial protection: The planned corrosion of a metal coating for the sake of protecting the substrate metal. For example, when zinc is in contact with a less reactive metal, such as steel, a galvanic cell is created. When an electrolyte such as atmospherically contaminated moisture is present, an electric current will flow in the cell and the zinc coating is sacrificed to protect the steel. See also galvanizing.

SAE: Abbreviation for Society of Automotive Engineers. The initials of this organization are used in its tests and specifications for motor oils, fuels, and steels.

SAE steel: See AISI/SAE steel designations.

safe edge: An edge of a file that is made smooth or uncut, so that it will not injure that portion or surface of the workplace with which it may come in contact during filing.

sal ammoniac: See ammonium chloride.

salt bath: A molten mixture of chemicals that may be as high as 2,399°F/1,315°C, used to harden and temper metals and for annealing both ferrous and nonferrous metals, used to enhance fatigue strength, wear, and corrosion resistance.

salt bath tempering: A heat-treatment process involving the use of molten salt baths for drawing or tempering operations that result in the relief of internal strain in hardened steel, and thus increases toughness. Tempering temperatures range from 300°F/145°C to 1,150°F/621°C. The temperature should be increased gradually, for example, from 300°F/149°C to 400°F/204°C up to the tempering temperature, which may range from 380°F/193°C to 600°F/316°C for most carbon-steel tools, and up to 1,050°F/566°C to 1,150°F/621°C for high-speed steel requiring extreme toughness and little hardness.

sampling length: A term related to the measurement of surface texture, the nominal spacing within which a surface characteristic is determined. Roughness is measured along the smallest sampling length of the workpiece; waviness is measured within the next higher level of the sampling length on the surface.

samarium: A rare-earth metal of the lanthanide group. Used in metallurgical research, permanent magnets; as a dopant for laser crystals, neutron absorber. An active reducing agent; keep away from oxidizers.

Symbol: Sm	**Melting point:** 1,962°F/1,072C°
Physical state: Silvery white	**Boiling point:** 2,966°F/1,630°C
Periodic Table Group: IIIB	**Source/ores:** Monozite
Atomic number: 62	**Oxides:** Sm_2O_3
Atomic weight: 150.36	**Crystal structure:** Rhom. (α); cubic (β)
Valence: 2,3	

sand: Small particulate material consisting mainly of such minerals as quartz, mica and feldspar. Occurring widely in nature, sand is formed by the disintegration of rocks from weathering. Molding sands usually contain about 85% silica; between 8 and 20% alumina (clay), and the balance magnesia and other minerals. Molding sand must possess six main characteristics: porosity, plasticity, adhesiveness, cohesiveness, refractoriness, and strength when heated. Few sands possess all these qualities in the proportions so it is usual to compensate for the deficiency of a sand in any particular characteristic by mixing it with other sands or substances which possess the needed characteristic to a high degree. The size and the shape of sand grains have a large bearing upon its strength and general character. Because round grains do not interlock or overlap each other, round grain sands are weaker than those that are sharp and irregular, especially when rammed together. Grain size determines the smoothness of the mold surface, and for that reason large-grained sands are generally unsuitable when castings with very smooth skins are required.

sand blasting: A process using a stream of sand that is propelled by high pressure using compressed air flowing through a hose. Used to clean or roughen castings, stonework, and remove paint, scale or corrosion from metal surfaces, etc., without removing significant amounts of metal.

sand casting: Castings are made from patterns which are exact facsimile of the article to be produced. The patterns are pressed into sand, and when removed leave their impression. Into this sand impression, or mold, molten metal is poured and allowed to cool. When it is removed it will be the same shape as the mold, only slightly smaller owing to the contraction of the metal. Molds may be poured while moist, or they may be dried out in an oven before the metal is cast. These are known respectively as *green-sand* and *dry-sand* molds, and the sand mixtures used vary considerably.

saponification number (SAP No.): A lubricant term indicating the amount of fatty material present in an oil. The SAP No. can vary; an oil containing no fatty material will be SAP No. 0 (zero), an oil containing 100% fatty material will be SAP No. 200.

SAW: See submerged arc welding.

sawing machine: See hacksaw, power; bandsaw, circular saw, cutoff wheel.

Saybolt universal viscosity rating: One of the standards for evaluating the internal friction or stiffness of a liquid The Saybolt viscometer measures the flow rate in Saybolt universal seconds (SUS) of a standard 60 ml of sample flowing through a calibrated Universal orifice in a Saybolt viscometer under specified conditions. A similar rating, the Furol viscosity test measures the in Saybolt Furol seconds (SFS) the same sample and conditions flowing through a calibrated Furol orifice in a Saybolt viscometer. Furol viscosity is approximately 1/10 of Saybolt Universal viscosity and is used for relatively high viscosity materials and fuel oils. Typical ratings are expressed in seconds and temperature such as "Saybolt viscosity 90-50 seconds @100°F/37.8°C" for refined mineral oil. Other types of viscometers include the Engler, Redwood, Brookfield, and Krebs-Stormer. See also viscometer.

Sb: Symbol for antimony.

scale: (1) Thin, flaky leaflets; (2) an incrustation, as in the formation of a layer of oxides on metal surfaces during corrosion processes; or, the formation of insoluble salts caused by the evaporation of hard water, such as boiler scale; (3) a series of graduation markings (at regular intervals) used for measurement or computation on drawings, instruments, graphs, etc; (4) a balance instrument used for weighing.

scalper: A coarse screen used to protect fine screens, especially from abnormally large particles.

scarf: A name for the bevel edge formed on a piece of metal which is to be lap-welded.

scavenger: Also known as a "getter." A substance added to remove impurities, or to overcome undesirable effects of the one or more of the substances contained in a mixture. In metallurgy, an active metal added to a molten metal or alloy that combines with oxygen or nitrogen in the melt and triggers its removal into the slag. Vanadium is a well-known scavenger for steel, being added to remove nitrogen.

Scav-Ox®: A registered trademark of the Olin Corp., Inc. for a hydrazine oxygen scavenger used to prevent corrosion in boilers and oil-wheel casings.

scheelite: [$CaWO_4$] A natural calcium tungstate and a chief native ore of tungsten.

sclerometer: An apparatus for determining the hardness of a material by measuring the pressure on a standard point that is required to scratch the surface of the material.

scleroscope: An apparatus for determining the hardness of a material by measuring the rebound of a standard diamond pointed hammer dropped from a fixed height through a guiding glass tube. The harder the steel, the higher the hammer will rebound. The scleroscope is portable and can be used to check work that are too big to place on the anvil or other machines such as the Rockwell or brinell tester.

scleroscope test: The scleroscope test is the same as the Shore test.

scrapers: A hand tool for removing a very small amount of metal and used to correct the irregularities of a machined work surface so that the finished surface is a plane surface. Scrapers may be classed as flat, hook (left or right hand), half round, triangular or three cornered, two-handled, and bearing.

scratch gage: A gage containing a graduated rule and lockable slide used for scratching a line at a given distance from one side of a workpiece.

screw flat: See flat.

screw pitch: The distance from the center of one screw thread to the center of the next. In screws with a single thread, the pitch is the same as the lead, but not otherwise.

screw-pitch gage: Also known as a thread gage. A tool for determining the number of internal or external threads per inch or *pitch* of a screw or nut. It consists of a holder containing a series of thin steel or stainless steel blades or "leaves," somewhat like a pocket knife. The blades are notched to create a profile and are used for comparing and measuring pitch of inside threads or inside holes and nuts as well as external nuts. Each individual blade has a standard pitch thread shape cut into it, and may be obtained in all standard and special shapes of threads. Each blade is stamped on the side with the pitch per inch, and sometimes with information about the type of thread, and the decimal equivalent of the pitch.

screw thread: A continuous and projecting helical ridge usually of uniform section on a cylindrical or conical surface.

scriber: A tool with long, slender, sharp points of hardened steel used to scratch or mark lines on metal. Some scriber have a single point while others have a straight point and a point bent at 90° on the other. The bent point is used to mark lines where the straight end cannot reach, such as the inside of cylindrical objects. Used for measuring or layout work.

Se: Symbol for the element selenium.

sealed bearing: An antifriction rolling bearing fitted with a seal on one or both sides of the bearing.

seam: (1) The fold or ridge formed at the juncture of two pieces of sheet material; (2) on the surface of a metal, an unwelded crack, fold, or lap that has been closed but not welded; usually resulting from a defect resulting from casting or in working.

seaming: The process of using multiple bends to join the edges of metal sheets.

secondary hardening: A hardening effect which occurs in previously hardened high-speed and other steels that retain some austenite during the quenching process. When reheated and cooled from temperatures in the tempering range, quenching strains are relieved and any retained austenite is converted to the harder martensite. This effect is also described as the precipitation of further carbides. The resulting hardness is greater than that obtained by tempering the same steel at some lower temperature for the same period of time. See also precipitation hardening.

second-cut file: A file having approximately 40 teeth per inch; a grade of file coarseness between bastard and smooth of American pattern files and rasps.

segregation: A term used to describe the nonheterogeneous structure of and ingot caused by nonuniform distribution or concentrations of impurities, alloying elements, or microphases (such as entrapment of liquid zones) that arise during solidification.

selective quenching: Quenching only certain portions of a workpiece.

selenium: A nonmetallic element. Occurs in several allotropic forms: monoclinic (red) (*α*-Se); (*β*-Se), hexagonal [known as metallic selenium (gray)], amorphous (dark red to black powder). The latter form is the most stable. Used as a steel and copper degasifier and to improve machinability.

Symbol: Se
Physical state: See above
Periodic Table Group: VIA
Atomic number: 34
Atomic weight: 78.96
Valence: 2,4,6

Density (g/cc): 4.42 (mono); 4.84 (hex); 4.28 (amorphous)
Melting point: 291°F/144°C (mono-); 428°F/220°C (hex-)
Boiling point: $1,265\pm2$°F/685 ± 1°C
Source/ores: clausthalite, silver selinide; by-product of copper refining
Oxides: SeO_2
Crystal structure: monoclinic (α, β); hexagonal (gray); cubic (α', β')

semi-killed steel: Steel partially deoxidized by the addition of small, controlled amounts of silicon or other deoxidizers, but containing sufficient dissolved oxygen that reacts with carbon in the steel, producing enough carbon monoxide [CO] during solidification to offset shrinkage in the mold.

semi-steel: An indefinite trade name for a gray iron product near the dividing line between steel and cast iron. A high-grade, high-strength cast iron with lower carbon, better physical structure, and fewer impurities than generally found in ordinary cast iron, made by adding 20% or more mild steel scraps to the pig iron.

semisynthetic cutting fluids: See synthetic and semisynthetic cutting fluids.

Seqlene®: A registered trademark of C. P. Co. Hall for chelating agents. Used in derusting and descaling processes for metal; concrete admixtures to retard set and reduce water required; textile applications; aluminum etching; and caustic bottle washing.

serial taps: Made for progressive cutting of internal threads in very tough metals. These taps are usually made in sets of three. The top (handle) end of the shank contains identifying marks such as one, two, or three circumferential rings to identify them as part of a set. This is important because they resemble other common tap types, but differ in both major and pitch diameter. Each tap starting with No.1 (marked with a single circle) is used in succession to start-, rough-, and final-cut the thread to its correct size.

set: (1) A file term describing a process to blunt the sharp edges or corners of file blanks before and after the overcut is made, in order to prevent weakness and breakage of the teeth along such edges or corners when the file is put to use; (2) a saw term used to describe the pattern of teeth and their offset. The most generally used saw tooth set pattern is the raker set. See also raker set.

setover screws: Screws found on a lathe's tailstock, used for aligning it with the headstock.

setup point: A numerical control term use to describe the beginning and ending point of a program. Also known as "home."

sfm or sfpm or sf/min: Abbreviations for surface feet per minute. To convert surface feet per minute (sfm) to meters per second (m/s): m/s=sfm/197. To convert meters per second (m/s) to surface feet per minute (sfm): sfm=m/s x 197.

SFSA: Abbreviation for Steel Founders' Society of America.

shaft basis: The system of fits where the maximum shaft size is basic. The fundamental deviation for a shaft basis system is *h*.

shank: A drill term used to describe the part of the drill by which it is held and driven.

shaper: A planing machine in which the work remains stationary except for cross feed, the single-point cutting tool having a reciprocating motion is forced across the work by means of a rigid arm or ram moving in a horizontal direction, and cutting on the forward stroke only. There is also a special type of shaper chiefly used for heavy work known as the *draw cut machine* in which the cutting is done on the return stroke. In other words, the work is drawn in rather than thrust out. The shaper uses tools similar to those of the planer but differs from the planer in that the tool moves while the work is stationary, whereas the planer tool is stationary and the work moves.

shaping: An operation used for producing flat surfaces and has a similarity to planing, but differs from it in the following respects. Shaping is only carried out on small areas of metal, usually less than 1 ft. in length, and mostly much less. It is, in short, used for parts not bulky enough for the planing, milling machines, and machining centers.

sharp crest: Crest apex. The apex formed by the intersection of the flanks of a thread when extended, if necessary, beyond the crest.

sharp root: Root apex. The apex formed by the intersection of the adjacent flanks of adjacent threads when extended, if necessary, beyond the root.

shear: (1) The effect of external forces acting so as to cause adjacent sections of a member to slip past each other; (2) an inclination between two cutting edges; (3) a tool for cutting metal and other material by the closing motion of sharp, closely adjoining edges; (4) to cut by shearing dies or blades; (5) the ratio between a stress applied laterally to a material and the strain resulting from this force.

shearing: The process of cutting by means of blades, one of which may be immovable.

shears: Generally used to describe scissors or large cutting machines resembling scissors, used for cutting sheet metal into smaller pieces for machine operations such as punching, blanking, and forming.

shear strength: Resistance to slicing, side cutting, or lateral motion forces.

shedder: A general term for various devices used to release or eject parts, blanks, or adhering scrap from punch, die, or pad surfaces. These can be a pin, ring, rod, or plate, activated by mechanical means or compressed air.

sheet metal: A general term used to describe metal products that are thinner than plates, thicker than foil, and wider than strips.

shell end mill: An end milling tool having teeth on the circumference and the end, and a hole through its center for mounting the cutter on a stub arbor. Cutters of this type are attached to the arbor with a nut or set screw. The teeth are usually helical but can be straight. Made in sizes larger the solid shank-type end mills; normally they are available in diameters from 1-1/4 to 6 in.

shell reamer: A fluted machine (chucking) tool with straight or helical flutes and tapered hole which fits snugly on a special arbor that can hold different sizes of this type of reamer. Generally used for reaming large holes. Shell reamers perform essentially the same work as ordinary machine reamers and are often used for economic reasons; they can be changed quickly to accommodate different size holes and are usually discarded and replaced with new ones when they wear out.

sherardizing: A dry galvanizing process for covering the surface of relatively small articles made of ferrous materials with a thin, tight, protective, and corrosion-resistant iron-zinc alloy coating. The objects are coated with zinc dust or powder at a temperature of about 1472°F/800°C in a closed vessel.

Sherbrite®: A registered trademark of Aqualon Co. for brightening agents for nickel plating.

shielded-arc welding: An arc welding technique using a gaseous atmosphere is used to protect the arc and the weld metal. See also: MIG (metal inert gas) welding, CO_2 welding, carbon-arc welding.

shim: A thin piece of metal used between mating parts to adjust or improve fit.

Shore scleroscope test: Also called the Shore scale. A hardness test for metals that measures the rebound, or loss in kinetic energy, of a falling hammer with a standard diamond point called a tup which is dropped from a fixed height. The harder the metal, the higher the tup will rebound which is read directly on a dial or scaled vertical column.

shortness: A form of metallic brittleness.

shot-blasting: The process of blasting the surface of a metal with tiny pieces of angular, sharp, abrasive metal propelled by a blast of compressed air, used for cleaning metals.

shot metal: Lead alloys nominally containing arsenic, usually less than 3%.

shot peening: The process of blasting the surface of a metal with tiny, hard steel balls propelled by a blast of compressed air or thrown at the surface by centrifugal force. Used on metals to clean castings, remove scale, harden the surface layers, improve fatigue resistance, and to produce complex shapes.

shotting: A casting process for making tiny steel balls from molten metal. The molten metal is dropped from a predetermined height into a cold quenching vat of water.

shoulder or heel: See heel or shoulder, file.

shrinkage: A term used to describe the ratio of the dimension of the molding to the corresponding dimension of the mold. The mold maker uses this ratio to determine mold cavity measurements that will produce a part of the required dimensions.

shrink fit (FN): See fits, description: *force fit (FN)*.

shrink rule: See rules.

shut height: A term applied to power presses, indicates the die space when the slide is at the bottom of its stroke and the slide connection has been adjusted upward as far as possible. The shut height is the distance from the lower face of the slide, either to the top of the bed or to the top of the bolster plate, there being two methods of determining it; hence, this term should always be accompanied by a definition explaining its meaning. According to one press manufacturer, the safest plan is to define shut height as the distance from the top of the bolster to the bottom of the slide, with the stroke down and the adjustment up, because most dies are mounted on bolster plates of standard thickness, and a misunderstanding that results in providing too much die space is less serious than having insufficient die space. It is believed that the expression shut height was applied first to dies rather than to presses, the shut height of a die being the distance from the bottom of the lower section to the top of the upper section or punch, excluding the shank, and measured when the punch is in the lowest working position.

Si: Symbol for the element silicon.

side fit, involute splines: In the side fit, the mating members contact only on the sides of the teeth; major and minor diameters are clearance dimensions. The tooth sides act as drivers and centralize the mating splines.

side-milling cutters: A term describing several types of milling tools that are similar to plain milling cutters; but, with cutting teeth on the periphery of the cutter; and, depending on the type, may have teeth on one or both sides. The teeth may be straight, staggered, or helical. These cutters are used for side-milling, slotting, grooving, and *straddle milling*. There are three basic kinds of side-milling cutters in common use: *plain side-milling cutters* have straight teeth on the periphery and on both sides; *half side-milling cutters* have helical teeth on the periphery and on one face only; *staggered tooth side-milling cutters* are narrow cutters with teeth that alternate on opposite sides, thus providing space for chip removal and less dragging and scoring. The plain side and half side cutters are recommended for medium and heavy-duty face milling and straddle milling; the staggered tooth cutter is recommended for heavy-duty cutting of keyways, slots, and grooves.

siderite: Chalybite [$FeCO_3$] A native ore of iron.

sigma ($\Sigma, \varsigma, \sigma$): The eighteenth letter of the Greek alphabet. Mathematical symbol for summation (Σ). Symbol for surface tension, millisecond, electrical conductivity (σ).

silicate bond wheels: Also called semi-vitrified or water glass wheels. Bonded abrasive wheels made from natural and synthetic abrasive grains that are mixed with sodium silicate [$O_3Si\cdot2Na$], also known as water glass (the simplest form of glass), and baked in an electric furnace. The final product is hard and can be made thinner in width than fully-vitrified bond wheels. Not recommended for heavy-duty grinding, used primarily for finishing fine edges on tools and knives. See also vitrified bond wheels, bonded abrasive wheels.

silicomanganese: Alloys nominally containing 60-75% manganese, 20-25% silicon, 5-24% iron, and some carbon. Used for springs and high-strength structural steels. See also ferromanganese.

silicon: CAS number: 7440-21-3. A nonmetallic element. Exists in four varieties: Amorphous, crystalline, graphitoidal, and adamantite. Silicon is an important element in iron because it promotes the formation of graphite. Thus by changing white iron to gray iron, it makes possible the production of gray iron. In itself, silicon may not be considered an alloying element of tool steels, but it is needed as a deoxidizer, and improves the hot-forming properties of the steel. In combination with certain alloying elements, especially in combination with manganese, the silicon content is sometimes raised to about 2% to increase the strength and toughness of steels used for tools that have to sustain shock loads. Used in magnetic sheet steels, where it aids in crystallization and increases electrical resistance.

Symbol: Si	**Density (g/cc):** 2.00 (amorphous), 2.42 (crystalline)
Physical state: Gray solid	**Melting point:** 2,570°F/1,410°C
Periodic Table Group: IVA	**Boiling point:** 4,496°F/2,480°C
Atomic number: 14	**Source/ores:** Quartz and other silicates.
Atomic weight: 28.086	**Oxides:** SiO_2
Valence: 4	**Crystal structure:** cubic, diamond
	Mohs hardness: 7

silicon-aluminum bronze: Alloys containing nominally 90-91% copper, 7.0% aluminum, and 2-3% silicon, and having moderate machinability.

silicon brass: Alloys nominally containing 81-83% copper, 14-15% zinc, and 3-4% silicon. Alloy 12B contains 81% copper, 4% silicon, 15% zinc.

silicon bronze: Alloys of copper, tin, and silicon. Alloy 12A contains 87% copper, 4% silicon, 1% tin, 4% zinc, 2% iron, 1% aluminum, 1% manganese. Another alloy contains 97.3% copper, 1% tin, 1% zinc, and 0.7% silicon.

silicon carbide: [SiC] A black crystalline solid nearly as hard as diamond. Used as an abrasive, generally used for cast iron and many grinding processes and cutting off. Also known as Carborundum, and known by other trade and chemical names including Annanox® CK; Betarundum®; Carbofrax® M; Carbolon®; carbon silicide; Crystar®; Crystolon®; green densic; silicon monocarbide; silundum; tokawhisker.

silicon-copper: Also known as copper silicide. Alloys of copper nominally containing up to 10% silicon.

silicone: Polymers that may be liquids, resins, semisolids, or solids with unique elastomeric properties, used as adhesives, mold release agents, and making greases, oils, rubbers, water repellency agents, and anti-foaming agents.

silicon gold alloy: See gold silicon alloy.

siliconizing: The process of dispersing silicon into solid metal, usually steel, at an elevated temperature.

silicon steel: Steel alloys nominally containing 0.40-4.50% silicon.

silicotungstic acid: [$H_4SiW_{12}O_{40} \cdot 5H_2O$] A white crystalline solid used in electroplating.

Sil-Trode®: A registered trademark for silicon bronze electrodes and filler rod for use in inert-gas welding.

Silumin®: Aluminum-silicon alloys nominally containing 12% silicon.

silundum: Silicon carbide

silver: CAS number: 7440-22-4. A metallic element having the highest electrical and thermal conductivity of all metals. It has many uses.

Symbol: Ag	**Density (g/cc):** 10.49
Physical state: Soft metal	**Melting point:** 1,761°F/961°C
Periodic Table Group: IB	**Boiling point:** 3,542°F/1,950°C
Atomic number: 47	**Source/ores:** Argenite, pyragyrite, chloragyrite; also
Atomic weight: 107.8682	by-product of copper and other metal production
Valence: 1,2	**Oxides:** Ag_2O, Ag_2O_2
	Crystal structure: f.c.c.

silver bell metal: An alloy containing 58-60% tin and 40-42% copper.

silver brazing alloys: Alloys containing variable amounts of silver, copper, zinc and sometimes cadmium and tin, and having liquidus ranges from approximately 1,150°F/620°C to 1,550°F/845C or higher.

silver chloride: [AgCl] A white powder used in silver plating, and the production of pure silver

silver cyanide: [CagN] A white powder used in silver plating. To clean up waste cyanide salts from case hardening of steel, react the salts at 1,202-1,292°F/650-700°C with waste ferric hydroxide [$Fe(OH)_3$] sludges available from various operations.

silver glance: Another name for argenite, an ore of silver.

Silver Institute: 1112 16th St., N.W., Ste. 240, Washington, DC 20036. Telephone: 202/835-0815. FAX: 202/835-0155. WEB: http://www.silverinstitute.org

silver lead solder: Also known as B32-60T, and Alloy 1.5S. A lead alloy containing 97.5% lead, 1.5% silver, and 1% antimony.

Silver-Lume® A,B: A registered trademark of M&T Chemicals, Inc. for a bright silver electroplating process for use by silversmiths and electronics manufacturers.

silver-magnesium-nickel alloy: Alloys containing 0.22-0.28% magnesium and 0.2% nickel. Type A contains approximately 0.28% magnesium and Type B contains approximately 0.22% magnesium. Increased magnesium content increases hardness, tensile strength. These alloys are slightly harder than fine silver.

silver metal: Alloys nominally containing 33.5% silver and 66.5% zinc.

silver nitrate: [AgNO$_3$] A colorless crystalline solid used in silver electroplating.

silver potassium cyanide: [KAg(CN)$_2$] A white crystalline solid used in silver electroplating.

Silver Users Association: 1717 K St. N.W., Ste. 911, Washington, DC 20006. Telephone: 202/ 785-3050. FAX 202/659-5760.

single cut: A file tooth arrangement where the file teeth are composed of single unbroken rows of parallel teeth formed by a single series of cuts. Refers to the tooth-forming cut found on the faces and edges of various kinds of files. A single cut file has one series of rows of cuts running across the file face at an angle varying from 45 to 85 degrees with the axis of the file. This angle depends upon the form of the file and the nature of the work for which it is intended. Various kinds of hand files have single cuts on their edges. The single cut file is customarily used with a light pressure to produce a smooth finish. See also double cut.

sine: A trigonometric term defined as the ratio of perpendicular to hypotenuse.

sine bar: A tool that utilizes trigonometrical ratios in conjunction with a height gage and accurate surface plate to determine the value of an angle to a high degree of accuracy. The sine-bar consists of a hardened, ground, and lapped steel bar with two very accurate, hardened, and ground cylindrical plugs of the same (and equal) diameter attached to or fitting into a notch at each end. The edges of the bar are parallel with the line of the plug centers. The standard center-to-center distance between the plugs, measured along the bar, is either 5, 10 or 20 in. The sine bar is always used in conjunction with an accurate surface plate or master flat in order to form the base from which the vertical measurements are taken.

single-row bearing: A bearing having only one row of rolling elements.

Sinimax®: An alloy nominally containing 54% iron, 43% nickel, and 3% silicon. having high permeability at low field strength and high electrical resistance.

sintered carbide: See cemented carbide.

sintering: The coalescence by heat of powdered metals into a solid mass. The bonding of powdered metals by firing a mixture of them at a temperatures below the melting point. Sintering is used to increase strength, conductivity, and density and the major powder metallurgy operation.

siserskite: A name used to describe the alloy osmiridium or iridosmine when it contains a high content of the element osmium. See also osmiridium.

Skamex®: A registered trademark of DuPont for a fluorocarbon plastic used as a metallurgical additive to ferrous and nonferrous metals, used for the removal of excessive quantities of dissolved hydrogen or in casting molds to create protection against the effects of absorbed oxygen.

skelp: The strip of preformed metal blanks used to produced pipe or tubing. The skelp is shaped to correct size in a rolling mill and the edges are subsequently joined by welding, riveting, etc.

skull: The layer of dross or solidified metal still clinging to the wall of a crucible following the pouring of metal.

slack quenching: The incomplete hardening of steel due to quenching from the austenitizing temperature at a rate slower than the critical cooling rate for the particular steel, resulting in the formation of one or more transformation products in addition to, or instead of, martensite.

slag: By-product of metal smelting and refinement; the medium by means of which impurities may be separated from fused metals during the melting of ores.

sliding clearance fit: See fits, description.

sliding fit (RC): See fits, description.

sliding hook rule: See hook rule.

slimicide: A general name for any substance that is toxic to the types of bacteria and fungi that are characteristic of aqueous slimes that may be found in some cutting fluids. Examples are chlorine and its compounds, phenols, and related substances. See also biocide.

slip renewable bushings: See renewable bushings.

slitting saw: Also called metal-slitting cutters. A thin, saw-like plain milling cutter having fine pitch teeth on the periphery only and sides that are slightly tapered toward the center or hole (creating a dish-effect) to provide side relief and prevent binding. Used for common cutoff processes and for cutting narrow slots, these cutters are available in narrow widths from 0.020 in (0.5 mm) to 3/16 in (4.8 mm) and in diameters from 2 ½ in (63.5 mm) to 8 in (203 mm).

slotter: Also called a vertical shaper. A machine containing a vertically traveling tool. Ssed to plane vertical surfaces or cutting slots.

slotting: (1) A milling operation using a slotting attachment mounted on a milling machine spindle that produces slots, grooves, keyways, dovetails, etc. The slotting attachment converts the rotary motion to a reciprocation action; (2) a punch-press operation for cutting slots and other rectangular shapes.

Sm: Symbol for the element samarium.

SME: Society for Mining, Metallurgy, and Exploration, Inc.

smelting: A process used to separate metal from its ore or gangue by roasting in the presence of air or oxygen in contact with a fluxing lime [calcium oxide(CaO)].

smooth cut: An American pattern file and rasp cut that is smoother than second cut.

Sn: Symbol for the element tin.

snagging: A term used in casting or forging to describe the removal of casing imperfections such as flash with a grinder.

snap gage: A fixed gage with inside measuring surfaces for calipering outside diameters, lengths, thicknesses of parts. Similar to and often called a caliper gage. *See also* caliper snap gage.

snap temper: A precautionary interim stress-relieving treatment applied to high hardenability steels immediately after quenching to prevent cracking because of delay in tempering them at the prescribed higher temperature.

soaking: The process of holding a metal to a prescribed temperature for a given period of time so that transformational (internal structural) changes can be completed. Generally, the soaking period is about one hour per inch of metal thickness.

Society of Automotive Engineers (SAE): 400 Commonwealth Drive, Warrendale, PA 15086. Telephone: 724/776-4841; 800/TEAM SAE. FAX: 724/776-5760. WEB: http://www.sae.org An engineering society established for the advancement and diffusion of knowledge of the arts and sciences, standards, and engineering practices related to the design, construction, and utilization of self-propelled vehicles, their components, and related materials and equipment.

Society of Manufacturing Engineers (SME): 1 SME Drive, P.O. Box 930, Dearborn MI 48121-0930. Telephone: 313/271-1500; 800/723-4763. FAX 313/271-2861. WEB: http://www.sme.org

Society for Mining, Metallurgy, and Exploration, Inc. (SME): 8307 Shaffer Parkway, PO Box 625002, Littleton, CO 80162-5002. Telephone: 303/973-9550. FAX: 303/973-3845. WEB: http://www.smenet.org

Society of Plastics Engineers (SPE): 14 Fairfield Dr., Brookfield, CT 06804-0403. Telephone: 203/775-0471. FAX: 203/775-8490. WEB: http://www.4spe.org An engineering society devoted primarily to the application of engineering principles to the manufacture and use of plastics. Publishes a monthly journal and technical books on all aspects of plastics technology.

Society of Plastics Industry (SPI): 1801 K Street, NW. Washington DC 20006. Telephone: 202/974-5200. FAX 202/296-7005. WEB: http://www.plasticsindustry.com

Soda Sil®: A registered trademark of Kaiser Chemicals for sodium silicofluoride. Used in aluminum metal refining.

sodium: Soft, ductile, malleable, silver-white, metallic element. Violent reaction with water. Must be stored in airtight bottles, in naphtha or other liquid that does not contain water for free oxygen. A key industrial chemical.

Symbol: Na	**Density (g/cc):** 0.97
Physical state: Silver metal	**Specific heat:** 0.253
Periodic Table Group: IA	**Melting point:** 208°F/98°C
Atomic number: 11	**Boiling point:** 1616°F/880°C
Atomic weight: 22.99	**Source/ores:** rock salt (halite), cryolite, trona;
Valence: 1	seawater
	Oxides: Na_2O, Na_2O_2
	Crystal structure: hexagonal (α); b.c.c. (β)

sodium acetate: [$NaC_2H_3O_2$] Colorless crystalline solid used in electroplating.

sodium bicarbonate: [$NaHCO_3$] Also known as baking soda. A white powder used in gold and platinum electroplating.

sodium bisulfide: [NaHSO$_3$] A white crystalline powder used in brass and copper electroplating.

sodium-p-chloro-m-cresolate: A preservative used in some water-soluble cutting oils.

sodium chloroplatinate: [Na$_2$PtCl$_6$·4H$_2$O] A yellow powder used in electroplating.

sodium citrate: [C$_6$H$_5$O$_7$·3Na] White crystalline solid or powder used in electroplating.

sodium copper cyanide: [NaCu(CN)$_2$] White crystalline solid used in copper electroplating.

sodium cyanide: [NaCN] A white crystalline powder used in electroplating, agent for extraction of gold and silver from ores, case-hardening of steel. Highly toxic. See MSDS. To clean up waste cyanide salts from case hardening of steel, react the salts at 1,202-1,292°F/650-700°C with waste ferric hydroxide [Fe(OH)$_3$] sludges from various operations.

sodium fluoride: [NaF] Clear crystalline solid or white powder used in electroplating. Highly toxic. See MSDS.

sodium hydroxide: Also known as lye and caustic soda. [NaOH]. Electrolytic extraction of zinc; in electroplating, metal etching.

sodium molybdate: [MoO$_4$·2Na] White crystalline solid used as brightening agent in zinc plating and corrosion inhibitor.

sodium phosphate, monobasic: [H$_2$O$_4$P·Na] A white crystalline powder used in electroplating.

sodium pyrophosphate, acid: Also known as SAPP. [Na$_2$H$_2$P2O$_7$·6H$_2$O] A white crystalline powder used in electroplating.

sodium stannate: [Na₂Sn(OH)₆] White to tan crystalline solid used as a source of tin for electroplating and immersion plating.

sodium thiocyanate: [NaSCN] Colorless crystalline solid or white powder used in electroplating.

soft hammer: Describes hammers with heads made from copper, lead, babbitt, rubber, plastic material, or rawhide. Used on a finished surface, to drive a mandrel, to seat work in a machine vise, and in similar operations where a steel hammer might cause marks or injury to the workpiece surface. Some models have renewable faces inserted into a metal head while others have heads made entirely of metal, fiber, or other material. These are available in various types and sizes.

soft solder: Also known as eutectic solder. Alloys nominally containing 60%-70% tin and 30-50% lead. Typically used for joining metals and for electrical work.

soft soldering: The process of joining metals by employing a nonferrous metal or metallic alloy filler whose melting point is lower than the arbitrarily set temperature of 800°F/427°C.

solder: (1) Used as a verb to describe the joining of metals using a nonferrous filler metal having a melting point that is lower than the metals being joined; (2) used as a noun to describe a metal or metal alloy of relatively low melting point used in fused form (when melted) to join metallic surfaces or components. Solder is often classified as *soft* to *hard*, depending its melting point, which some experts arbitrarily set as 800°F/427°C. *Soft solder* is usually an alloy of lead and tin, with bismuth and cadmium frequently included to lower the melting point (approx. 360-415°F); or zinc, aluminum and copper (aluminum solder, with approximate melting point of 658°F). Electrician's solder is usually 60-65% tin and freezes rapidly, while plumber's solder is 30-35% tin and has a long freezing range to permit cleaning of the joint. So-called "medium" solder (actually a kind of soft solder) usually contains equal parts of lead and tin. Hard solders (often called spelter), used for brazing, may contain equal parts of copper and zinc (approx. m. p., 1580-1598°F); or silver, copper zinc and cadmium (silver brazing solder, approx. m.p.,

1148-1184°F). High-temperature brazing alloys are usually composed of 96% silver and 4% manganese, or 64% silver, 33% palladium and 3% manganese. See also solder, ultrasonic fluxless soldering.

soldering: The process of joining metals by employing a nonferrous metal or metallic alloy filler whose melting point is lower than that of the base metal and in all cases below 800°F/427°C. The filler is applied as a thin layer, in a molten state, and drawn into the space between them by capillary action, and allowed to cool. The term brazing or hard soldering is used for a similar process when the temperature exceeds an arbitrary temperature of 800°F/427°C. Soldering is used to provide a convenient joint that does not require any great mechanical strength.

solid drilling: The most common drilling method where the hole is drilled in solid material, to a predetermined diameter, in a single operation.

solidus: The highest temperature at which metal or alloy is completely solid.

solubility: The property of a substance describing the degree to which one material may be completely mixed or dissolved in another material. The degree of solubility of most substances increases with the rise in temperature; however, in the case of organic salts of calcium the substance may be more soluble in cold than in hot solvents.

soluble-oil cutting fluids: Also known as emulsifying oils. These metal-cutting and boring lubricants, when mixed with water, produce milky solutions with large emulsions or amber-colored, transparent solutions with fine emulsions.

soluble oils: Types of cutting oils that are suspensions of mineral oil droplets in water and generally contain additives of animal and vegetable oils (polar oils) and chemicals such as chlorine, sulfur, and phosphorus. Soluble oils offer certain advantages including cooling and lubricating properties, cleanliness, good operator acceptance, economy of use (since water dilution lowers cost), and improved health benefits because they are less prone to bacteria growth. The disadvantages of soluble oils include less rust control and a tendency to foam during use. See also emulsifiable mineral oil. *Note:* Do not use water-based cutting fluids on magnesium.

solution heat treatment: A treatment in which an alloy is heated to a suitable temperature and held at this temperature for a sufficient length of time to allow a desired constituent to enter into solid solution, followed by rapid cooling to hold the constituent in solution. The material is then in a supersaturated, unstable state, and may subsequently exhibit age hardening.

solvent: A chemical liquid, capable of dissolving another substance. An industrial term generally used to describe organic solvents.

spacing: A term related to the measurement of surface texture, the distance between specified points on the profile measured parallel to the nominal profile.

spade drill: A drill bit containing a replaceable, flat end-cutting blade containing two cutting lips, attached to a rigid shaft, used for drilling large holes, beyond the limits of twist drills. The normal size ranges from 1 in (25.4 mm) to 5 in (127 mm), but specialized sizes, both smaller and larger, are available.

spalling: The cracking and flaking of portions of the surface of solids resulting from internal or mechanical stresses. This can possibly happen on the surfaces of rollers and bearings.

spatter: The metal particles expelled during fusion welding that do not form a part of the weld.

SPC: Abbreviation for statistical process control.

SPE: Abbreviation for Society of Plastics Engineers.

specific gravity: (1) The ratio of the mass of a given volume of a material to the mass of the same volume of water, both measured at 73°F/23°C. The ratio is without dimension so it is useful for comparing different materials, and is used in cost estimating and quality control. Water is the standard for solids and liquids, air or hydrogen for gases; (2) the weight of a material compared to the weight of an equal volume of water is an expression of the density (or heaviness) of a material. Insoluble materials with specific gravity of less than 1.0 will float in (or on water). Insoluble materials with specific gravity greater than 1.0 will sink in water. Most (but not all) flammable liquids have specific gravity less than 1.0 and, if not soluble, will float on water, an important consideration for fire suppression.

speculum metal: Alloys nominally containing 64-66% copper, 32-34% tin and possibly 4% nickel, with trace of arsenic. Melting point: 1,382°F/750°C.

spelter: A name used to describe zinc in ingots, crude zinc, brazing brass, and sometimes to the relatively pure zinc used in galvanizing.

sphalerite: [ZnS] Natural zinc sulfide. The most important ore of zinc and a source of cadmium.

spheroidal: Shaped by a sphere.

spheroidizing: An annealing process of heating and cooling steel just below A_{c1} long enough to form cementite (iron carbide [Fe_3C]) resulting in the formation of relatively large rounded (spheroidal) or globular form of carbide in the structure of the steel. Used to improve machinability.

SPI: Abbreviation for Society of Plastics Industry.

spicular: Shaped like a needle.

spindle speed (n): The speed at which the main movement takes place and is expressed in revolutions per minute.

spiral flute taps: Also called helical flute taps. Taps having spiral flutes similar to a drill, used where chip disposal is a problem. Spiral fluted taps have improved chip drawing action and permits the bridging of gaps or slots inside the hole. They are available in high-angle and low-angle types. Hi-angle taps can be used when tapping relatively deep, blind holes in tough steel alloys. Low-angle taps are often used for machine tapping of ductile materials such as aluminum, brass, bronze, copper, magnesium, leaded steels, and die cast metals.

spiral gear: Gears in which the center line of gears is not parallel, but at an angle to each other. As the contact between the spiral gears is limited to a single point instead of a line the entire width of the gears, their capacity to carry a load is less and the wear greater.

spiral point (chip driver): A supplementary angular fluting cut in the cutting face of the land at the chamfer end. It is slightly longer than the chamfer on the tap and of the opposite hand to that of rotation.

spiral saw band: See band saw.

spiral point taps: Also known as gun taps, these solve the problem of tap breakage by pushing or shooting the chips ahead of tap. Contains concentric threads with no pitch diameter relief and two, three, or four straight flutes, depending on size.

spline: (1) A machine element consisting of integral keys (spline teeth) or keyways (spaces) equally spaced around a circle or portion thereof; (2) another name for a *feather*, a sliding key. See also feather.

splined shaft: A shaft having a series of parallel keys formed integrally with the shaft and mating with corresponding grooves cut in a hub or fitting; this arrangement is in contrast to a shaft having a series of keys or feathers fitted into slots cut into the shaft. The latter construction weakens the shaft to a considerable degree because of the slots cut into it and consequently, reduces its torque-transmitting capacity. Splined shafts are most generally used in three types of applications: (1) for coupling shafts when relatively heavy torques are to be transmitted without slippage; (2) for transmitting power to slidably-mounted or permanently-fixed gears, pulleys, and other rotating members; (3) for attaching parts that may require removal for indexing or change in angular position.

sponge: As used in the metals industry, a porous and finely divided form of metal, frequently employed as catalytic agents in chemical reactions, or pressed into metal ingots.

spot-facing: The operation of using a pilot-guided spot facing tool (spotfacer) to create a bearing surface area around the top of a hole by making it smooth, flat, and square for the head of a nut or a cap screw. If the depth to be drilled is shallow a counterbore or boring tool might be used.

spot-finishing: An ornamental metal finish created by placing a hardwood dowel (about 2 in long and ½ in diameter or less, as desired for spots) in a drill press and bringing the revolving dowel in contact with the surface which has been treated with a fine abrasive suspended in oil or valve grinding compound.

spot welding: A welding process for joining relatively thin and overlapping metal parts in which the weld cross section or fusion point is confined to a relatively small, approximately circular area. It is generally resistance welding, but may also be other processes such as one of the gas- shielded arc methods (gas-metal arc, gas-tungsten arc, etc.), or submerged arc welding (SAW).

spray: An application and distribution of small droplets of a liquid throughout a gas.

spray-mist cooling: A method of cooling an object in a fine spray of liquid.

spray quenching: Quenching in a fine spray of liquid.

springs: See compression springs, leaf springs, spiral springs, tension springs, torsion spring.

sprocket gear: A chain-driven wheel such as those found on the driving wheel of a bicycle. Sprocket gears can have regular teeth that can move in either direction or hook teeth that can run one way only.

sprue: A term used in casting to describe the vertical channel through which molten metal flows.

SPFA: Abbreviation for Steel Plate Fabricators Association.

square: See steel square.

square file: A file that is double cut on each face and may be parallel square or tapered for the last third of its length. Used for enlarging square holes and filing square corners, keyways, splines, and slots.

squareness: The degree of conformity of an axis or plane to a corresponding axis or perpendicular plane.

stabilizer: A substance that renders another substance or system more stable, or resistant to change in temperature, pH, or other conditions. For example *antioxidants* stabilize the product in which they are contained by rendering it more resistant to oxidation, because they act as negative catalysts for reaction of the other components with oxygen, or because the stabilizers undergo preferential oxidation.

stabilizing treatment: A treatment applied to stabilize the dimensions of a workpiece or the structure of a material such as (1) before finishing to final dimensions, heating a workpiece to or somewhat beyond its operating temperature and then cooling to room temperature a sufficient number of times to ensure stability of dimensions in service; (2) transforming retained austenite in those materials that retain substantial amounts when quench hardened (see cold treatment); (3) heating a solution-treated austenitic stainless steel that contains controlled amounts of titanium or niobium plus tantalum to a temperature below the solution heat-treating temperature to cause precipitation of finely divided, uniformly distributed carbides of those elements, thereby substantially reducing the amount of carbon available for the formation of chromium carbides in the grain boundaries on subsequent exposure to temperatures in the sensitizing range.

stainless iron: Alloys nominally containing 3-38% chromium, with or without traces of nickel, essentially magnetic and ferritic in character. High chromium irons are brittle after welding. Most popular composition for fabrication is 15-18% chromium, 0.1% C (max).

stainless steel: High-alloy steels contain relatively large amounts of chromium and possessing high strength and superior corrosion and oxidation resistance when compared to the carbon and conventional low-alloy steels. Most stainless steels contain at least 10% chromium and few contain more than 30% chromium or less than 50% iron. However, in the United States the stainless steel classification includes those steels containing as little as 4% chromium. The standard stainless steels can generally be divided into three groups based on their structures: austenitic, ferritic, and martensitic. Austenitic grades are nonmagnetic in the annealed condition, although some may become slightly magnetic after cold working. They can be hardened only by cold working, but not by heat treatment. They have outstanding corrosion and heat resistance with good mechanical properties over a wide temperature range. Ferritic grades are magnetic and contain 12-17% chromium but no nickel. They can be hardened to some extent by cold working, but not by heat treatment. They combine corrosion and heat resistance with moderate mechanical properties. Martensitic grades are magnetic and can be hardened by quenching and tempering. They nominally contain 12% chromium and, with few exceptions, no nickel. The martensitic grades are excellent for service in mild environments such as the atmosphere, freshwater, steam, and weak acids, but are not resistant to severely corrosive solutions.

standard: Any established measure of extent, quantity or quality of values.

standard conditions: For a solid, the allotropic form in which it most commonly occurs, at ordinary temperatures, and at one atmosphere of pressure. for a gas, a temperature of 32°F/0°C and a pressure of 760 mm of mercury (Hg).

standard fits designations: Standard fits are designated by means of the following symbols which facilitate reference to classes of fit for educational purposes. The symbols are not intended to be shown on manufacturing drawings; instead, sizes should be specified on drawings.

The letter symbols used are as follows:

> RC = Running or Sliding Clearance Fit
> LC = Locational Clearance Fit
> LT = Transition Clearance or Interference Fit
> LN = Locational Interference Fit
> FN = Force or Shrink Fit

These letter symbols are used in conjunction with numbers representing the class of fit; thus FN 4 represents a Class 4, force fit. Each of these symbols (two letters and a number) represents a complete fit for which the minimum and maximum clearance or interference and the limits of size for the mating parts. See also fits, description.

standard gold: A legally adopted alloy for coinage of gold in the United States. An alloy containing 10% Cu. Gold coins have not been made in the United States since 1933, when the country went off the gold standard.

standard (main) pressure angle, involute spline ($ö_d$): The pressure angle at the specified pitch diameter.

standing balancing: See static balancing.

standoff: A thread term used to describe the axial distance between specified reference points on external and internal taper thread members or gages, when assembled with a specified torque or under other specified conditions.

stannous: A compound containing divalent tin, as tin(II), or tin(2+).

stannum: Latin name for tin.

start: A term used in thread design to describe the entry point of one thread on a workpiece. Thread starts can be *single start*, or multi-starts, as *two start*, *three start*, etc., depending on the number of beginnings or entry points on a workpiece.

static balancing: Also known as standing balancing. Distributing the weight of rotating parts such as pulleys, shafts, or flywheels so that, when placed on knife-edge bearings, they will stand in any position. If unbalanced, the heavy side rolls downward.

stator: The stationary part of a machine (i.e., motor, turbine, dynamo, etc.) around which a rotor revolves.

steam: (1) Water vapor, especially when at temperature at or above the boiling point of water; (2) the vapor of any liquid at or above the boiling point of that liquid.

steel: An alloy of iron and carbon and other elements including chromium, silicon, phosphorus, sulfur, manganese, aluminum, vanadium, and nickel. The terms *carbon steel* or *plain carbon steel* are used to describe steel that contains no other alloying element other than carbon. See also low carbon steel, medium carbon steel, high carbon steel.

Steel Founders' Society of America (SFSA): 205 Park Avenue, Barrington IL. Telephone: 847/382-8240. FAX: 847/382-8287. Web: http:///www.sfsa.org An association of companies engaged in the manufacture of steel castings. Publishers of technical books and bulletins including the *"Steel Castings Handbook"* and the *"Journal of Steel Castings Research."*

Steel Manufacturer's Association (SMA): 1730 Rhode Island Avenue, NW, Suite 907, Washington DC. Telephone: 202/296-1515. FAX: 202/296-2506. WEB: http://www.steelnet.com

Steel Plate Fabricators Association (SPFA): 15 Spinning Wheel Road, Hinsdale, IL 60521. A non-profit association of metal plate fabricators.

steel rule: Rules made of tempered steel and containing graduations cut with great accuracy. The rules are made in various lengths and a great variety of graduations. The thickness of steel rules vary from 1/64 to 1/20 in., depending upon length.

steel square: A precision measuring instrument that consists of a thick stock or *beam* and a thin blade set exactly (or as near as possible) at 90°of one another. The stock and edges are hardened and accurately ground to insure straightness and parallelism. The stock is made thick so that it will easily stand on a flat surface, and to also create a bearing surface when it is pressed against the edge of the work. The accuracy of the angle of the blade and beam are affected by care and abuse. See also try square.

Stellite®: Proprietary trade names of Haynes International, Inc. for a family of cast alloys nominally containing 11-32% chromium, 20-68% cobalt, 4.5-14% tungsten, 1-2.5% carbon and sometimes 1-11% iron, 0.5-19% nickel, 0.5-22.5% molybdenum, 9% vanadium, 1% manganese, or 0.4-1.4% silicon. These alloys retain their harness at red heat and are highly resistant to corrosion, used for making high speed cast toolbits, valve seats, etc.

Stellite® 8, Stellite® 21, Stellite® 23, Stellite® 25, Stellite® 27, Stellite® 30, Stellite® 36, Stellite® 8a, Stellite® C: Proprietary trade names of Haynes International, Inc. for a family of cast cobalt-chromium based alloys used as brazed tips on tool shanks, as removable tool bits, as inserts in toolholders and in milling cutters.

Stellite® 31, Stellite® x40, Stellite® 31 x 40: A proprietary trade name of Haynes International, Inc. for a family of cobalt-chromium-nickel-tungsten alloys.

step drill: A multifunction or combination tool used for drilling and reaming or drilling and counterboring or drilling and countersinking, or other function in a single operation. For example, the *center drill* can accomplish drilling and countersinking at the same time; the *combination drill and reamer* makes it possible to drill and ream a hole in one operation. See also center drill and sub-land drill.

sterling silver: Alloys that must contain at least 92.5% silver, the remaining 7.5% metal is unrestricted and often unspecified, but is usually copper.

Stillson wrench: See pipe wrench.

stibium: Latin name for the element, antimony.

stibnite: The most important ore of antimony. Also known as antimonite [Sb_2S_3]. Density 4.52-4.62, Mohs hardness 2.

stiffness: A property of materials, defined as their resistance to deformation.

stp: Abbreviation for standard temperature and pressure [0°C, and 1 atmosphere (atm) pressure].

straddle milling: A milling operation using two side-milling cutters mounted on a milling arbor to simultaneously machine two parallel sides of the same workpiece. The required distance between the cutters is established with arbor spacers between them.

straight carbon steel: See carbon steel.

straight-flute drill: Also called the Farmer drill for its inventor. Used for soft metals such as brass, bronze, copper. This type of drill has straight flutes running along its body parallel to its axis, and may contain oil holes. A gun drill is a variation of the straight-flute drill, used for drilling very deep holes. See gun drill.

straightness: The degree of conformity of and axis along its length to a standard.

straight oil: See mineral oil.

straight-peen hammer: A common hammer used by machinists. The head is made of tool steel having one face that is flat, and the other is a wedge-shaped straight-peen positioned with the edge parallel to the handle. The flat face is used for driving center punches, chisels, and for various other general purposes; the wedge-shaped end is used for purposes such as hand swaging or peening. The cross-peen hammer is similar, with the edge of the wedge-shaped end positioned perpendicular to the handle.

straight set: A saw tooth set pattern having one tooth bent to the right followed by one tooth bent to the left, etc. No longer used in metalworking.

straight thread: A screw thread projecting from a cylindrical surface.

strain: The resistance to deformation of a material subjected to the application of a static (constant) or dynamic (increasing at a uniform rate) external load in one or more directions. The types of deformation produced under strain are compression, elongation, shear, torsion, and bending. Strain increases as a function of stress and at rupture it represents the maximum strength of the material, usually measured in pounds per square inch or kilograms per square centimeter. Unit strain is the amount of deformation per unit length. Compare to stress.

strain hardening: See work hardening.

Straits tin: A name for 99.895% pure tin.

strap wrench: A tool using a fabric or leather strap and a lever to create torque to revolve any workpiece on which the surface finish must be protected, used for turning cylindrical parts, plated or polished pipes, removing bezels, etc.

strength: Power to resist force; solidity or toughness; the quality of material by which they may endure the application of force without breaking or yielding.

stress: The force applied per unit area that tends to deform a body. The load may be static (constant) or dynamic (increasing at a uniform rate). In either case, it induces a strain in the material that results in rupture if the deforming force exceeds its strength. There are three kinds of stress: tensile, compressive, and shear. Unit stress is the amount of load per unit area. Compare to strain.

stress relieving: A heat-treatment process used to reduce internal residual stresses in metals by heating the object to a suitable temperature, below the phase transformation range, and holding long enough to reduce residual stresses, followed by slow cooling. This treatment is used to relieve stresses induced by casting, quenching, normalizing, machining, cold working, or welding. In general there is a slight increase in ductility and a negligible loss of strength and hardness.

stretch forming: The shaping of a metal sheet by stretching it over a preformed shape.

stretchout: A sheet metalworking term used to describe a pattern displaying the size and shape of the flat sheet required to make an object.

stripper: A tool used to remove or strip the workpiece or part from the punch.

strontium: A metallic element used in alloys. Will burn in air; contact with water causes decomposition.

Symbol: Sr	**Density (g/cc):** 2.54
State: Silvery, soft metal	**Melting point:** 1,386°F/752°C
Periodic Table Group: IIA	**Boiling point:** 2,534°F/1,390°C
Atomic Number: 38	**Ores:** Strontianite, celestite; strontianite
Atomic Weight: 87.62	**Crystal structure:** f.c.c. (α); h.c.p. (β); b.c.c. (γ)
Valence: 2	

strontium chromate: [$SrCrO_4$] A yellow crystalline solid or powder used for metal protective coatings to prevent corrosion; in aluminum flake coatings, to control sulfate concentration of solution in electroplating baths.

stub pitch, involute spline (P_s): is a number used to denote the radial distance from the pitch circle to the major circle of the external spline and from the pitch circle to the minor circle of the internal spline. The stub pitch for splines in this standard is twice the diametral pitch.

stud: Also called a stud bolt. A headless bolt that is threaded at both ends. One threaded end may be permanently installed in a fixture to receive a removable part, such as a cover and a fastening device such as a nut is affixed to the other threaded end. Other kinds of studs used in metal working include (a) those used to strengthen welded or brazed joints; (b) those used in foundry work to support cores during pouring operations; and (c) those used for carrying gears, rocker levers (collar stud), etc.

sub-land drill: A kind of *step drill* used to drill several diameters in a single operation and are available with four flutes.

submerged arc welding (SAW): A widely-used arc welding process in which the arc and filler metal are both covered with a mound of sand-like flux. Both the flux and filler wire are automatically fed from feed tubes, resulting in a uniform weld with little or no spark, smoke, or spatter.

substrate: Any solid surface on which a coating or layer of a dissimilar material is deposited.

sulfamic acid: [H_3NO_3S] A white crystalline solid used in metal cleaning and electroplating.

sulfur: CAS number: 7704-34-9. Nonmetallic element. Exists in five solid allotropic forms: Amorphous (soft), amorphous (yellow), α- (rhombic), β- (monoclinic), and plastic. In cast iron sulfur makes it harder, tending to produce unsound castings. In plain carbon steels it is usually held to a maximum of 0.05% because excess amounts increases *red shortness*, increasing brittleness at high temperatures, and interfering with shaping and forging. However, carefully controlled amounts of sulfur in amounts from 0.08% to 0.33% may be added to certain grades of steel to improve machinability. These steels are called resulfurized carbon steels and they may be heat treated. This element makes cast iron harder and more brittle, tending to produce unsound castings. Powders and finely divided forms are a fire and explosion risk.

Symbol: S
Physical state: Yellow solid
Periodic Table Group: VIA
Atomic number: 16
Atomic weight: 32.066
Valence: 2,4,6

Density (g/cc): 2.046
Melting point: 240°F/115°C
Boiling point: 833°F/445°C
Source/ores: Native sulfur; from pyrites and other sulfide ores
Oxides: SO_2, SO_3, S_2O_3, S_2O_7
Crystal structure: monoclinic (β; γ-S_8) rhombohedral (ϵ-S_6), and orthorhombic (α-S_8)

sulfuric acid: [H_2SO_4] A clear, colorless, oily liquid when pure, and brownish when impure. It has many uses in industry including for etching metals; in electroplating baths. Highly corrosive. Wear protective equipment and see MSDS.

Sulfur Institute: 1140 Connecticut Ave., NW, Washington DC 20036.

sulfurized and chlorinated mineral oil: A type of mineral oil used for cutting tough metals and for severe operations that put a heavy strain on machine tools. Sulfurized oils, with or without chlorine, are recommended for tapping., broaching, drilling, reaming, threading turning, and for milling carbon steels, malleable iron, wrought iron, stainless steels, tool steel, and high-speed steels.

superabrasives: Ultrahard, ultra-wear-resistant materials such as natural diamond, and polycrystalline diamond (PCD), cubic boron nitride (CBN), polycrystalline cubic boron nitride (PCBN), etc., used for making abrasives and cutting tool inserts. The man-made materials are compressed at pressures up to 10^6 psi at high temperature in electric furnaces, making them as hard or nearly as hard as diamond, with excellent heat-shock resistance.

superalloy: Alloys developed for combined very high-temperature mechanical properties, service where relatively high stresses (tensile, thermal, vibratory, and shock) are encountered, creep resistance to an unusual degree, and where oxidation resistance is frequently required. These alloys are often iron-, cobalt-, or nickel-based and may contain be composed as follows : Cobalt-base: 0-26% nickel, 0-26% chromium, 0-15% tungsten, balance cobalt. Iron-base: 10-45% nickel, 13-19% chromium, 1.3-6% molybdenum, balance iron. Nickel-base: 55-75% nickel,

10-20% chromium, 0-6% aluminum, 0-5% titanium. All contain less than 0.5% carbon, plus other special ingredients. Superalloys can be used up to 2,500°F/1,371°C. Used in extreme high-temperature applications.

super-austenitic steel (25Ni/20Cr/6.5Mo): See WN 1.4529.

super bronze: Alloys nominally containing 57% copper, 21-37% zinc, 3-3.2% manganese, 1.3-2% iron, and 1.2-5.1% aluminum.

superficial Rockwell hardness test: A less harmful form of the Rockwell hardness test used for determining the surface hardness of delicate small parts or thin sections. This test involves the use of relatively light loads which produce minimum damage or penetration. See also Rockwell hardness test.

Superfine®: A registered trademark of Stauffer Chemical for a flour sulfur used for casting magnesium and aluminum.

Superglue®: See cyanoacrylate adhesives.

superheat: A casting term used during pouring of a metal to describe the degrees of heat above the melting point.

Superlume®: A registered trademark for a superleveling, bright-nickel electroplating process on steel stampings, brass, copper, zinc die castings, etc. The materials used are nickel sulfate, nickel chloride, boric acid, and addition agents.

Supermalloy®: An alloy nominally containing 79% nickel, 18% iron, and 3% molybdenum, having high permeability at low field strength and high electrical resistance.

supplementary angle: Any two angles whose sum is 180°, whether they are adjacent or not, are called supplementary.

surface: (1) The boundary which separates one object from another object, substance or space; (2) Mathematically, a two-dimensional entity, a figure having length and width but no thickness.

surface gage: An adjustable type gage used for many purposes including testing the accuracy or parallelism of planed surfaces, and for gaging the height between a flat surface and some point on the work. The gage can also be used for the scribing of lines at a given height from some or all faces of the work. It consists of a heavy, flat, steel base with a vertical post containing a pivoted scriber, the latter being pinched by a screw and adjustable at different angles and heights. It has a V-groove at the bottom of the base and at one end for use with round work.

surface grinding: The process of producing and finishing flat surfaces by means of a grinding machine that uses a revolving abrasive wheel as the cutting tool.

surface plate: A large flat surface of iron or granite used for layout of a workpiece.

surface speed: The rotation speed of a workpiece measured at its circumference.

surface texture: The repetitive or random deviations from the nominal surface that form the three-dimensional topography of the surface. Surface texture includes roughness, waviness, lay, and flaws.

Sur-Gard®: A registered trademark of Nalco Chemical co. for chemicals used to remove oxygen from boiler water and to inhibit scale and corrosion.

swaging: (1) Changing the shape of a piece of metal by rolling, hammering, bending, or otherwise forcing a change in shape without cutting; (2) a hand- or machine-process used to stretch, spread, or compress the diameter of a rod, tube, or wire, reduce wall thickness, or form tapers, points, and other shapes. The work, supported on a mandrel, sustains many successive and rapid blows by a pair of dies, special hammers, or swage blocks of desired shape. See also peening.

sweating: Another name for soldering.

swing: A term used to denote the maximum diameter of work that can be machined in the lathe. The rated swing of a lathe is referred to the lathe bed and not the tool carriage. The actual swing is the radial distance from the center axis to the bed.

Swiss-type file: Also called Swiss pattern file. See dead smooth file.

synergistic effect, steel: The addition of more than one element to a steel often produces what is called a synergistic effect. Thus, the combined effects of two or more alloy elements may be greater than the sum of the individual effects of each element.

synthetic and semisynthetic cutting fluids: Synthetics are solutions compounded from chemicals such as sulfur, chlorine, and phosphorus rather than oils. Semisynthetic fluids contain small amounts of mineral oil as well as chemical additives. Advantages include lighter film residues and good workpiece visibility, better tank life, better mixtures than soluble oils and less tendency to foam than soluble oils. Disadvantages include less rust control, irritation of the operator's skin, and inferior lubricating properties. *See also* additives, cutting oils.

system, stable: A system that can undergo considerable variation in external conditions, such as temperature, pressure, etc., without fundamental change.

T

T: The abbreviation for absolute temperature. The symbol for critical temperature (T_c).

Ta: Symbol for the element tantalum.

tachometer: An instrument used to measure angular velocity, as of a shaft, either by registering directly the number of rotations during a specified period of time, as revolutions per minute.

tack welds: Small welds made to hold parts of a weldment in proper alignment until the final welding is finished.

tailstock: Also referred to as dead-center. A movable fixture opposite the headstock on a lathe. The tailstock is movable and can be clamped along the bed of the lathe. It contains a spindle that does not turn, but is adjustable (it can be moved in and out) and is used to support one end of a workpiece. The tapered hole of the tailstock also accepts tools for drilling operations including drill chucks, tapered shank drills, and reamers.

tailstock setover: The amount that the tailstock or dead-center on a lathe is offset from the headstock during the turning of tapers.

tall oil: Also known as liquid rosin. Used in cutting oils, lubricants, and greases. Obtained from the treatment of the spent liquor when pine wood is made into pulp. Combustible.

tang: (1) A file term use to describe the narrow, tapered, or pointed end of a file that fits into a wooden handle. While the blade is hard and brittle, the tang is tempered to be soft and tough as it would otherwise be easily broken where the handle meets the blade; (2) a drill term used to describe the flattened end of a taper shank, intended to fit into a drill press spindle or in a drill sleeve.

tang drive: A drill term used to describe two opposite parallel driving flats on the end of a straight shank.

tangential hook angle: A tap term used to describe the angle between a line tangent to a hook cutting face and the cutting edge and a radial line to the same point.

tangent plane, gear: A plane tangent to the tooth surfaces at a point or line of contact.

tannic acid: [$C_{76}H_{52}O_{46}$] A yellowish-white to light brown powder, flakes, or spongy masses used in electroplating.

tantalum: CAS number: 7440-25-7. A rare metal of the vanadium family. Corrosion resistant. A substitute for platinum. Used in high-speed tools, electrical capacitors, getter alloys in electronic tubes, catalyst for synthetic diamonds. Surgical instruments and body implants; may be heat-sterilized without losing hardness.

Symbol: Ta	**Density (g/cc):** 16.6
Physical state: Gray solid	**Specific heat:** 0.033
Periodic Table Group: VB	**Melting point:** 5,425°F/2,996°C
Atomic number: 73	**Boiling point:** 9,797°F/5,425°C
Atomic weight: 180.9479	**Source/ores:** Columbite, tantalite, yttrotantalite,
Valence: 2,3,5	fergusonite; also a by-product of tin production
	Oxides: TaO_2, $Ta_2 O_5$
	Crystal structure: b.c.c.

tantalite: The most important ore of tantalum.

tantalum carbide: [TaC] a hard, heavy, crystalline solid. Used in cutting tools, especially cemented carbide tools, and dies.

tantiron: Acid resistant alloy containing 85% iron, 13.5% silicon, 1% carbon, 0.4% manganese, 0.18% phosphorus, 0.05% sulfur. Another alloy is 84% iron, 15% silicon, and 1% carbon.

Tantung®: A proprietary name for a nonferrous cast alloy used as brazed tips on tool shanks, as removable tool bits, as inserts in toolholders and milling cutters.

tap: A tool for cutting internal threads in metal. A short length of cylindrical, hardened, and tempered tool steel with a straight or slightly tapered thread at one end and a square shank on the other. The threaded portion has flutes cut in it parallel with the axis, which form the cutting edges and allow space for the shavings. Taps are obtainable in three forms: the taper tap, the plug tap, and the bottoming tap. Taps are used for cutting internal or female threads. See also taper tap.

Tap Aid®: A proprietary trade name of Doyle Specialties, Inc for an oil made from 1,1,1-trichloroethane, used for tapping small holes, grinding, and wire drawing. ASTM S-215 oil.

tap and drill gage: A gage consisting of plate containing a series of holes of diminishing diameters and a series of numbers and designed to enable the user to easily select the correct sized drill to suit machine screw taps most commonly used, which will leave the right amount of stock (just enough) to allow the tap to cut as near a full thread as is practicable without breaking it. The first row of figures of a tap and drill gage may contain the numbers 2-56-50-44-228. the number 2 is the number of the tap; 56, the pitch of the thread; 50, the size of the drill to use which will leave just enough stock for the proper thread, and the number 44 is the size of the drill to use to let the tap, screw, or bolt through freely. The figure 228 designates the size of the hole in thousandths of inches.

tap extractor: Also called a tap remover. A tool designed to dislodge and remove broken and tightly jammed taps. It contains a handle at one end of a shaft containing slender steel fingers or prongs at the other end. These fingers fit into the flutes and around the center section of the tap. Once penetrating oil has been applied, a twisting force can be applied to the handle to remove the broken tap. For a right-hand tap, the extractor is gently turned counterclockwise.

tap drill: A drill used to make a hole prior to tapping.

taper gage: A gage consisting of a strip of tool steel containing a graduated taper for measuring *tapered thickness* used for gaging slots and bearing work and *tapered width* used as a tubing gage, for gaging the width of slots and sizes of holes in nuts drilled for tapping. It can also be used for setting calipers.

taper-pin: Standard dowel pins 3/4 in. to 6 in. long, manufactured with a taper of 1/4 in. to the foot, made to fit into reamed holes. See also taper-pin reamer.

taper-pin reamer: A hand reaming tool used to ream predrilled holes for standard taper pins. These reamers are available in a range of 17 standard sizes which are designated by numbers, with straight- and left-hand helical flutes, and in various steels. Each size overlaps the next smaller size by about ½ in.

taper plug gage: See plug gage.

taper tap: A tap having 7 to 10 threads at the front end of the land. These threads are tapered to allow for an easier start and to distribute the cutting action over several teeth while the internal threads are being cut. Also, the taper makes it easier to keep the tap straight as the cut is started. See also chamfer tap.

taper thread: A screw thread projecting from a conical surface.

tapping: An operation using a manual tap or tapping attachment in a drill press to cut internal threads in a previously drilled hole having a diameter smaller than that of the tap.

tapping attachment: A drill press accessory containing a collet chuck in which only spiral ground and helical-fluted taps can be mounted for power tapping a predrilled hole. Tapping attachments have various features to minimize tap breakage and enhance operation. Drill presses equipped with *reversing spindles* can utilize a reversing tap driver, allowing the tap to be extracted. Drill presses not equipped with reversing spindles must use a clutch-activated *nonreversing spindle* that has a right-hand rotation when pressure is applied downward on the feed handle, stops rotating (although the spindle is rotating) when pressure is released, and extracts the tap from the hole by rotating in a left-hand direction when upward pressure is exerted on the drillpress feed handle.

tap wrench: Sometimes called a T-handle wrench or a T-handle tap wrench. A hand tool used for holding and turning small taps and hand reamers. The T-handle tap wrench has two jaws inserted in an adjustable chuck and fits the square end of the tap. The handle with tap attached can be turned gradually into the hole and cuts threads of an interior surface. Available in various sizes, each T-handle tap wrench can hold several size taps. See also adjustable tap wrench.

tarnish: A reaction product that occurs readily at room temperature between metallic silver atmospheric sulfur or sulfur dioxide, forming silver sulfide. Gold can also tarnish in the presence of a high concentration of environmental sulfur.

tartaric acid: Dihydroxysuccinic acid. A chemical used to color metals.

tau (T or τ): The nineteenth letter of the Greek alphabet. Symbol for time, transmittance, and unit vector tangent to path (τ).

TD: Abbreviation for tool diameter in inches.

TDA: Abbreviation for Titanium Development Association.

Te: Symbol for the element tellurium.

telescoping gage: A T-shaped gage used for measuring inside diameters and the widths of slots and grooves. The gage is equipped with a plunger located at 90° to the knurled handle. The tool is inserted into a part with the plunger retracted; the plunger is released, the knurled nut on the handle is tightened to lock the plunge, and the gage is removed from the part. The distance across the top of the "T" which includes the plunger are measured with a micrometer caliper.

tellurium: CAS number: 13494-80-9. A nonmetallic element with metallic characteristics. Silvery -white in bulk; dark gray semi-metal. An alloying element. Used in alloys including stainless steel, iron, lead, and stainless steel castings; in thermoelectric devices. See also Selenium-Tellurium Development Association.

Symbol: Te
Physical state: Gray powder
Periodic Table Group: VIA
Atomic number: 52
Atomic weight: 127.60
Valence: 2,4,6

Density (g/cc): 6.00 (amorphous); 6.25 (crystalline)
Melting point: 842°F/450°C
Boiling point: 1,814±6.8°F/990±4.8°C
Source/ores: Sylvanite, tetradymite, tellurite; by-product of copper refining.
Oxides: TeO, TeO_2, TeO_3
Mohs hardness: 2.3
Crystal structure: hexagonal

tellurium lead: An acid-resistant alloy nominally containing 0.05% tellurium.

Tempaloy®: A proprietary trade name for alloys nominally containing approximately 95% copper, 4% nickel and, 1% silicon.

temper: (1) To increase the hardness and strength of a metal by quenching or heat treatment. In tools steels, temper is sometimes used, but inadvisably, to denote the carbon content.

temperature color scale: The relationship between the temperature of an incandescent substance and the color of light emitted. Thus, when carbon steels are heated, a very faint yellow glow becomes perceptible at 420°F/215°C, or slightly higher, becoming red at around 1,095°F/590°C, and reaching white heat at about 2,250°F/1,230°C.

temper brittleness: Brittleness that results when certain low-alloy steels are held within, or are cooled slowly, through a certain range of temperatures below the transformation range. The brittleness is revealed by notched-bar impact tests at or below room temperature.

temper carbon: The free or graphitic carbon that is precipitated out of solution from iron-carbon alloys usually in the form of rounded nodules in the structure during graphitizing or malleablizing.

tempering: Also called drawing. (1) In general, the process of rendering a material more suitable for its purpose; (2) a process of heating or reheating a metal, followed by cooling, at predetermined rates and temperatures (but in no case above the lower critical temperature) to remove internal stresses, reduce brittleness, and obtain a desired degree of toughness and hardness. The tempering process consists in heating the steel by various means to a certain temperature followed by quenching in air, water, or oil. When steel is in a fully hardened condition, its structure consists largely of martensite. On reheating to a temperature of from about 300 to 750°F/149 to 399°C, a softer and tougher structure known as troostite is formed. If the steel is reheated to a temperature of from 750-1,290°F/399-699°C a structure known as sorbite is formed that has somewhat less strength than troostite but much greater ductility; (3) in forging, mixing the green sand for molding with water so that it will pack properly.

tempering oil: A heavy oil used for cooling metals during tempering. According to *Machinery's Handbook* the following tempering oil specifications delivers satisfactory results: 94% mineral oil, 6% saponifiable oil with a specific gravity= 0.920; flash point = 550°F/288°C; fire test = 625°F/329°C.

template: An outline used for laying out the shape of an article to be constructed or for replication of shapes such as arcs, holes, and other features. Usually made from sheet metal, plastic or other material; used as a pattern.

tensile strength: Also known as *ultimate strength*. (1) The maximum longitudinal stress in tension, measured in pounds or kilograms, that a cross-sectional area of 1 in^2 (645.2 mm^2) of a material can withstand before rupturing or tearing apart; (2) In tensile testing, the ratio of maximum load to specimen cross-sectional area (usually less than one square inch) before the application of the strain; (3) the cohesive power by which a material resists an attempt to pull it apart in the direction of its fibers, this bearing no relation to its capacity to resist compression.

tensile stress area: An arbitrarily selected area for computing the tensile strength of an externally threaded fastener so that the fastener strength is consistent with the basic material strength of the fastener. It is typically defined as a function of pitch diameter and/or minor diameter to calculate a circular cross section of the fastener correcting for the notch and helix effects of the threads.

tereplate: A dull finish alloy of lead (approx. 75%) and tin (approx 25%) used for coating steel or iron used for stamped metal products.

Termalloy®: A proprietary trade name for a nickel alloy nominally containing 67% nickel, 30% copper, and 2% iron and some carbon, manganese, and tungsten.

Termalloy® A: A proprietary trade name for a nickel alloy nominally containing about 68% nickel, 30% copper, and 0.15% carbon, and 0.15% silicon.

Termalloy® B: A proprietary trade name for a nickel alloy nominally containing about 58% nickel, 40% copper, and 0.15% carbon, and 0.15% silicon.

terne metal: Alloys nominally containing 80% lead, 18% tin, and 2% antimony.

Terneplate: A lead-tin alloy composed of 75% lead and 25% tin, used as a stamping metal and for coating iron or steel for roofing.

Th: Symbol for the element thorium.

thallium: CAS number: 7440-28-0. Siver-gray, soft metal. Used in mercury and fusible alloys; in low temperature switches. An alloying element; lowers the freezing point of certain metals. Highly toxic.

Symbol: Tl	**Density (g/cc):** 11.85
Physical state: Gray metal	**Specific heat:** 0.0326
Periodic Table Group: IIIA	**Melting point:** 577°F/303°C
Atomic number: 81	**Boiling point:** 2,655°F/1,457°C
Atomic weight: 204.383	**Source/ores:** In certain iron and copper pyrites,
Valence: 1,3	crooksite, lorandite, potash, feldspar, pollucite;
	by-product of lead and zinc smelting.
	Oxides: Tl_2O, Tl_2O_3
	Crystal structure: hexagonal (α); cubic (β)

thermal: Of, or pertaiing to, heat, as thermal conductivity, thermal capacity.

thermal conductivity: The ability to conduct heat in an unequally heated system from a region of high to one of lower temperature. The coefficient of thermal conductivity is the time rate of heat conduction per unit area per unit temperature gradient.

thermal damage: Damage resulting from heat.

Thermalloy: See Termalloy

thermal spraying: A group of processes for resurfacing metal parts, involving the simultaneous melting and accelerating of an atomized spray of particles, usually a metal or ceramic, onto the surface of the part to be coated, forming a solid coating.

thermal stability: Sometimes termed "thermostable." Neither destroyed nor altered by moderate heating.

Thermenol®: An proprietary trade name for an alloy nominally containing 81% iron, 16% aluminum, and 3% molybdenum.

Thermit®: A proprietary trade name for a mixture of ferric- or manganese-oxide and powdered aluminum, used for welding. Also, an ingredient in thermite (incendiary bombs). Thermite is very dangerous because they provide their own oxygen supply; and, once started, are extremely difficult to stop.

thermit: A bearing metal containing 78-79% lead, 19-20% antimony, and small quantities of copper, nickel, and tin.

thermodynamics: That branch of physics which deals with the relationships between heat and other forms of energy and laws that govern their interconversion.

theta (Θ, θ): The eighth letter of the Greek alphabet. Symbol for time, thermodynamic temperature, plane angle, glancing angle, angle of contact, angle of diffraction, etc.

thickness gage: See feeler gage.

thorium: CAS number: 7440-29-1. A radioactive metallic alloying element. The powder form is flammable and explosive.

Symbol: Th	**Density (g/cc):** 11.3
Physical state: Silvery metal	**Melting point:** 3,182°F/1,750°C
Periodic Table Group: IIIB	**Boiling point:** 7,952°F/4,400°C
Atomic number: 90	**Source/ores:** Monazite, thorite, thotianite
Atomic weight: 232.0381	**Oxides:** ThO_2
Valence: 4	**Crystal structure:** f.c.c. (α); b.c.c. (β)

thorite: A natural thorium silicate ($ThSiO_4$) and a source of thorium.

thread: A thread is a portion of a screw thread encompassed by one pitch. On a single-start thread it is equal to one turn. See also threads per inch and turns per inch.

thread axis: See axis if thread.

thread form: Defines the basic thread shape (V, square, buttress, etc.) and specification of radii or flats at he crest or root of a thread.

thread gage: A tool containing a number of blades secured in a holder, each blade is accurately shaped to the profile of a standard thread form (having same number of notches per inch) as the thread it represents. Made for different kinds of threads in various forms. Each blade is stamped with the number of threads per inch that it represents and also, in some makes, with the double depth of thread expressed in thousandths of an inch. The diameter of the screw has no connection with the application of the gage. See also screw thread gage.

threading die: Carbon or high-speed steel tools used to cut external or male threads on round stock or bolts. Designed with female or internal threads similar to a nut, they can be solid or adjustable. There are a great number of die types available to meet varied conditions.

thread relief: The clearance produced by removal of metal from behind the cutting edge. When the thread angle is relieved from the heel to cutting edge, the tap is said to have eccentric relief. If relieved from heel for only a portion of land width, the tap is said to have co-eccentric relief.

thread runout: See vanish thread.

thread series: Groups of diameter/pitch combinations distinguished from each other by the number of threads per inch applied to specific diameters.

thread shear area: The total ridge cross-sectional area intersected by a specified cylinder with diameter and length equal to the mating thread engagement. Usually the cylinder diameter for external thread shearing is the minor diameter of the internal thread and for internal thread shearing it is the major diameter of the external thread.

threads per inch: The reciprocal of the axial pitch in inches. The number of threads in one inch of length parallel to the centerline of the workpiece.

three corner file: Contains faces at an angle of 60° to each other and tapers to a blunt point.

three-square parallel file: A triangular file that does not taper. Used for finishing square corners and filing internal angles.

three-square tapered file: A triangular file that tapers. Used for finishing corners and internal angles less than 90°.

three-wire system: A system used to measure the pitch diameter of threads. On one side of a thread, two wires are laid in adjacent grooves, and one wire is laid in a groove on the opposite side and measured with an outside micrometer over the three wires. The three wires have the same diameter within 0.00003 in. and a common diameter equal to 0.57735 divided by the number of threads per inch. However, the best-size wires can be found on readily available tables.

thrust bearing: A rolling bearing designed to support primarily axial loads, having a nominal contact angle of greater than 45° up to and including 90°. Principal parts are shaft washer, housing washer, and rolling element with or without a cage. Thin, disklike thrust bearings are called thrust washers.

Ti: Symbol for the element titanium.

TIG (tungsten inert gas welding): Also known as gas tungsten arc welding (GTAW). A welding technique using a non-consumable tungsten electrode and a shield of inert gas containing argon or helium. When used for steel, carbon dioxide gas may replace the argon or helium shield.

time quenching: Interrupted quenching in which the duration of holding in the quenching medium is controlled.

tin: CAS number: 7440-31-5. An alloying element. Used for tin foils, plating and corrosion-resistant coatings, anodes, electro-tinning, solders, dental alloys, manufacture of tin salts.

Symbol: Sn (from stannum) **Density (g/cc):** 5.75 (á); 7.31(â) [20°C]
Physical state: silver metal **Melting point:** 449°F/232°C

Periodic Table Group: IVA
Atomic number: 50
Atomic weight: 118.710
Valence: 2,4

Boiling point: 4,118°F/2,270°C
Source/ores: cassiterite
Oxides: SnO, SnO$_2$
Crystal structure: cubic (α-gray)

tin babbitt: Also known as tin-base babbitt alloys. Alloys nominally containing 65-90% tin, 4.5-15% antimony, 2-8% copper and sometimes 10-15% lead. Alloy 1 (ASTM B23-49) contains 91% tin, 4.5% antimony, and 5.5% copper. Alloy 2 (ASTM B23-49) contains 89% tin, 7.5% antimony, and 3.5% copper. Alloy 3 (ASTM B23-49) contains 84% tin, 8% antimony, and 8% copper. Alloy 4 (ASTM B23-49) contains 75% tin, 12% antimony, 10% lead, and 3% copper. Alloy 5 (ASTM B23-49) contains 65% tin, 18% lead, 15% antimony, and 2% copper.

tin bronze: Alloys nominally containing 88-89% copper, 8-11-12% tin, 4% zinc, and sometimes 0.35% phosphorus, and having moderate machinability. Tin bronze (1A) contains 88% copper, 10% tin, and 2% zinc. Tin bronze (1B) contains 88% copper, 8% tin, and 4% zinc. Tin bronze (SAE 65) contains 89% copper, 11% tin.

tin bronze, high-leaded: Alloys nominally containing 70-80% copper, 7-10% tin, 10-15% lead, 0.75% nickel, and having good machinability. High leaded tin bronze (3B) contains 83% copper, 7% tin, 3% lead, and 3% zinc.

tin bronze, leaded: Alloys nominally containing 87-88% copper, 6-8% tin, 1-2% lead, 4% zinc, and having moderate machinability. Leaded tin bronze (2C) contains 87% copper, 10% tin, 1% lead, and 2% zinc.

tin foil: Either pure tin or the more commonly used alloys containing 92% tin and 8% zinc; or, 91.85% tin, 8% zinc, and 0.15% nickel; or, 90% tin, 5% lead, 3% antimony, 2% copper.

tinplate: The process of utilizing the remarkable corrosion-resistance of tin by surface coating steel, iron, or other metal with a layer of tin. The base metal is covered by electroplating, dipped in the molten metal, or by immersion in solutions containing components that deposit tin by chemical action. This term is also applied to electroplated tin-zinc, a sacrificial coating on steel and tin-nickel plate which is virtually tarnish-proof in the atmosphere.

tin-silver solder: Alloys nominally containing 95% tin and 5% silver. Used for coating and joining metals; electrical soldering, high temperature service . See also antimonial tin solder.

tinstone: See cassiterite.

tin sulfate: [$O_4S \cdot Sn$] White to yellowish crystalline solid used in tin electroplating.

tin-zinc alloy plate: Alloys of up to 50-50% tin-zinc may be deposited from an alloy anode; however, the plate is normally 75-80% tin and 20-25% zinc.

Tin Research Institute (TRI): 1353 Perry St., Columbus, OH 43201. Telephone: 614/424-6200. FAX: 614/424-6924.

tip relief, gear: An arbitrary modification of a tooth profile whereby a small amount of material is removed near the tip of the gear tooth.

titanium: CAS number: 7440-32-6. A hard, lustrous, silver metal. An alloying element with copper, bronze, iron and other metals. As ferrotitanium, a good cleaner and deoxidizer in steel; prevents grain growth at high temperatures in stainless steels; reduces hardenability in medium chromium steels. Used in alloys, structural materials, surgical instruments, orthopedic appliances, cermets; metal-ceramic brazing, in nickel-cadmium batteries for space vehicles, coatings on metals.

Symbol: Ti	**Density (g/cc):** 4.5
Physical state: Silver metal.	**Melting point:** 3,020°F/1,660°C
Periodic Table Group: IVB	**Boiling point:** 5,949°F/3,287°C
Atomic number: 22	**Source/ores:** Ilmenite, rutile, titanite
Atomic weight: 47.88	**Oxides:** TiO_2, TiO_3, Ti_2O_3, TiO
Valence: 2,3,4	**Crystal structure:** h.c.p. (α-); b.c.c. (β)

titanium boride: [TiB$_2$] Used as metallurgical additive; a cermet component, coatings resistant to attack by molten metals, aluminum manufacture; making super alloys.

titanium carbide: [TiC] As an additive to tungsten carbide, this extremely hard material is used for making cutting tools and other parts subjected to thermal shock, important industrial cermets, arc-welding electrodes, etc.

Titanium Development Association (TDA): 4141 Arapahoe Ave., Ste. 100, Boulder, CO 80303; Telephone: 303/443-7515; FAX: 303/443-4406.

titanium disulfide: A chemical used as a solid lubricant.

titanium nitride: [TiN]. Used in cermets and alloys.

Titanox®: A registered trademark of Velsicol Chemical Corp., Inc. for a series of white pigments comprising titanium dioxide in both anatase and rutile crystalline forms and also extended with calcium sulfate (titanium dioxide-calcium pigment). Noncombustible. Used in welding rod coatings.

Tl: Symbol for the element thallium.

Tobin bronze: A special bronze alloy nominally containing from 59 to 63% of copper, from 0.5-1.5% tin, about 40% zinc and possibly small amounts of iron. It has a specific gravity of 8.4, the weight per cubic inch being 0.304 pound. Tensile strength is about 60,000 lb/in^2. Its tensile strength and its resistance to the corrosive action of sea water, makes it ideal for use in marine engineering. See also delta metal.

tolerance: (1) The total amount by which a specific dimension or surfaces of machine parts is permitted to vary. The tolerance is the difference between the maximum (upper) and minimum (lower) limits of and specified dimension. See also limit; (2) the maximum error allowable in the graduation of instruments.

tolerance class, metric: The combination of a tolerance position with a tolerance grade. It specifies the allowance (fundamental deviation), pitch diameter tolerance (flank diametral displacement), and the crest diameter tolerance.

tolerance grade, metric: A numerical symbol that designates the tolerances of crest diameters and pitch diameters applied to the design profiles.

tolerance limit: The variation, positive or negative, by which a size is permitted to depart from the design size.

tolerance position, metric: A letter symbol that designates the position of the tolerance zone in relation to the basic size. This position provides the allowance (fundamental deviation).

tong hold: That protruding section at one end of a forging billet that is gripped by the operator's tongs and subsequently removed from the workpiece at the end of the forging operation. Commonly found on drop-hammer and press-type forgings.

tool bits: Small pieces of hardened and sharpened material which are held, welded, or cemented in the tool holder and do the actual cutting in the lathe.

tool holder, lathe: A steel device with a rectangular body which is clamped into the tool post of a lathe, designed to rigidly and securely hold small pieces of tool steel or the working end of a cutting tool bit in a desired position. The cutting tool bit can be removed from the tool holder for sharpening or replacement without disturbing the holder. Used for the various lathe operations: turning, cutting off, boring, threading, and knurling. The *straight tool holder* holds the bit in line with the body of the holder, and tipped up at and angle of about 20°. The *right-hand offset* turns the tool bit to an angle of about 30° to the left of the body of the holder, when looking from the rear to the front of the holder. It also tips the tool bit up, as does the straight type. The *left-hand offset* is similar to the right-hand type, except that the tool is turned to the right.

toolmakers' clamps: See parallel clamps.

toolmakers' hand vise: A small, portable steel vise used by toolmakers for bench work such as tapping and drilling. This vise comes with two interchangeable blocks whose use depends on the size of the workpiece held in the jaws of the vise.

toolpost: That part of the lathe compound rest used to mount the toolholder.

tool steel: Any high carbon steel that is suitable for blanking and forming tools. Tool steel contains a sufficient carbon content (usually .60-1.50) that will allow hardening if heated above a certain temperature and rapidly cooled.

tool steel classifications: Steels for tools must satisfy a number of different, often conflicting requirements. The need for specific steel properties arising from widely varying applications has led to the development of many compositions of tool steels to meet specific needs. The resultant diversity of tool steels makes it extremely difficult for the user to select the type best suited to his needs, or to find equivalent alternatives for specific types available from particular sources. As a cooperative industrial effort under the sponsorship of AISI and SAE, a tool classification system has been developed in which the commonly used tool steels are grouped into seven major categories. These categories, several of which contain more than a single group, are listed in the following with the letter symbols used for their identification. The individual types of tool steels within each category are identified by suffix numbers following the letter symbols.

Category Designation	Letter Symbol	Group Designation
High-Speed Tool Steels	MT	Molybdenum types Tungsten types
Hot-Work Tool Steels	H1-H19H20-H39H40-H59	Chromium types Tungsten types Molybdenum types

Category Designation	Letter Symbol	Group Designation
Cold-Work Tool Steels	DAO	High-carbon, high-chromium types Medium-alloy, air-hardening types Oil-hardening types
Shock-Resisting Tool Steels	S	. .
Mold Steels	P	. .
Special-Purpose Tool Steels	LF	Low-alloy types Carbon-tungsten types
Water-Hardening Tool Steels	W	. .

Source: Machery's Handbook, 25th Edition, used with permission.

tooth pitch: A measure of the spacing between saw blade teeth; expressed as number of teeth per inch or millimeter.

tooth set: Describes the convention of bending saw teeth to one side or the other in order to increase the width of the kerf or saw cut. See also kerf.

Tophet® C: An alloy nominally containing 57-62% nickel, 22-28% iron, 14-18% chromium, 0.8-1.6% silicon, 0-1% manganese, and 0-0.2 carbon. It is used for electrical resistance metals and offers good resistance to seawater and wet, sulfurous environments.

torque (moment of force): A measure of the tendency of the force to rotate the body upon which it acts about an axis. The magnitude of the moment due to a force acting in a plane perpendicular to some axis is obtained by multiplying the force by the perpendicular distance from the axis to the line of action of the force. (If the axis of rotation is not perpendicular to the plane of the force, then the components of the force in a plane perpendicular to the axis of rotation are used to find the resultant moment of the force by finding the moment of each component and adding these component moments algebraically.) Moment or torque is commonly expressed in pound-feet, pound-inches, kilogram-meters, etc. The metric SI unit is the newton-meter.

torque wrench: A special wrench that measures the exact amount of twisting or turning force being applied to a fastening device such as a bolt or nut.

torsional strength: The ability to resist twisting forces, at right angle to the length of a shaft

total face width, gear: The actual width dimension of a gear blank. It may exceed the effective face width, as in the case of double-helical gears where the total face width includes any distance separating the right-hand and left-hand helical teeth.

total index variation, involute spline: is the greatest difference in any two teeth (adjacent or otherwise) between the actual and the perfect spacing of the tooth profiles.

total thread: includes the complete and all the incomplete thread, thus including the vanish thread and the lead thread.

total tolerance, involute spline: is the machining tolerance plus the variation allowance.

toughness: (1) A measure of the ability of a material to absorb mechanical energy (to withstand shock) without cracking or breaking. Tough material can absorb mechanical energy with either elastic or plastic deformation; (2) having the quality of flexibility without brittleness; capable of resisting great strain; able to sustain hard usage; not easily separated or cut. Toughness is usually measured by the energy absorbed in a notch impact test, but the area under the stress-strain curve in tensile testing is also a measure of toughness.

toxin: (1) Any poisonous substance; (2) a bacterial toxin, i.e., a poisonous substance produced by microorganisms, as in cutting fluids.

trammels: A divider-like layout tool composed of two moveable legs having steel points and that are attached to a beam along which the legs can travel. This tool can be used to measure distances or scribing large circles and arcs on metal surfaces.

transfer punch: A kind of pin punch containing a sharp point in the center of its straight end like a center punch. It is used to transfer the location of the center of a hole from one part to another. These punches come in various sizes. See also pin punch.

transformation: A term having the general meaning of a change in structure or arrangement.

transformation hardening: The process of hardening metal by heating to the critical temperature range or transformation range (the point at which phase changes occur), quenching it, and solidifying it with uniform distribution of its carbon content.

transformation, order-disorder: A change in the arrangement of the atoms in the lattice of certain alloys from a random distribution of the atoms of the metals to an ordered arrangement in which certain regularly chosen positions are regularly occupied by the atoms of each metal. The random arrangement results on quenching, whereas annealing causes the transformation to the ordered arrangement.

transformation ranges: Also called transformation temperature ranges. Those ranges of temperature within which there is a change from one form of iron to another. During heating (e.g. Ac_1 to Ac_3) austenite disappears, transforming to ferrite during cooling (e.g. Ar_3 to Ar_1). See also transformation temperature.

transformation temperature: The temperature at which allotropic changes, or changes in phase take place. The term is sometimes used to denote the limiting temperature of a transformation range. The following symbols are used for iron and steels:

A_{ccm} = In hypereutectoid steel, the temperature at which the solution of cementite in austenite is completed during heating

Ac_1 = The temperature at which austenite begins to form during heating

Ac_3 = The temperature at which transformation of ferrite to austenite is completed during heating

Ac_4 = The temperature at which austenite transforms to delta (ä) ferrite during heating

Ae_1, Ae_3, A_{ecm}, Ae_4 = The temperatures of phase changes at equilibrium

A_{rcm} = In hypereutectoid steel, the temperature at which precipitation of cementite starts during cooling

Ar_1 = The temperature at which transformation of austenite to ferrite or to ferrite plus cementite is completed during cooling

Ar_3 = The temperature at which austenite begins to transform to ferrite during cooling

Ar_4 = The temperature at which delta ferrite transforms to austenite during cooling

M_s = The temperature at which transformation of austenite to martensite starts during cooling

M_f = The temperature, during cooling, at which transformation of austenite to martensite is substantially completed

All these changes except the formation of martensite occur at lower temperatures during cooling than during heating, and depend on the rate of change of temperature.

transition fit: (1) One having limits of size so specified that either a clearance or an interference may result when mating parts are assembled; (2) the relationship between assembled parts when either a clearance or an interference fit can result, depending on the tolerance conditions of the mating parts.

transition temperature: The arbitrarily defined temperature at which there is a transition from ductility to brittleness in the fracture of material.

transverse pitch: In helical gear design, the distance between similar flanks of adjacent teeth, measured along the pitch circle. See also circular pitch.

transverse plane, gear: A plane perpendicular to the axial plane and to the pitch plane. In gears with parallel axes, the transverse plane and the plane of rotation coincide.

traversing length: A term related to the measurement of surface texture, the length of profile which is traversed by the stylus to establish a representative measurement.

trepanning: A boring operation for solid material used principally for large hole diameters where a solid cylindrical shaped core or plug is left at the center of the hole instead of all the material being removed in the form of chips. Trepanning is carried out in one operation, although subsequent counterboring can be used to improve surface quality or the tolerance, and it is used for through-hole applications only. The power requirements for trepanning is lower than for solid drilling. Shallow trepanning, known as face grooving, can be performed on a lathe using a single-point tool that is similar to a grooving tool but has a curved blade.

TRI: Abbreviation for Tin Research Institute.

triangles: If all three sides of a triangle are of equal length, the triangle is called equilateral. Each of the three angles in an equilateral triangle equals 60°. If two sides are of equal length, the triangle is an isosceles triangle. If one angle is a right or 90° angle, the triangle is a right or right-angled triangle. The side opposite the right angle is called the hypotenuse. If all the angles are less than 90°, the triangle is called an acute or acute-angled triangle. If one of the angles is larger than 90°, the triangle is called an obtuse-angled triangle. Both acute and obtuse-angled triangles are known under the common name of oblique-angled triangles. The sum of the three angles in every triangle is 180°.

triangular file: Files having the shape of an equilateral triangle. They are usually single or double cut and tapers to a point from about two-thirds of their length from the tip. The type of triangular file which does not taper is known as three-square parallel. A triangular file used for saw sharpening has edges that are slightly rounded with teeth cut on them.

tributyl phosphite: [$(C_4H_9O)_3P$] A water-white chemical liquid. Combustible. Used as an additive for greases and extreme-pressure lubricants.

trichloroethane: [$C_2H_3C_{l3}$] A colorless liquid used as a coolant and lubricant in metal cutting oils.

tricresyl phosphate: [$(CH_3C_6H_4O)_3PO$] A chemical additive made from cresol and phosphorus oxychloride. Used as an additive in extreme pressure lubricants. Combustible.

tripoli: A silica also known as "rottenstone" used as an abrasive powder for buffing and polishing.

trituration: The process of fragmenting a material into powder by grinding, pounding, pulverizing, rubbing, etc.

trochoid, gear: The curve formed by the path of a point on the extension of a radius of a circle as it rolls along a curve or line. It is also the curve formed by the path of a point on a perpendicular to a straight line as the straight line rolls along the convex side of a base curve. By the first definition, the trochoid is derived from the cycloid; by the second definition, it is derived from the involute.

trootsite: A structureless micro-constituent of steel tempering occurring in the transformation of austenite, the stage before sorbite, and following martensite.

true involute form diameter, gear: The smallest diameter on the tooth at which the involute exists. Usually this is the point of tangency of the involute tooth profile and the fillet curve. This is usually referred to as the *tif diameter*.

true position: The theoretically exact location of a feature established by basic dimensions.

try square: A term used to describe any square used to test or try the accuracy of work. A square designed for this sole purpose would not necessarily require graduations on the blade. However, many try squares are manufactured with graduated blades that are not hardened and held in a stock containing a special nut and bolt arrangement that allows the tool to be taken apart so the worn or "out of square" blade and/or stock can be re-ground or lapped. See also combination square.

T-slot milling cutter: A special milling machine end-mill tool used to make T-slots such as those found on the tables of milling machines, shapers, drill presses, and other machine tools. The T-slot is produced by first cutting the narrow portion of the slot with an end mill or side cutter followed by machining the wide portion with the T-slot cutter.

tube: A hollow cylinder.

tungsten: CAS number: 7440-33-7. Tungsten is one of the important alloying elements of tool steels. In many of its alloying effects, tungsten is similar to molybdenum. Tungsten increases the density of alloys to which it is added. It improves "hot hardness," that is, the resistance of the steel to the softening effect of elevated temperature, and it forms hard, abrasion-resistant carbides, thus improving the wear properties of tool steels. Used in high-speed tool steel, ferrous and nonferrous alloys, ferrotungsten products, hot-working die steels, welding electrodes, heating elements in furnaces, and vacuum-metallizing equipment, aerospace applications, shell steel, high-speed rotors as in gyroscopes, solar energy devices (as vapor-deposited film that retains heat at 500°C), metallic filaments in electric light bulbs, and electron tubes.

Symbol: W (from wolfram)　**Density (g/cc):** 19.3
Physical state: Gray powder　**Specific heat:** 0.036
Periodic Table Group:　**Melting point:** 6,165°F/3,407°C
Atomic number: 74　**Boiling point:** 10,215°F/5,657°C
Atomic weight: 183.85　**Source/ores:** Scheelite, wolframite, tungstite
Valence: 2,4,5,6　**Oxides:** WO_2, WO_3
　Crystal structure: b.c.c.

tungsten bronze: Alloys containing 90-95% copper, 3% tin and 2-10% tungsten.

tungsten carbide (WC): The strongest of all structural materials. Used in cemented carbide, dies and cutting tools, wear-resistant parts, cermets.

tungsten carbide, cemented: A mixture consisting of tungsten carbide and cobalt. Tungsten carbide is bonded with cobalt at 2,550°F/1,399°C. From 3 to 25% cobalt is used, depending on the properties desired. Used in machine tools and abrasives, for machining and grinding metals and other materials.

tungsten-molybdenum base steels: A series of high-speed tool steels containing tungsten, molybdenum, chromium, and vanadium. One common type (M2) contains 6% tungsten, 5% molybdenum, 4% chromium, and 2% vanadium., the balance being iron with a small amount of carbon.

tungsten steel: High strength, hard alloys used for high-speed tool steels. One common type is the T1 or 18-4-1 steel containing 18% tungsten, 4% chromium, and 1% vanadium., the balance being iron and a small amount of carbon. Other types nominally contain 6.75-91% iron, 5-8% tungsten, 1.25% carbon, and sometimes 3-4% chromium.

turning: An operation performed on a lathe using a single-point cutting tool that removes metal as it moves longitudinally along the workpiece diameter which is revolving. Turning is used to machine many different straight or tapered cylindrical shapes.

turret drill press: A production drill press equipped with a rotating, multiple-spindle turret that can hold 6,8, or 10 different drilling and finishing tools, used to quickly perform multiple drill press operations. Although most of the large production machines perform automatic drilling operations, smaller, hand-operated presses are also available. Compare to gang drill.

turret lathe: A production lathe equipped with multiple tools mounted on a revolving turret that takes the place of the tail stock, used for performing multiple operations.

twist drill: A rotary end-cutting tool for cutting holes. Contains two helical grooves or flutes along its length for the purpose of forming the cutting edge and clearing itself from the waste material, the borings passing up the grooves as the drill is fed into the work. See also drill.

type metal: A variable alloy nominally containing 50-93% lead, 2.5-30% antimony, 2-40% tin, 0-5% copper, and 0-29% bismuth.

U

Udimet® 700: A nickel-based cobalt alloy containing 47-59% nickel, 17-20% cobalt, 13-17% chromium, 4.5-5.7% molybdenum, 3.7-4.7% aluminum, 3-4% titanium, 0-1 iron, and 0-0.1% carbon.

UEL: Abbreviation for upper explosive limit.

UL: Abbreviation for Underwriters Laboratories, Inc.

ultimate strength: The maximum unit stress (tensile, compressive, or shear) that a material can withstand before or at rupture.

Ultimet®: A trademark of Haynes International, Inc. for a line of cobalt-based alloys having high-strength, excellent resistance to corrosion, and exceptional wear resistance. Used in welding electrodes and wire.

ultrahard abrasive: Another name for superabrasives.

ultrasonic frequency (ultrasonics): Describing high-frequency sound vibrations beyond the limit of human hearing of audible frequencies (about 20 kHz= 20,000 cycles per second). Used for descaling and cleaning metal parts, friction welding and fluxless soldering, degassing and solidification of molten metals, electroplating, drilling very hard materials, nondestructive testing of materials.

ultrasonic waves: Waves or vibrations of high-frequency sound. They include longitudinal, transverse, surface, and standing waves.

ultrasonic welding: A rapid, fluxless welding process using high-frequency (20-40 kHz) sound waves (ultrasonic vibrations) to create vibrational energy that is directed to the interfaces to be joined. This focused energy creates friction caused by molecular excitation, becoming the heat source that causes the plastics to melt. Thus, it is a kind of friction welding. Metals that can be soldered by this method include aluminum, brass, copper, germanium, magnesium, silicon, and silver.

undercut, gear: A condition in generated gear teeth when any part of the fillet curve lies inside of a line drawn tangent to the working profile at its lowest point. Undercut may be deliberately introduced to facilitate finishing operations, as in preshaving.

Underwriters Laboratories, Inc. (UL): A non-profit organization that performs testing of devices, systems, and materials. Also UL publishes safety standards for many items including welding torches and machines, regulators, gages, acetylene generators. Located at 207 East Ohio Street, Chicago, IL 60611.

Unichrome®: A registered trademark of M&T Chemicals for a copper pyrophosphate used for electroplating.

unified form thread: See Unified Screw Thread.

Unified Screw Thread: The basic standard for fastening types of screw threads in the United States, Great Britain, and Canada. This standard was agreed upon in 1948 to attain screw thread interchangeability among these nations. In relation to previous American practice, Unified threads have substantially the same thread form and are mechanically interchangeable with the former American National Screw Threads of the same diameter and pitch. Also known as Unified Form Thread.

Unified and American (National) screw threads (UN): The common thread encountered in the United States for fastening threads and given in the standards,

ASA B1.1-1949 and ASA B1.1-1960, published by the America Society of Mechanical Engineers (ASME), New York, NY. The UN thread is comprised of a 60° included angle. The UN thread is generally identified on part drawing as UNC, UNF, or UNEF. The UNC (Unified Nation Coarse) is the coarse pitch version of the standard UN thread for the particular size or diameter of the thread, UNF (United National Fine) is the fine pitch version, and UNEF (United National Extra-Fine) is the extra-fine version.

unilateral tolerance: A tolerance in which variation is permitted in only *one* direction from the specified dimension or design size.

unilateral tolerance system: A design plan that uses only unilateral tolerances.

United States Cutting Tool Institute (USCTI): 1300 Sumner Avenue, Cleveland OH 44115-2851. Telephone: 216/241-7333. FAX: 216/241-0105. WEB: http://www.taol.com/uscti

United Steelworkers of America (USWA): 5 Gateway Center, Pittsburgh, PA 15222. Telephone: 412/562-2400. FAX: 412/562-2445.

universal spiral milling attachment: An milling machine accessory for making helical (spiral) grooves and surfaces, screw threads, worms, helical gear teeth, etc., using a plain horizontal milling machine. The attachment is mounted on the column and driven by the machine spindle.

upcut, file: The series of teeth superimposed on the overcut, and at an angle to it, on a double-cut file.

up-milling: See conventional milling.

upper deviation (ISO term): In thread design, the algebraic difference between the maximum limit of size and the basic size. It is designated ES for internal and es for external thread diameters. See deviation and deviation, upper.

Upper Explosive Limit (UEL): The highest concentration (highest percentage of the substance in air) of a vapor or gas that will produce a flash of fire when an ignition source (heat, arc, or flame) is present. At higher concentrations, the mixture is too "rich" to burn. Also see Lower Explosive Limit.

upsetting: Another name for machine forging by push or squeezing pressure, as opposed to the impact pressure in drop forging.

upsilon (Υ or υ): The twentieth letter in the Greek alphabet.

uranium: CAS number: 7440-61-1. A hard, heavy nickel-white metal. It is radioactive and the isotopes ^{238}U and ^{235}U occur in nature. Forms alloys for nuclear reactors with molybdenum, niobium, titanium, and zirconium. Uranium powder can ignite spontaneously in air. Highly toxic.

uranvitrol: A copper-containing mineral. also called johannite.

USCTI: Abbreviation for United States Cutting Tool Institute.

USWA: Abbreviation for United Steelworkers of America.

u.v.: UV. Abbreviation for ultraviolet.

V

V: Symbol for the element vanadium. Abbreviation for volume (*V* or *v*). Symbol for the shearing force in beam (*V*). Symbol for velocity (*v*).

vacuum deposition: Also known as vacuum coating and vacuum metallizing. The process adding a protective coating, often aluminum, to a base material, usually another metal or a plastic, by evaporating a metal under high vacuum and condensing it on the surface of the base material. The coatings obtained range in thickness from 0.01 to as many as 3 mils. The process is used for inexpensive jewelry, electronic components, etc.

vacuum melting: A metal process involving the melting of metals in a vacuum to prevent contamination from air and remove gases already dissolved in the metal. Following melting the solidification may also take place in a vacuum or at low pressure.

valley: A term related to the measurement of surface texture, the point of maximum depth on that portion of a profile that lies below the centerline and between two intersections of the profile with the centerline.

valve, ball: A check valve having a control element incorporating a ball seated in the opening in the direction of flow. Reversal of flow seats the ball and clogs the flow.

valve, check: A valve designed to permit flow in one direction.

valve bronze: Alloys nominally containing 83-89% copper, 3-6% lead, 4-5% tin, and 3-7% tin.

valve, needle: A valve having a needle, or semi-pointed control element that seats in a small orifice in the line of flow.

vanadium: CAS number: 7440-62-2. An alloying element in steel; contributes to the refinement of the carbide structure and thus improves the forgeability of alloy tool steels. Vanadium has a very strong tendency to form a hard carbide, which improves both the hardness and the wear properties of tool steels. However, a large amount of vanadium carbide causes low grindability.

Symbol: V	**Density (g/cc):** 6.15
Physical state: Silver (pure)	**Melting point:** 3,429°F/1,887°C
Periodic Table Group: VB	**Boiling point:** 6,111°F/3,377°C
Atomic number: 23	**Source/ores:** Patronite, vanadinite, roscoelite,
Atomic weight: 50.941	carnotite, and phosphate rock.
Valence: 2,3,4,5	**Oxides:** V_2O, V_2O_2, V_2O_3, V_2O_4, V_2O_5
	Crystal structure: b.c.c.

vanadium brass: Alloys nominally containing 70% copper, 29.5% zinc, and 0.5% vanadium.

vanadium bronze: Alloys nominally containing 61% copper, 38.5% zinc, and 0.5% vanadium.

vanadium carbide: A crystalline solid used as a steel additive and making alloys for cutting tools. Mohs hardness = 2,800 kg/mm^2.

vanadium-molybdenum steel: Alloys nominally containing 92-99.28% steel, 0.1-1.0% vanadium, 0.1-1.0% carbon, and 0.52-6.0% molybdenum.

vanish thread: Also known as partial thread, washout thread, or thread runout. That portion of the incomplete thread which is not fully formed at the root or at crest and root. It is produced by the chamfer at the starting end of the thread forming tool.

valence: The property of an atom or radical to combine with other atoms or radicals in definite proportions, or a number representing the measure of reactivity of given atom or radical to combine. The standard of reference is hydrogen, which is assigned a valence of 1, and the valence of any given atom or radical is then the whole number of hydrogen atoms, or their equivalent, with which the given atom or radical combines (or which it will displace) with other atoms. Many elements have several valences, and their compounds are classified and designated accordingly. A molecule is always electrically neutral; therefore, the sum of the valences of atoms in a molecule must equal zero.

vapor density: Indicator of the number of times that the vapors of a substance are heavier or lighter than air. Vapor density measurement is taken at the boiling point. If the vapor density is greater than 1, the vapor will tend to collect at floor level. If the vapor density is less than 1, the vapor will rise in air.

vaporization: The change from the liquid to a gaseous state without a change to the molecular composition, usually by the application of heat. Sometimes called volatilization.

vapor pressure: When a substance evaporates, its vapors create a pressure in the surrounding atmosphere; therefore, vapor pressure is a measurement of how readily a liquid or a solid mixes with air at its surface. This measurement is expressed in millimeters of mercury (mm of Hg), at 68°F/20°C and normal atmospheric pressure (760 mm Hg, equaling 14.7 psi, or 1 atmosphere). A vapor pressure above 760 mm (1 atmosphere) indicates a substance in the gaseous state. The higher a product's vapor pressure, the more it tends to evaporate, resulting in a higher concentration of the substance in air and therefore increases the likelihood of breathing in it. See also volatiles.

variation allowance (λ), involute spline: The permissible effective variation. The effect of individual spline variations on the fit (effective variation) is less than their total, because areas of more than minimum clearance can be altered without changing the fit. The variation allowance is 60 percent of the sum of twice the positive profile variation, the total index variation, and the lead variation for the length of engagement. The variation allowances are based on a lead variation for an assumed length of engagement equal to one-half the pitch diameter. Adjustment may be required for a greater length of engagement.

V-block: A workpiece-holding tool made of steel and containing a 90° V-shaped groove, used to support cylindrical work on which a layout is being made, or which is being machined or inspected. The standard v-blocks are usually made in matched pairs to exact dimensions, and are often fitted with a clamp which aids in holding the work in the groove.

velocity: The time-rate of change of distance and is expressed as distance divided by time, that is, feet per second, miles per hour, centimeters per second, meters per second, etc.

vent: A small hole in a punch, or die for admitting air to prevent suction holding, or in a mold to allow pockets of trapped air, steam, or gasses to escape.

verdigris: The greenish atmospheric corrosion compound, principally basic copper sulfate, formed on copper, brass, or bronze. This is often referred to as antique patina and is desired in many articles, especially those found out-of-doors. A metal coloring solution can be made as follows: *Important:* Wear safety glasses and protect skin.

cream of tartar	240 g
ammonium chloride	80 g
cupric nitrate	600 g
table salt	240 g
water	1000 ml

Heat water in a non-aluminum pot to boiling, dissolve all the ingredients in it. Apply the hot solution to the metal with a swab or paint brush. When the desired color has been obtained, wash and dry.

vernier: A scale that has been added to a measuring device, used to make very fine measurements.

Vicalloy®: A proprietary trade name for 35-62% cobalt and 5-17% vanadium.

Vickers hardness test: Used for measuring the hardness of a material, obtained by measuring the resistance to indentation of a standard diamond penetrator at a standard pressure. The standard penetrator is a square-based pyramid having an included point angle of 136°; usually applied under a load of 5,10, 20, 30, 50, or 120 kg, usually applied for 30 seconds. The indentation, irrespective of load, is always the same shape, and the numerical value of the hardness number equals the applied load in kilograms, divided by the area of the pyramidal impression. Where P=load in kg, d=diagonal of impression in millimeters.

Vienna cement: An imitation gold. An amalgam of 86% copper and 14% mercury.

virtual diameter: See pitch diameter, functional diameter.

viscometer: Also refered to as a viscosimeter. An instrument for measuring the internal friction or viscosity of a liquid by the flow rate of the test liquid through an orifice of standard diameter or the rate of flow of a metal ball through a column of the liquid, or from the rate of revolution of a rotating spindle or vane immersed in the test liquid in comparison with its rate of revolution or speed in water. See also Saybolt universal viscosity rating.

viscosification: Increasing the viscosity of a liquid.

viscosity: The measure of internal fluid friction; the property of thickness, stiffness or pourability of a fluid such as an oil. Oil is available in a variety of *weights*; the higher the weight, the greater the viscosity of the oil. Viscosity is measured by a viscometer under specified conditions; the ratio of shear stress to rate of shear of a fluid. See also viscometer.

viscosity index improver: VI improver. An additive to lubricating oil that increases the viscosity of the oil in such a way that it becomes greater at increased temperature. VI improvers usually contain complex polymers, and are primarily used in engines.

vise: A workpiece-holding device that can be mounted on a bench for benchwork, or used as an accessory on various machines to hold work between jaws that can be manually adjusted by a screw mechanism attached to a handle, or sometimes with a toggle or lever, or can be adjusted hydraulically. The jaws can be fixed or changeable. There are many types available to the machinist including plain, swivel-base, universal, etc.

vitrified bond wheels: Nonelastic grinding wheels containing natural or synthetic abrasive grains of abrasives that are mixed with a clay material and baked in an oven at high temperatures (about 3,000°F/1,649°C), producing a hard, glasslike material that is impervious to acids, water, oil as well and heat and cold. Nearly all grinding wheels produced today are vitrified or semivitrified. See also bonded abrasive.

vixen file: A single-cut file with large, curved rows of teeth running across its surface, used for filing soft materials such as lead and babbitt. A similar file with curved teeth is the *lead float.*

volatiles: A substance, usually a liquid, that has a high vapor pressure and easily vaporizes or evaporates to form a gas or vapor.

W: Symbol for the element tungsten (wolfram).

Wagner's alloy: An alloy containing 85% tin, 10% antimony, 3% zinc, 1% copper, and 1% bismuth.

W5 Alloy: Alloy containing 95.7% nickel, 4% silicon, 0.3% manganese, used for high temperature applicatons.

Warne's metal: Alloys nominally containing 37% tin, 26% nickel, 26% bismuth, and 11% cobalt.

washout thread: See vanish thread.

watchmaker's alloy: A brass containing 59% copper, 40% zinc, and 1% lead.

water absorption: The amount of increase in weight of a material due to absorption of water, expressed as a percentage of the original weight. Standard test specimens are first dried for 24 hr, then weighed before and after immersion in water at 73.4°F/23°C for various lengths of time. Water absorption affects both mechanical and electrical properties and part dimensions. Parts made from materials with low water absorption rates tend to have greater dimensional stability.

water-miscible fluids: Fluids that easily mix with water.

water quenching: A process for hardening plain carbon steel by immersion in a bath of fresh water. By itself, water allows an insulating vapor barrier of gas bubbles to form on the surface of hot steel, and cooling is slowed. The result is uneven cooling and sometimes excessive strains which may cause cracking. In order to secure more even cooling, reduce danger of cracking, and remove scale from the steel, either rock salt (8 or 9%) or caustic soda (3 to 5%) may be added to the bath to promote rapid early cooling by making better contact with the surface. Brine or water should be at a temperature of about 60-70°F/16-21°C. After immersion, the part to be hardened should be agitated in an up-and-down or figure-eight movement in the solution. A more uniform rate of cooling is obtained by the agitation, and the formation of a vapor coating on the surface is avoided.

water soluble oil: See soluble oils and emulsifiable mineral oil. *Note:* Do not use water-based cutting fluids on magnesium.

wave: An undulation, vibration, or rhythmic motion of the particles of a solid, liquid, or gas; a form of regular movement by which all radiant energy of the electromagnetic spectrum is believed to travel.

wave set: A saw tooth set pattern having multiple teeth bent to the right followed by and equal number of teeth bent to the left, etc., creating a wavelike pattern when viewed from above the teeth. Recommended for cutting irregular structural shapes and work of varying thickness.

waviness (W): A term describing the more widely spaced component of surface texture. Unless otherwise noted, waviness is to include all irregularities whose spacing is greater than the roughness sampling length and less than the waviness sampling length. Waviness may result from such factors as machine or work deflections, vibration, chatter, heat-treatment or warping strains. Roughness may be considered as being superposed on a micro-geometric *wavy* surface.

waviness height: A term related to the measurement of surface texture, the peak-to-valley height of the modified profile from which roughness and flaws have been removed by filtering, smoothing, or other means. The measurement is to be taken normal to the nominal profile within the limits of the waviness sampling length.

waviness sampling length: A term related to the measurement of surface texture, the sampling length within which the waviness height is determined.

waviness spacing: A term related to the measurement of surface texture, the average spacing between adjacent peaks of the measured profile within the waviness sampling length.

wavy set: A tooth set containing teeth are set to the right and left alternatively.

ways: The precision-machined guiding or bearing surfaces (tracks) on which moving parts slide, as a carriage or tailstock of a lathe plane or milling machine.

wear: See also adhesive wear, abrasive wear, and pitting.

web: A drill term used to describe the central portion of the body that joins the end of the lands. The thin, extreme end of the web forms the chisel edge at the cutting end on a two-flute drill.

web thickness: A drill term used to describe the thickness of the web at the point unless another specific location is indicated.

web thinning: A drill term used to describe the operation of reducing the web thickness at the point to reduce drilling thrust.

weld: The union made between two metals that may be similar or dissimilar made by welding.

weldability: The suitability of a metal for welding under specific conditions. The nearer the properties of the material, after being welded, are to what they were before being heated and welded, the more weldable it is.

welding: A term used to describe all those processes using heat, pressure, or a combination of both to cause metals to form a joint having continuous crystallization and as homogenous as possible. There are two groups of weld types, groove and fillet. In general, if addition metal (filler) is added to a joint, it is composed of a similar material to that of the original metals being joined.

There are approximately 100 welding and allied welding processes, but the four manual arc welding processes are: gas metal arc welding (GMAW), which is also commonly known as metal inert gas welding (MIG); flux-cored arc welding (FCAW); shielded metal arc welding (SMAW); gas tungsten arc welding (GTAW). These four account for over 90% of the arc welding used in production, fabrication, structural, and repair applications. Flux-cored arc welding and shielded metal arc welding use fluxes to shield the arc, and flux-cored arc welding uses fluxes and gases to protect the weld from oxygen and nitrogen. Gas metal arc welding and gas tungsten arc welding use mixtures of gases to protect the weld. The two most cost-effective manual arc welding processes are gas metal arc welding and flux-cored arc welding. These two welding processes are used with more than 50% of the arc welding consumable electrodes purchased. Gas metal arc welding modes extend from short-circuit welding, where the consumable electrode wire is melted into the molten pool in a rapid succession of short circuits during which the arc is extinguished, to pulsed and regular spray transfer, where a stream of fine drops and vaporized weld metal is propelled across the continuous arc gap by electromagnetic forces in the arc.

Welding Institute: Abington Hall, Cambridge, England. A professional institute devoted to research, education, and dissemination of technology knowledge through professional meetings, publications, information services, its library and courses in its School of Welding Technology.

Welding Research Council (WRC): 3 Park Avenue, New York, NY 10016-5902.

Telephone: 212/591-7956. FAX: 212/591-7183. A nonprofit association devoted to cooperative research, education, and dissemination of technology in the welding field.

wetting: A surface phenomenon involving such intimate contact between solids and liquids that the adhesive force between the two phases is greater than the cohesive force within the liquid. Addition agents, such as soaps, detergents, alcohols, and fatty acids can induce wetting by surface tension reduction of the liquid. Foreign substances such as fats, oils, grease, and waxes may prevent wetting.

wetting agents: A general name for surface-active, addition agents used to lower the surface tension and reduce cohesion within water, alcohol, and other solutions, suspensions, and pastes, giving them greater penetrating power, ease in mixing, and improved surface flow. Soaps, detergents, fatty acids, alcohols, and polyoxyethylene ethers of higher aliphatic alcohols are examples of wetting agents. Some may act as stabilizers, flotation agents, mold lubricants, etc.

wheel balancing stand: A grinding wheel accessory used to test for balance and dangerous cracks. See also ring test.

wheel dresser: See grinding wheel dresser.

wheel flanges: Also called safety flanges. Large metal washers that are bolted to the sides of grinding wheels mounted on the machine spindle or shaft. They support the sides of the wheel and tends to hold the wheel together in case of a break. These flanges are usually mounted on the grinding wheel along with safety washers, made of soft material slightly larger than the flanges, that locks into the pores of the wheel.

whiskers: Single, axially-oriented crystal fibers up to 2 in. long, with diameter of 10 microns, having extremely high tensile strength. They are made from aluminum, cobalt, nickel, iron, tungsten, rhenium, tantalum, and other metals; aluminum oxide

[Al$_2$O$_3$], sapphire, silicon carbide [SiC], and other refractory materials; carbon, boron, etc., and are used largely in composite structures with glass, graphite, and plastics for specialized functions.

white alloy: Nickel-silver alloys nominally containing 80% zinc, 10% copper, and 10% cast iron. This term is also used to describe alloys nominally containing approximately 50% copper, 23% nickel, 23-24% zinc, and 2% iron.

white brass: Alloys nominally containing 30.75-39.75% copper, 60-69% zinc, 0.25% phosphorus. Another formula calls for 75% zinc, 10% copper, 10% aluminized zinc, 4% tin, and 1% phosphorus.

white cast iron: Alloys nominally containing approximately 97% iron and 2-2.6% carbon and graphite. When nearly all of the carbon in a casting is in the chemically combined or cementite form, it is known as white cast iron. It is so named because it has a silvery-white fracture. White cast iron is very hard and also brittle. Its ductility is practically zero. Castings of this material need particular attention with respect to design since sharp corners and thin sections result in material failures at the foundry. These castings are less resistant to impact loading than gray iron castings, but they have a compressive strength that is usually higher than 200,000 pounds per square inch as compared to 65,000 to 160,000 pounds per square inch for gray iron castings. Some white iron castings are used for applications that require maximum wear resistance, but most of them are used in the production of malleable iron castings.

white copper: A nickel-silver alloys nominally containing approximately 70% copper, 18% zinc, and 12% nickel.

white gold: Alloys containing gold, nickel, copper and zinc or palladium. Some alloys contain: 90% gold and 10% palladium, or 41% gold and 59% nickel, or 58% gold, 17% copper, 17% nickel, 7% zinc. Also 18-, 14-, and 10-karat alloys are widely used. Following are typical compositions of white golds: 18-K, 75% gold,

2.23% copper, 17.3% nickel, and 5.47% zinc. 14-K, 58.3% gold, 23.5% copper, 12.2% nickel, and 6% zinc or 58.3% gold, 22.5% copper, 10.8% nickel, and 8.7% zinc. 10-K, 41.7% gold, 32.8% copper, 17.1% nickel, and 8.4% zinc or 41.7% gold, 30.8% copper, 15.2% nickel, and 12.3% zinc.

white iron: See white cast iron.

white lead: A compound of white lead carbonate [$2PbCO_3 \cdot Pb(OH)_2$] a toxic material, that is mixed with powder and linseed oil for making high-pressure lubricants. Often used on the tailstock or dead-center of metal lathes when work is turned between centers.

white metal: A term used to describe any of a group of alloys containing mainly zinc and tin, or zinc, tin, and lead. One commercial alloy called 92-8 white metal contains 92% tin and 8% antimony. Casting temperature is usually in the 600-625°F/316-329°C range. Used for die casting decorative objects and inexpensive jewelry. Type metal, Babbitt, pewter, and Britannia metal are of this group.

white rust: See rust, white.

white solder: A soldering alloy nominally containing 45% zinc, 45% copper, and 10% nickel.

white spirit: A petroleum-based having similar physical properties to turpentine and used as a substiture.

white tellurium: A mineral containing tellurides of 25-29% gold, 3-15% silver, and 2.5-20% lead.

whole depth, gear: The total depth of a tooth space, equal to addendum plus dedendum, also equal to working depth plus clearance.

Widmanstatten structure: A characteristic structural formation distinguished by a geometrical pattern of a parent solid solution resulting from the formation of a new solid phase along certain crystallographic planes of the original crystals. This results in the formation of a geometric pattern of parallel plates, mutually inclined to each other at equal angles. The structure was originally observed in iron-nickel meteorites but is readily produced in certain steels, brasses, and other alloys such as aluminum bronzes, and in precipitation hardening a component in a supersaturated solid solution. These formations can be seen with the unaided eye on a polished and etched metal surface or under a microscope at low magnification, up to about 10 diameters.

wire feed speed: The rate of speed at which wire is consumed in arc welding.

wire gage: A tool used for gaging metal sheets and plates, and for measuring the diameter of round wire generally along the circumference of a round graduated plate containing a series of slots of diminishing widths. The article being gaged is measured between the sides of the opening, not in the holes. The U.S. gage is based on weight in ounces per square foot. Each gage has the decimal equivalent of the gage number stamped on the reverse side.

WN 1.4529®: A registered trademark of Wiggin Alloy Products Ltd. for a fully-austenitic, super-corrosion resisting steels nominally containing 24-26% nickel, 19-21% chromium, 6-7% molybdenum, 0.50-1.50% copper, and possibly a maximum amount of the following: 0.02% carbon, 1.0% silicon, 2.0% manganese, and 0.015% sulfur, and the balance iron.

wolfram: Alternate name for tungsten. Although tungsten is the preferred name in the United States, wolfram is an official name used in international trade. See tungsten.

wolframite: One of the two chief ore sources of wolfram or tungsten.

Woodruff key: A remountable machinery part having the shape of a half circle, which, when assembled into key-seats, provides a positive means for transmitting torque between the shaft and hub.

Woodruff key-seat cutter: Also called key-seat cutter. A milling machine end-mill or arbor-type cutter used for cutting key seats for standard Woodruff keys. The end mill type is available in diameters from 1/4 in (6.35 mm) to 1 ½ in (38 mm) and the arbor-type, in diameters from 2 1/8 (54 mm) to 3 ½ (90 mm).

Woodruff key number: An identification number by which the size of key may be readily determined.

Woodruff keyseat-hub: An axially located rectangular groove in a hub. This has been referred to as a keyway.

Woodruff keyseat milling cutter: An arbor type or shank type milling cutter normally used for milling Woodruff keyseats in shafts.

Woodruff keyseat-shaft: The circular pocket in which the key is retained.

Woods metal: Alloys with the nominal composition of 50% bismuth, 10% cadmium, 13.3% tin, 26.7% lead. Used in sprinkler systems; melts at 158°F/70°C.

wootz: A ferroalloy containing 1-1.6% carbon produced by direct reduction of bloomery iron to steel in a primitive charcoal oven.

word address form: And N/C computer language which uses numerical data followed by letters that identify different machine functions.

work: (1) Objects that are undergoing treatment, or have been treated; (2) in mechanics, is the product of force times distance and is expressed by a combination of units of force and distance, as foot-pounds, inch-pounds, meter-kilograms, etc. The metric SI unit of work is the joule, which is the work done when the point of application of a force of one newton is displaced through a distance of one meter in the direction of the force.

work hardening: The hardening that occurs in metal during cold working.

workhead: See headstock.

working depth, gear: The depth of engagement of two gears, that is, the sum of their addendums. The standard working distance is the depth to which a tooth extends into the tooth space of a mating gear when the center distance is standard.

working gage: A gage used by the manufacturer to check the workpiece as it is produced.

workpiece burning: See burning

worm gear: A gear used to connect shafts whose center lines do not meet. Nearly all worm gears work with shafts at 90°, although they are sometimes made for shaft angles of about 70°. Especially useful when a big ratio of speed reduction is required, for it can be made to give a ratio of 70 to 1, whereas the other types of gear cannot easily give more than about 6 to 1 in one pair.

WRC: Abbreviation for the Welding Research Council.

wringing: A term used to describe the manual action of attaching rectangular or angle gage blocks to each other by squeezing them together with a twisting motion, forcing all air from between the blocks. The flat, lapped surfaces of these precision tools when thoroughly cleaned and slid one on the other with slight inward pressure, will strongly adhere to each other and have been reported in various text books to sustain a considerable direct weight without coming apart.

wrought iron: A low-carbon content steel containing a considerable amount (1-4%) of slag fibers (iron silicates) entrained in a ferrite matrix. A commercial, virtually pure form of iron made from pig iron containing about 0.035% carbon. Wrought iron is tough, malleable, and has good ductility, weldability, and corrosion resistance. Also, it is relatively soft, lacking the strength of most steels, and is expensive to produce. Wrought iron has a tensile strength of approximately 50,000 pounds. Wrought iron softens and welds at 1,600°F/871°C, and can be forged at still lower temperatures. Wrought iron nominally contains 0.035% carbon, 0.06% manganese (max.), 0.10% to 0.15% phosphorus (which adds about 1000 psi for each 0.01% above 0.10%), 0.006% to less than 0.015% sulfur, 0.075% to 0.15% silicon, and less than 0.05% chromium, cobalt, copper, molybdenum, and nickel.

wrought metal: A general term applied to metal that has been subjected to one or more forms of *hot* or *cold* mechanical working such as drawing, extruding, forging, pressing, spinning, and swaging. Often used to differentiate worked metals, with higher tensile strength and ductility, from cast metals.

wulfenite: [$PbMoO_4$] Chief ore source of molybedenum, sometimes containing calcium, chromium, vanadium. Mohs hardness 2.75-3.0.

X

x: (1) Symbol for an unknown quantity; (2) rectangular coordinate. See x-axis.

X-alloy: An alloy containing 93.1-94.1% aluminum, 3-4% copper, 1% iron, 0.7% silicon, 0.7% nickel, 0.5% magnesium. The name is also used for a bearing alloy containing 93.5% aluminum, 6% iron, and 0.5% magnesium.

x-axis: The horizontal axis on the Cartesian coordinate system, or coordinate plane.

Xe: Symbol for the element xenon.

xenon: A noble gas element. Colorless, odorless.

Symbol: Xe	**Density:** 5.8971 g/L @ STP
Physical state: Gas	**Melting point:** $-170°F/-112°C$
Atomic number: 54	**Boiling point:** $-161°F/-107°C$
Atomic weight: 131.29	**Source:** liquid air
Valence: 2,4,6,8	**Crystal structure:** f.c.c.

XH: Symbol for "extra heavy."

Xi (Ξ, ξ): The fourteenth letter of the Greek alphabet and symbol for x.

x-ray(s): Electromagnetic radiation ranging in wavelength from 0.06-20 Å commonly produced by directing cathode rays upon a metal target.

Y

Y: Symbol for the element yttrium.

Y-alloy: An aluminum alloy used for casting. Contains aluminum, 4% copper, 2% nickel, 1.5% each of magnesium, silicon, and iron.

y-axis: The vertical axis on the Cartesian coordinate system, or coordinate plane.

Yb: Symbol for the element ytterbium.

yellow brass: (1) A general name for brass castings; (2) alloys nominally containing 65-67% copper, 34-37% zinc. Melt at 1639°F/893°C; having good ductility, high strength, corrosion resistance, and can with stand severe cold-working.

yellow brass, leaded: Alloys nominally containing 67-70% copper, 25-29% zinc, 3% lead, 1% tin, and good machinability.

yellow gold: Alloy nominally containing 53% gold, 25% silver, and 22% copper.

yield point: (1) The point at which stresses and trains become equal, so that deformation occurs. The point at which the stresses equal the elasticity of a test piece; (2) the initial stress in a material, usually less than the maximum attainable stress at which an increase in strain occurs without an increase in stress. Only certain metals exhibit a yield point. If there is a decrease in stress after yielding, a distinction may be made between upper and lower yield points.

yield strength: The stress at which a material exhibits a specified deviation from proportionality of stress and strain. The deviation is expressed in terms of strain. An offset of 0.2% is generally used for many metals.

Yoloy®: A steel alloy containing 96.7% iron, 1% copper, 2% nickel, 0.15-0.2% carbon, and small amounts of phosphorus.

Young's modulus: See modulus of elasticity.

ytterbic: Containing ytterbium.

ytterbite: Another name for gandolinite, a source of ytterbium.

ytterbium: CAS number: 7440-64-4. A trivalent metallic element of the lanthanide (rare earth) group. Used in lasers, source of portable x-rays.

Symbol: Yb	**Melting point:** 1,515°F/824°C
Physical state: Silvery metal	**Boiling point:** 2,786°F/1,530°C
Periodic Table Group: IIIB	**Source/ores:** Gadolinite, euxenite, xenotime
Atomic number: 70	**Oxides:** Yb_2O_3
Atomic weight: 173.04	**Crystal structure:** f.c.c. (α); b.c.c. (β)
Valence: 2,3	

ytterbium oxide: [Yb_2O_3] Also called yttria. Used for making special alloys.

yttria: Another name for gandolinite, a source of ytterbium.

yttric: Containing yttrium..

yttrium: CAS number: 7440-65-5. A rare-earth metallic element of the aluminum group. Alloying element in super alloys. Used as deoxidizer to remove dissolved gasses and improve ductility in non-ferrous metals including vanadium. also used in atomic reactors, propulsion systems, and missiles. This material reacts slowly with cold water forming flammable hydrogen. Dust may be flammable or explosive, especially in moist air.

Symbol: Y

Physical state: Gray metal

Periodic Table Group: IIIB

Atomic number: 39

Atomic weight: 88.90585

Valence: 2,3

Density (g/cc): 4.47

Boiling point: est. 5,800°F/3,200°C

Melting point: 2,748°F/1,509°C

Source/ores: Gadolinite, thalenite, xenotime

Oxides: Y_2O_3

Brinell hardness: 30-45

Crystal structure: h.c.p. (α); b.c.c (β)

Z

z: (1) Symbol for compression factor; (2) symbol for a rectangular or cylindrical coordinate.

Z-alloy: Alloys containing 93% aluminum, 6.5% nickel, 0.5% titanium. A bearing metal.

Zamak-3®: A proprietary trade name for a die-casting zinc alloy containing 95.96% zinc, 4% aluminum, and 0.04% magnesium.

Zamak-5®: A proprietary trade name for a die- and sand-casting zinc alloy containing 95.96% zinc, 4% aluminum, 1% copper, and 0.04% magnesium.

zelco: An alloy containing 83% zinc, 15% aluminum, and 2% copper, used as a solder for aluminum and its alloys.

zero: The complete absence of a particular quantity. By convention zero (or null) is considered a *natural* number.

zero-defect: A term used to describe improvements in the manufacturing process or operation whose goal is to produce products with no defects.

zeta (Z or ζ): The sixth letter of the Greek alphabet.

Zilloy®-15: A proprietary trade name for an alloy containing 98.99% zinc and 1% copper, and 0.010% magnesium, used for structural material.

Zilloy®-40: A proprietary trade name for an alloy containing 99% zinc and 1% copper.

zimalloy: An alloy containing cobalt and chromium.

zinc: CAS number: 7440-66-6. Used as an alloying element in brass, etc. It is brittle at ordinary temperatures. Between 212-300°F/100-149°C it is ductile and malleable, and becomes brittle again at about 410°F/210°C. Cast zinc's tensile strength is about 5,000-6,000 lb/in². Used for as an alloying element in brass, bronze, and die-casting alloys, galvanizing and hot-dipping iron and other metals, electroplating, metal spraying, automotive parts, roofing, gutters, engravers plates; in fungicides. A reducing agent. The dust is flammable. Dry material may ignite spontaneously in air. See also American Zinc Association.

Symbol: Zn	**Density (g/cc):** 6.7-7.14
Physical state: Bluish metal	**Melting point:** 787°F/420°C
Periodic Table Group: IIB	**Boiling point:** 1,663°F/906°C
Atomic number: 30	**Source/ores:** Franklinite, hydrozincite, zincite,
Atomic weight: 65.38	calamine (Smithsonite), sphalerite, zinc blende
Valence: 2	**Oxides:** ZnO, ZnO₂
	Crystal structure: h.c.p.

zincaloy: An alloy of zirconium with small amounts of iron, chromium, nickel, and not more than 0.5 ppm boron.

Zincalume®: A registered trademark of M & T Harshaw (division of Atochem) for a bright zinc electroplating process containing zinc cyanide, sodium cyanide, and additional agents.

zinc ammonium chloride: A chemical used in welding, as a soldering flux, and for galvanizing metals.

zincate treatment: A process for preparing the surfaces of light alloys to improve the electroplate bond. The treatment involves cleaning and acid pickling followed by immersion in a solution of sodium zincate.

zinc-base slush-casting alloy: Alloys containing 94.5-95% zinc, 4.75-5.5% aluminum, 0.25% copper.

zinc chloride: [$ZnCl_2$] A white crystalline solid or powder used as a soldering flux, especially on tin, copper and brass. Also known as "killed" muriatic acid and chloride of zinc, it is prepared by slowly adding small scraps of zinc to muriatic (hydrochloric) acid until the acid fails to "eat" the zinc and becomes a saturated solution. The escaping fumes from this chemical preparation are dangerously corrosive to eyes, skin, and respiratory tract, and require the wearing personal protection equipment, including NIOSH-approved respirator. The fumes also will cause ferrous metal tools and machines to rust. Zinc chloride is usually diluted with 50% alcohol for use on tin. Where zinc chloride is used as a flux, the part should be cleaned after soldering to prevent corrosion of the metal.

zinc chromate: [Cr_2O_4Zn] A yellow-orange pigment found in paint primers, imparts some sacrificial protective action to metals found in marine environments.

zinc coating: A sacrificial, corrosion-resistant coatings used on steels. The coating can be applied by hot-dipping, electroplating, etc. See sacrificial protection, galvanizing, galvanized iron.

zinc cyanide: [$Zn(CN)_2$] A white powder used in electroplating. Poisonous. *Warning:* Contact with magnesium; may cause a violent reaction.

zinc fluoride: [ZnF_2] Colorless or white crystalline solid used for galvanizing steel and in electroplating baths.

zinc fluoroborate: [$Zn(BF_4)_2$] A colorless liquid used in electroplating.

zincic-: Containing zinc.

Zinc Omadine®: A registered trademark of the Olin Corporation for zinc pyrithione. Used as an additive to inhibit bacterial and fungal growth in metalworking fluids.

zinc phosphate:[$Zn_3(PO_4)_2$] A white powder used for conversion coating of aluminum, steel, and other metal surfaces.

zinc stearate: [$Zn(C_{18}H_{35}O_2)_2$] A white powder used as a mold release agent. A nuisance dust that may cause lung effects; pulmonary fibrosis.

zinc sulfate monohydrate: [$ZnSO_4 \cdot H_2O$] A white powder used in electroplating.

Zin-O-Lyte®: A registered trademark of DuPont for a series of electroplating products for use in zinc cyanide plating baths.

Zippo®: A proprietary trade name for an aluminum solder. Used to join aluminum to itself or to brass, copper, tin, or zinc.

Zircal®: A registered trademark for alloys nominally containing 7-8.5% zinc, 1.75-3.0% magnesium, 1-2% copper, 0.1-0.4% chromium, 0.1-0.6% manganese, 0.7% iron and silicon.

Zircaloy®: A registered trademark for an alloy nominally containing zirconium and small amounts of antimony, chromium, nickel, and iron.

Zircaloy®-2: A trademarked name for an alloy nominally containing 98.25% zirconium, 1.5% tin, 0.1% iron, 0.1% chromium 0.05% nickel, used in water reactors.

zirco-: Containing zirconium.

zircon: A naturally occurring zirconium silicate and the source of metallic zirconium and used as an abrasive.

zirconic: Pertaining to or containing zirconium.

zirconium: CAS number: 7440-67-7. A hard, lustrous, silver colored rare-earth metal and metallic element. Used to make corrosion and heat resistant alloys; welding fluxes; deoxidizer and scavenger in steel manufacturing. Powder and dust are flammable and potentially explosive. A possible human carcinogen. In storage powder should be kept wet and protective clothing should be worn.

Symbol: Zr	**Density (g/cc):** 6.51-6.53
Physical state: Silvery metal	**Melting point:** 3,366°F/1,852°C
Periodic Table Group: IVA	**Boiling point:** 7,911°F/4,377°C
Atomic number: 40	**Source/ores:** zircon, baddeleyte (zirconia)
Atomic weight: 91.22	**Oxides:** ZrO_2 Zr_2O_5
Valence: 4, 3, 2	**Crystal structure:** h.c.p. (α); b.c.c. (β)

zirconium carbide: [ZrC] A gray, crystalline solid used as an abrasive; in cermets, metal cladding, cutting tool component.

zirconium disilicide: [$ZrSi_2$] Used for making specialized alloys.

zirconium disulfide: [ZrS_2] A solid lubricant.

zirconium nitride: [ZrN] Used for cermets.

zircono-: Pertaining to or containing zirconium.

zircon sand: A sand containing considerable zirconium, titanium, and related metals, used for foundry casting of alloys.

Zirmax®: A trademarked name of Magnesium Elektron Ltd. for a zirconium hardener.

Zirmel®: A trademarked name of Magnesium Elektron Ltd. for a series of powder and alloy products used in the metalworking trade.

Ziscon: An alloy nominally containing 60% aluminum, and 40% zinc.

Zn: Symbol for the element zinc.

Z-nickel: An alloy containing 93.85% nickel, 4.5% aluminum, 0.5% silicon, 0.4% iron, 0.4% titanium, 0.3% manganese, 0.05% copper.

Zr: Symbol for the element zirconium, taken from the Greek word zircon.

Appendix

Manufacturer's Directory

Air Products and Chemicals Inc. 7201 Hamilton Blvd., Allentown PA 18195. Telephone: 215/481-4911; 800/345-3148. FAX: 215/481-5900.

Alox Corporation. 3943 Buffalo Ave. Niagara Falls, NY 14302. Telephone: 716/282-1295. FAX 716/282-2289.

Ampco Metal, Inc. 1745 S. 38th St. Box 2004 Milwaukee, WI 53215. Telephone: 800/844-6008. FAX: 414/645-6466.

Aqualon Co. (Division of Hercules, Inc). PO Box 15417, 2711 Centerville Rd. Wilmington, DE 19850-5417. Telephone: 302/996-2000; 800/345-8104. FAX: 302/996-2049. Canada: 5407 Eglington Avenue. West Suite 103, Etobicoke, ON, M9C 5K6. Telephone: 416/620-5400. *UK:* Genesis Centre, Garrett Field, Birchwood, Warrington, Cheshire, WA3 7BH. Telephone: (Country code 44) 925 83 00 77.

Atlas Minerals & Chemicals Co. PO Box 38, 1227 Valley road, Mertztown, PA 19539. Telephone: 800/523-8269. FAX: 610/682-9200.

Atochem North America Inc. Three Parkway, Philadelphia 19102. Telephone: 215/587-7000; 800/225-7788. FAX 215/587-7591.

Barium and Chemicals Inc. County Rd. 44, Box 218 Steubenville, OH 43952. Telephone: 740/282-9776. Web: http://www.bariumchemicals.com

Cabot Corp. **(1)** 300 Holly Rd. Boyertown, PA 19512-1807. **(2)** P.O. Box 188, Tuscola IL 61953. Telephone 217/253-3370; 800/222-6745. FAX: 217/253-4334. *UK:* Carbon Division: Lees Lane, Stanlow, South Wirral, Merseyside, L65 4HT. Telephone: (Country code 44) 051 355 36 77. FAX 051 356 07 12.

Calgon Corp. PO Box 1346, Pittsburgh, PA 15230. Telephone: 412/777-8000. FAX 412/777-8927.

The Carborundum Co. (subsidary of BP America, Inc.) PO Box 156, 345 Third Street, Niagara Falls NY 14302. Telephone: 716/278-2000. FAX 716/278-2900.

Chemtronics, Inc. 8125 Cobb Center Drive, Kennesaw GA 30144. Telephone: 800/645-5244. FAX 404/424-4267.

Cytec Industries, Inc., 5 Garret Mountain Plaza, West Patterson NJ 07424. Telephone: 973/357-3100.; 800/652-6013. FAX: 973/357-3065. Web: http://www.cytec.com Engineered Materials. 4300 Jackson Street, Greenville, TX 75401. Telephone: 903/457-8500. FAX: 903/457-8598. *Canada:* Garner Road, Niagra Falls, ON L2E 6T4. Telephone: 905/356-9000. FAX 905/374-5939.

Dow Chemical USA, 2020 Willard H. Dow Center, Midland, MI 48674. Telephone: 517/636-1000; 800/441-441-4369. Canada: 1086 Modeland Road, PO Box 1012, Sarnia, ON, N7T 7K7. Telephone: 519/339-3131; 800/363-6250. *UK:* Lakeside House, Stockley Park, Uxbridge, Middlesex UB10 1BE. Telephone: (Country code 44) 081 848 86 88. FAX: 081 848 54 00.

DuPont, E.I. DuPont de Nemours & Co., Inc., 1007 Market Street, Wilmington, DE 19898. Telephone: 302/774-7573; 800/441-9442. FAX: 302/774-7573. Canada: Box 2200, Streetsville, Mississauga, ON L5M 2H3. Telephone: 416/821-5612. *UK:* Wedgewood Way, Stevenage, Hertsfordshire, SG1 4QN. Telephone: (Country code 44) 0438 73 40 00. FAX: 0438 73 41 54.

Duriron Co. Inc. PO Box 1145 Dayton, OH 45401. Telephone: 937/476-6263. Web: http://www.duriron.com

Eastman Chemical Co. (division of Eastman Kodak Co. Inc.) PO Box 431, Kingsport, TN 37662. Telephone: 615/229-2000; 800/327-8626. FAX: 615/229-1064. *UK:* Hemel Hempstead, PO Box 66, Kodak House Station Road, Hertsfordshire, HP1 1JU. Telephone: (Country code 44) 0442-24 11 71. FAX: 0442 24 11 77.

Foseco (FS) Ltd/Metallurgical Division (subsidiary of Foseco PLC), Tamworth, Staffordshire, B78 3TL. Telephone: (Country code 44) 0827-28 99 99. FAX: 0827-250-806.

W. R. Grace, 55 Hayden Ave, Lexington, MA 02173. Telephone: 617/861-6600; 800/232-6100. FAX: 617/862-3869. *Canada:* 3455 Harvester Road, Unit No. 7, Burlington, ON L7N 3P2. Telephone: 416/681-0285. *UK:* **(1)** Northdale House, North Circular Road, London, NW10 7UH. Telephone: (Country code 44) 081 965 06 11; **(2)** Grace Dearborn Waterside Lane, Ditton, Widnes, Cheshire, WA8 8UD. Telephone: 051 424 53 51. FAX: 051 423 27 22.

Hall, C. P., Co. 7300 S Central Avenue, Chicago IL 60638-0428. Telephone: 708/594-6000; 800/321-8242. FAX: 708/458-0428.

Handy & Harman, Inc. 850 Third Avenue, New York, NY 10022. Telephone: 212/752-3400. FAX: 212/207-2614.

Haynes International, Inc. PO Box 9013, 1020 West Park Ave, Kokomo IN 46904-9013. Telephone: 800/342-9637; 800/354-0806. FAX: 317/456-6905. *UK:* PO Box 10, Parkhouse Street, Openshaw, Manchester M11 2ER. Telephone: (Country code 44) 061 23077 77. FAX: 061 223 24 12.

Inco Alloys International, Inc. PO Box 1958, 3200 Riverside Drive, Huntington WV 25720 Telephone: 304/526-5100. FAX: 304/526-5411. *UK:* 1-3 Grosvenor Place, London, SW1X 7EA. Telephone: (Country code 44) 071 235 20 40. FAX: 071 235 43 58.

International Rustproof Co. (division of Lubrizol Corp.) 29400 Lakeland Blvd., Wickliffe, OH 44092. Telephone: 216/943-4200. FAX: 216/943-5337. *UK:* Waldron House, 57-63 Old Church Street, London SW3 5BS. Telephone: (Country code 44) 351 33 11 20. FAX: 351 33 10.

Kolene Corp. 12890 Westwood Ave. Detroit, MI 48223. Telephone: 313/273-9220; 800/521-4182. FAX: 313/273-5207.

Magnesium Elektron Ltd., (subsidiary of British Alcan Aluminium PLC), Regal House, London road, Twickenham, Middlesex, TW1 3QA. Telephone: (Country code 44) 081/892 44 88. FAX: 081/891 57 44).

M & T Harshaw, (division of Atochem) 2 Riverview Drive, P.O. Box 6768, Somerset NJ 08875-6768. Telephone: 908/302-3500. FAX: 908/271-8960.

Nalco Chemical Co. Inc., One Nalco Center, Naperville IL 60563-1198. Telephone: 708/305-1000; 800/527-7753. *UK:* PO Box 11, Winnington Avenue, Northwich, Cheshire CW8 4DX Telephone: (Country code 44) 0606 74488. FAX: 0606 79557.

Occidental Chemical Corp. (OxyChem®), **(1)** 5005 LBJ Fwy, Dallas, TX 75244. Telephone: 214/404-3925 or 3800; 800/752-5151. FAX 214/404-3669. **(2)** 360 Rainbow Blvd., South, P.O. Box 728, Niagra Falls NY 14302. Telephone: 716/286-3000. FAX: 716/286-3441. *Australia:* Suite 15, Gateway Court, 81-91 Military Road, Neury Bay, NSW 2089.

Olin Corp., **(1)** PO, Box 586 350 Knotter Drive, Cheshire, CT 06410-0586. Telephone: 203/356-2000; 800/243-9171. **(2)** 120 Long Ridge Road, PO Box 1355, Stamford CT 06904. Telephone: 203/356-2000; 800/243-9171. *UK:* Suite 7, Kiddminster Road, Cutnall Green, Worcestershire WR9 0NS. *Australia:* 1-3 Atchison Street, PO Box 141, St. Leonards 2065, NSW.

Olin Hunt Specialty Products, Inc. 5 Garrett Mountain Plaza, West Patterson, NJ 07424. Telephone: 201/977-6195; 800/367-4868. FAX: 201/977-6110.

OxyChem®, See Occidental Chemical Corp.

PMC Specialties Group (division of PMC Inc.) **(1)** 20525 Center Ridge Rd, Rocky River, OH 44116. Telephone: 216/356-0700. FAX: 216/356-2787. **(2)** 501 Murray Road, Cincinnati OH 45217. Telephone: 513/242-3300; 800/543-2466.

PPG Industries Inc. **(1)** One PPG Place, Pittsburgh, PA 15272. Telephone: 412/434-3131; 800/2436-774. **(2)** 3938 Porett Drive, Gurnee IL 60031. Telephone: 708/244-3410; 800/323-0856. FAX: 708/244-9633. *Canada:* Two Robert Speck Parkway, Suite 750, Mississauga ON L4Z 1H8. Telephone: 416/848-2500. FAX: 416/848-2501. *UK:* Carrington Business Park, Urmston, Manchester M31 4DD. Telephone (Country code 44) 061 777-9203. FAX: 061 777-9064.

SCM Corp., 7 St. Paul Street, Suite 1010, Baltimore MD 21202. Telephone: 301/783-1120; 800/638-3234. FAX: 301/783-1087. *UK:* PO Box 26, Grimsby, South Humberside DN37 8DP. Telephone: (Country code 44) 0469 571000. FAX: 0469 571234. *Australia:* PO Box 465, Auburn 2144, NSW. Telephone: (Country code 61) 2-647-2566.

Stainless Foundry & Engineering, Inc, 5150 N. 35th Street, Milwaukee WI 53209. Telephone: 414/462-7400. FAX: 414/462-7303. Web: http://stainlessfoundry.com

Stauffer Chemical Co. Industrial Chemical Division (division of Rhône-Poulenc) Westport, CT 06880.

R.T. Vanderbilt, Inc. 30 Winfield Street, Norwalk CT 06856. Telephone: 800/243-6064. FAX: 203/853-1452. Web: http//:www.rtvanderbilt.com

Union Butterfield Inc. Crystal Lake IL 60039-9000. Telephone: 800/222-8665. FAX: 800/432-9482.

Uniroyal Chemical Co., World Headquarters, Benson Road, Middlebury, CT 06749. Telephone: 203/573-3880; 800/243-3024. FAX: 203/573-3393. *UK:* **(1)** Brooklands Farm, Cheltenham Road, Eversham, Worchester WR11 6LW. **(2)** Kennett House, 4 Langley Quay, Waterside Drive, Slough, Berkshire, SL3 6EH. Telephone: (Country code 44) 0753 580888. FAX: 0753 591352.

Velsicol Chemical Corp. 10400 W. Higgins Road, Rosemont IL 60068. Telephone: 800/843-7759. FAX: 312/298-9014. *UK:* Worting House, Basingstoke, Hampshire RG23 8PY. Telephone: (Country code 44) 0256 81 76 40. FAX: 0265 81 18 76.

Westinghouse Electric Corp. PO Box 355, Energy Center Complex, Pittsburgh, PA 15230-0355. Web: http://www.westinghouse.com

Wiggin Alloy Ltd. Ivy House Road, Hanley, Stoke-on-Trent ST1 3NR. Telephone: (Country code 44) 1782 220260. FAX: 01782 220250. Web: http://www.wigginalloy.com